GIS Applications for Enhancing Sustainable Operational Performance of Manufacturing Industries

Kamalakanta Muduli
https://orcid.org/0000-0002-4245-9149
Papua New Guinea University of Technology, Papua New Guinea

Sujoy Kumar Jana
https://orcid.org/0000-0003-0901-5316
Papua New Guinea University of Technology, Papua New Guinea

Rashmi Prava Das
CV Raman Global University, Odisha, India

Indrajit Pal
Asian Institute of Technology, Thailand

IGI Global
Scientific Publishing
Publishing Tomorrow's Research Today

Vice President of Editorial — Melissa Wagner
Director of Acquisitions — Mikaela Felty
Director of Book Development — Jocelynn Hessler
Production Manager — Mike Brehm
Cover Design — Jose Rosado

Published in the United States of America by
IGI Global Scientific Publishing
701 East Chocolate Avenue
Hershey, PA, 17033, USA
Tel: 717-533-8845 | Fax: 717-533-7115
Website: https://www.igi-global.com E-mail: cust@igi-global.com

Library of Congress Cataloging-in-Publication Data

Names: Muduli, Kamalakanta, 1977- editor | Jana, Sujoy Kumar, 1975- editor | Das, Rashmi Prava, 1979- editor | Pal, Indrajit editor
Title: GIS applications for enhancing sustainable operational performance of manufacturing industries / edited by Kamalakanta Muduli, Sujoy Kumar Jana, Rashmi Prava Das, Indrajit Pal.
Description: Hershey, PA : IGI Global Scientific Publishing, [2027] | Includes bibliographical references and index. | Summary: "This book explores how Geographic Information Systems can enhance sustainable manufacturing by integrating spatial intelligence into Industry 4.0 practices. Combining theory, decision frameworks, and case studies, it shows how GIS optimizes logistics, energy use, waste reduction, and environmental performance in pursuit of net-zero, resilient industrial systems"-- Provided by publisher.
Identifiers: LCCN 2026003279 (print) | LCCN 2026003280 (ebook) | ISBN 9798337379418 hardcover | ISBN 9798337379425 paperback | ISBN 9798337379432 ebook
Subjects: LCSH: Manufacturing processes--Environmental aspects | Manufacturing processes--Geographic information systems | Sustainable development | Sustainable engineering
Classification: LCC TS155.7 .G57 2026 (print) | LCC TS155.7 (ebook)
LC record available at https://lccn.loc.gov/2026003279
LC ebook record available at https://lccn.loc.gov/2026003280

British Cataloguing in Publication Data
A Cataloguing in Publication record for this book is available from the British Library.

Table of Contents

Detailed Table of Contents

sensing, the drone ensures robust defect detection under varying conditions. Onboard deep learning fusion, SLAM, and GPS-denied localization techniques enable seamless autonomous navigation in both indoor hangars and outdoor airfields. By combining sustainable propulsion with advanced AI-driven inspection capabilities, the system significantly improves inspection accuracy, reduces downtime, and enhances overall sustainability in aviation maintenance.

Manufacturing industries account for a significant share of global energy consumption and greenhouse gas emissions, making resource efficiency an urgent priority. This chapter examines the integration of Geographic Information Systems (GIS) with Industrial Internet of Things (IIoT) technologies, digital twins, and artificial intelligence to enable spatial energy mapping and resource efficiency in manufacturing environments. Energy consumption data collected from smart submeters and edge-computing platforms can be visualized as geospatial layers to identify energy hotspots, quantify waste sources, and prioritize intervention strategies.

The rapid advancements in Internet of Things (IoT) and Global Positioning System (GPS) technologies have revolutionized fleet management systems, enabling real-time monitoring and optimization of vehicle performance, fuel consumption, and overall fleet efficiency. This chapter explores the integration of IoT and GPS with fuel monitoring systems, highlighting their transformative impact on fleet operations. By leveraging IoT sensors and GPS tracking, fleet managers gain valuable insights

Alotau, the capital town of Milne Bay Province, is experiencing rapid urbanization, leading to significant pressure on its limited land resources and demanding improved planning strategies. The urban expansion from 2005 to 2025 was analysed using Geographic Information Systems (GIS), remote sensing techniques and land use land cover (LULC) maps, and found a significant improvement in urbanization i.e., ~148%. The expansion was predominantly confined to low-elevation, low-gradient terrain, resulting in the conversion of vegetated and barren land covers, and was strongly regulated by topographic constraints. The LULC maps with high classification accuracy (> 85%), providing reliable, data-driven evidence of Alotau's growth patterns. These findings offer critical insights for town planners and policy makers, underscoring the urgent need for strategic and sustainable urban planning to guide Alotau's future development via. manufacturing firms, logistics planning etc., in a manner that is both resilient and environmentally conscious.

The growing complexity and energy needs of the manufacturing sectors in developing countries like Papua New Guinea (PNG) require the incorporation of the modern geospatial and analytical technologies to improve operational efficiencies and sustainability. Python programming, integrated with Geographic Information Systems (GIS), provides an inexpensive and robust mapping, analysis and optimization of energy and resource flows in manufacturing operations. This paper discusses the application of GIS- python models of enhancing sustainable performance of operations within industrial sectors of the PNG and other Pacific Island states as well. It presents a theoretical framework of spatial analytics, energy flow mapping, and resource-optimization using Python automation. Based on case studies of the manufacturing clusters of Lae and Port Moresby (PNG), and the comparative analysis of Fiji and the Solomon Islands, the study draws measurable areas of efficiency improvements.

M. Gunasekar, Vel Tech Multi Tech Dr. Rangarajan Dr. Sakunthala Engineering College, India
S. Gunanandhini, SA Engineering College, India
S. Deepa, R.M.D. Engineering College, India
R. Vinoth, R.M.K. College of Engineering and Technology, India
P. Ranjidha, Panimalar Engineering College, India
M. G. Dinesh, Easa College of Engineering and Technology, India

The integration of Geographic Information Systems (GIS) has significantly transformed supply chain and logistics operations by enabling enhanced visibility, improved efficiency, and data-driven strategic decision-making. This chapter explores the fundamentals of GIS, its applications in real-time tracking, route optimization, warehouse management, risk assessment, and sustainability planning. It highlights how GIS integrates with emerging technologies such as IoT, AI, machine learning, blockchain, and digital twins to support predictive analytics and autonomous logistics systems. Through industry case studies and analysis of future trends, the chapter emphasizes the growing role of GIS in building resilient, agile, and environmentally sustainable supply chains.

Samrat Majumdar, KIIT (Deemed to be) University, India
Deepak Singhal, KIIT (Deemed to be) University, India
Sushanta Tripathy, KIIT (Deemed to be) University, India

The supply chains of packaged marine products are very perishable, geographically distributed and are under strict regulatory and cold-chain parameters. Geographic Information Systems (GIS) have good potentials to help overcome such challenges but few empirical studies have identified and ranked the major factors that drive the adoption of GIS in packaged marine products supply chains. The focus of this research is to fill this gap by determining and prioritizing the essential factors of GIS adoption based on the consensus of the experts. To begin with, the possible drivers of GIS adoption were established by a thorough literature review of works on the marine supply chain, cold-chain logistics, traceability, and spatial decision support. The Fuzzy Delphi Method was used in the second stage to assess and rank

Preface

The growing demand for sustainable industrial development has intensified the need for innovative technologies that enhance operational efficiency while minimizing environmental impact. Among these technologies, Geographic Information Systems (GIS) have emerged as a transformative tool capable of integrating spatial intelligence, real-time analytics, and decision-support mechanisms across manufacturing and industrial ecosystems. The book *GIS Applications for Enhancing Sustainable Operational Performance of Manufacturing Industries* presents a comprehensive exploration of how GIS, combined with emerging digital technologies such as Artificial Intelligence (AI), Internet of Things (IoT), edge computing, digital twins, and geospatial analytics, can contribute to resilient, efficient, and sustainable manufacturing operations.

This volume brings together contributions from researchers and practitioners across multiple countries, offering interdisciplinary perspectives on the evolving role of GIS in industrial sustainability. The chapters collectively demonstrate how spatial technologies can improve energy management, logistics planning, fleet operations, infrastructure monitoring, renewable energy deployment, and supply chain optimization. The integration of GIS with AI-driven systems, predictive analytics, and intelligent sensing technologies highlights the transition toward smart manufacturing environments aligned with the principles of Industry 4.0 and Industry 5.0.

Several chapters emphasize the importance of energy efficiency and environmental stewardship through applications such as spatial energy mapping, rooftop solar photovoltaic assessment, and intelligent fuel monitoring systems. Others explore how GIS-enabled logistics and supply chain systems can improve transparency, traceability, and operational responsiveness in complex industrial networks. The volume also showcases innovative implementations involving autonomous drones, real-time road damage detection, and urban expansion mapping, illustrating the versatility of geospatial technologies in solving contemporary industrial and infrastructural challenges.

A notable strength of this book lies in its practical orientation. Many contributions are supported by case studies from diverse regional contexts, including Papua New Guinea, India, and other developing economies, where sustainable industrial transformation remains both a challenge and a necessity. These studies demonstrate how cost-effective and scalable geospatial solutions can support policy development, infrastructure planning, resource optimization, and sustainable economic growth.

This book is intended for academics, researchers, industry professionals, policymakers, engineers, and students interested in GIS applications, sustainable manufacturing, industrial engineering, logistics, and smart infrastructure systems. By presenting both theoretical foundations and real-world implementations, the volume aims to inspire further research and encourage the adoption of geospatial technologies for sustainable operational excellence.

CHAPTER OVERVIEW

1. The opening chapter introduces an adaptive geo-intelligent framework that integrates GIS, AI, and IoT technologies to support sustainable smart manufacturing. It discusses spatial-temporal intelligence, predictive analytics, and environmental monitoring for enhancing operational efficiency and reducing waste.
2. The second chapter presents an autonomous drone-based aircraft inspection system powered by hybrid hydrogen-electric propulsion and multimodal AI vision technologies. The work highlights sustainable aviation maintenance through intelligent defect detection and autonomous navigation.
3. The third chapter examines energy mapping and resource efficiency using geospatial technologies integrated with Industrial IoT systems, digital twins, and AI-based analytics to identify energy hotspots and optimize industrial resource utilization.
4. The fourth chapter explores the implementation of IoT- and GPS-enabled fleet management systems with integrated fuel monitoring capabilities. It demonstrates how real-time vehicle tracking and predictive maintenance can improve operational efficiency and sustainability.
5. The fifth chapter focuses on rooftop solar photovoltaic potential mapping using UAV photogrammetry and GIS technologies. Through a university-based case study, it demonstrates the feasibility of renewable energy deployment and sustainable electrification planning.
6. The sixth chapter investigates urban expansion mapping using GIS, remote sensing, and land use analysis to support sustainable urban planning, logistics

development, and industrial infrastructure management in rapidly urbanizing regions.

7. The seventh chapter discusses the integration of GIS and Python programming for optimizing energy and resource flows within manufacturing industries. The chapter highlights automation-driven spatial analytics and sustainable operational improvements in developing economies.
8. The eighth chapter addresses supply chain and logistics optimization through GIS-enabled visibility, route planning, warehouse management, and sustainability assessment. It also examines the integration of GIS with emerging technologies such as AI, IoT, blockchain, and digital twins.
9. The ninth chapter explores the role of GIS in packaged marine product supply chains, particularly in ensuring cold-chain integrity, route optimization, traceability, and risk management for highly perishable products.
10. The final chapter presents a vehicle-mounted real-time road damage detection system using YOLOv8 and edge computing technologies. The system integrates GPS-enabled spatial mapping and automated defect detection to support efficient infrastructure maintenance and public reporting systems.

The editors sincerely hope that this book will contribute meaningfully to the advancement of sustainable industrial practices and encourage further innovation in the integration of GIS technologies within manufacturing and operational environments.

Chapter 1

An Adaptive Geo-Intelligent System Integrating AI and IoT for Sustainable Smart Manufacturing

Duggirala Aravind
http://orcid.org/0009-0005-4362-7684
ASET College of Science and Technology, India

N. V. Suresh
http://orcid.org/0000-0002-0393-6037
ASET College of Science and Technology, India

R. N. Ravikumar
http://orcid.org/0009-0009-3705-1681
Marwadi University, India

E. Eswara Reddy
http://orcid.org/0000-0003-0387-1437
CMR University, India

Vanathi GopiKrishna
http://orcid.org/0009-0009-0667-6507
AVICHI College of Arts and Science, India

Kasukurthi Aravind
http://orcid.org/0009-0007-3768-3422
ASET College of Science and Technology, India

ABSTRACT

The chapter introduces an adaptive geo-intelligent system that combines Geographic Information Systems (GIS), Artificial Intelligence (AI), and Internet of Things (IoT) in order to aid the spatially enabled decision-making in the current manufacturing context. The framework uses sensor data based on location, predictive analytics, and geo-AI models to track the movements of energy, emissions, materials and equipment performance. Through the generation of persistent spatial-temporal intelligence, it enhances the minimization of waste, mitigation of risks, supply chain

DOI: 10.4018/979-8-3373-7941-8.ch001

responsiveness, and compliance with the environment. The chapter also outlines the principles of architecture design, implementation journey, and application use cases of sustainable manufacturing operations. Future research topics lead to the opportunities of digital twins, autonomous optimisation, the creation of scalable geo-intelligent networks of Industry 5.0 ecosystems.

1. INTRODUCTION

The manufacturing systems are quickly changing towards data oriented, sustainable and spatially conscious systems. With the transition to sophisticated digital technologies in the industries, geo-intelligence in real-time is a key to better utilisation of resources, minimised emissions, and increased resilience of operations. The framework of this chapter includes a combination of GIS-AI-IoT to allow a sustained spatial-temporal sustainability intelligence of the next generation manufacturing.

1.1 Need for Real-Time, Spatially Enabled Intelligence in Manufacturing

Current smart manufacturing systems are in high dynamic spatial-temporal conditions where energy dynamic, material mobility, dispersion of emissions, and machine conditions are continuously changing throughout geographically dispersed sites. This real time spatial variability cannot be measured by traditional monitoring architectures that utilize isolated sensors or periodic reporting to capture spatial variations (Safari Bazargani et al., 2021). In the process of development of India towards Industry 5.0, manufacturing systems will need geo-intelligent infrastructures that can combine IoT-enabled field data together with Geographic Information Systems (GIS) to produce spatially intelligent analytics (Kamilaris & Ostermann, 2018). This type of intelligence allows the factories to analyze spatial mobility of inventory, geofenced equipment behaviour, and location-specific sustainability results.

Live geo-AI models continue to provide a higher predictive potential of equipment breakdowns, potential congestion and environmental variances. The manufacturers cannot operationalise the precision decision-making, minimise the carbon intensity, or optimises the resource cycles without spatial-temporal intelligence. Thus, a coherent, real-time geo-intelligent layer will be invaluable in resilient operations, sustainable manufacturing optimisation and nationally scalable industrial digitalisation in accordance with the advanced-manufacturing vision of India (Rozanec et al., 2022; San Emeterio de la Parte et al., 2023).

1.2 Gaps in Traditional Sustainability and Operational Monitoring

Traditional ways of monitoring sustainability in manufacturing have relied on detached environmental audit reports, monthly reports or non-spatial system dashboards that ignore geographical variation (Kumar & Mani, 2022). Such forms of legacy management view production units, supply networks, and resource systems as fixed entities leading to partial, slower, and usually inaccurate measurements. They are unable to detect micro-scale variations, including energy asymmetry at locations, emissions localisation point, material bottlenecks, in multi-plant or within-plant layouts. Moreover, classic MES/SCADA systems do not support geo-registration, which is an attribute of sensor outputs, and thus they do not provide spatial correlations between the activity and its impact on the environment (Nurmandi et al., 2025; Walzberg et al., 2021). The geo-tagged data lack of consecutive streams can also limit the predictive modelling of the sustainability deviations or hazard regions or space dependence within the multi-tier supply chains. With regulatory frameworks in India becoming more demanding in the granular accountability of the environment, these drawbacks of conventional monitoring are more noticeable and a shift to the spatial-temporal, AI-informed, real-time sustainability intelligence is required (Ncube & Ngulube, 2024; Olawade et al., 2024).

1.3 Purpose and Contributions of the Proposed Framework

This chapter presents a geo-intelligent framework that is adaptive and designed to combine and align GIS, AI and IoT to provide continuous, spatial-temporal operational intelligence to productive manufacturing. This is to provide an order of structure architecture that can absorb, digest, and scrutinize geo-referenced sensor data in real-time, to allow predictive sustainability operation, energy efficiency, and operational defence. The framework adds a multilayered architecture of the combination of, spatially aware analytics pipelines and AI-based environmental reasoning frameworks to the IoT streams of data. It enables automated warnings on emissions, geo-heating of resource wastage and geo-sensitive optimisation of supply chains and logistics. Moreover, the chapter adds a replicable implementation pathway in accordance to the growing Industry 5.0 objectives in India of digital sovereignty, sustainable production networks, of exercise-based industrial policy. The framework provides a holistic roadmap of space intelligence implementation as an essential facilitator of next-generation; sustainable manufacturing ecosystems through the integration of conceptual bearing, real-world applications.

2. BACKGROUND AND THEORETICAL FOUNDATIONS

The interplay between GIS, AI and IoT has seemingly changed the nature of industrial systems to dynamic, and spatially-aware ecosystems. Learning about their evolution and theoretical foundations, understanding their relevance in terms of sustainability is a key to creating geo-intelligent architectures that can support real-time optimisation and environmentally friendly manufacturing in Industry 4.0 and 5.0 settings

2.1 Evolution of GIS, AI, and IoT in Industrial Systems

The blending of GIS, AI, and IOTs into industrial systems has transformed solitary digital instruments in secluded forms to cyber-physical entanglements that allow spatially diligent manufacturing. Previously, GIS was utilized in facility mapping or site planning or merely in environmental assessment (Hasan et al., 2025). Within the past ten years, the IoT improved the creation of dense networks of sensors that measured energy, emissions, health of the machine, and the environment. Simultaneously, AI evolved into an executable framework of rules to corroborated machine learning and geo-AI solutions that extract insights using intricate spatial-temporal data. When the Industry 4.0 spurred automation in India, industries started implementing digital twin, predictive maintenance, and intelligent logistics, but they did not have spatial integration (Attaran et al., 2024; Bolbotinović et al., 2025). Sustainability, resilience, and human-centric manufacturing, which are emerging as the requirements of the Industry 5.0 event in the world, also require the unity of GIS-AI-IoT. Their intersection has become enabling of ongoing geo-tagged tracking, anticipative spatial modelling, and automated optimisation to create a fresh period of clever, resilient industrial intelligence structures (Meindl & Mendonça, 2021; Rahman et al., 2025).

2.2 Core Concepts of Geo-Intelligence and Spatial – Temporal Analytics

Geo-intelligence means the analytical combination of location-sensitive information, spatial interactions and temporal development to make decisions. Geo-intelligence has been used in manufacturing to map the energy flows, material trails, particle concentrations, and machine behaviour throughout both facility layouts or scattered industrial nodes, respectively (Kamilaris & Ostermann, 2018; Safari Bazargani et al., 2021). Spatial-temporal analytics extends this geospatial modelling of conditions changing through time and space creating insights that could not be acquired using time-series or non-spatial analysis. Spatial interpolation, geofenced

IoT triggers, spatial clustering, and dynamic heatmaps are some of the technologies enabling manufacturers to find anomalies, sustainability hotspots, and predict risk areas (Mkhitaryan et al., 2025). This is improved by geo-AI, which offers machine learning to spatial data so that predictive tasks, like forecasting emissions, or predicting downtime through probability maps, can be performed, or immediate hazards sensed. The two ideas in combination form a computation basis to temperate, real-time sustainability cognizance- fundamental to designate-precision optimisation in advanced manufacturing ecosystems (Park & Yang, 2020; Xu, 2025).

2.3 Sustainability Imperatives in Industry 4.0 and 5.0

Industry 4.0 was focused on automation, cyber-physical integration, and data-driven operations yet sustainability was a minor goal. Because with globalization regulations it gain stress on the environment, industries are no longer satisfied with the mere compliance to sustainability indices that demand sustained and measurable performance. The revolutionary change that Industry 5.0 brings about is that not only should manufacturing be intelligent, but also regenerative, resilient, and human-oriented (Aheleroff et al., 2022; Rame et al., 2024). This requires real time data regarding carbon emission, energy use, wastes and environmental responses in industrial spaces. The traditional sustainability reporting on the basis of static assessment is unable to fulfil these imperatives. Spatial intelligence can be highly important in monitoring environmental effects on a microsystem scale, simulating the carbon cycle, and facilitating recycling of resources (Cao et al., 2023). With the combination of GIS, AI and IoT, manufacturers can transform today reactive and compliance-based strategies into proactive and spatially optimised sustainability platforms that are consistent with national and global Industry 5.0 objectives (Ghobakhloo et al., 2024, Delabeye et al., 2021).

3. PROBLEM STATEMENT AND RESEARCH GAPS

Manufacturing systems produce both enormous amounts of operational and environmental data but the crucial spatial-temporal intelligence is not fully exploited. Current systems do not provide the possibility of integrating sensor streams with spatial analytics that leave significant gaps in the surveillance, anticipatory execution and real-time optimisation of operations.

3.1 Fragmented sensor data lacking spatial context

The purchasing machines installed in the majority of manufacturing facilities use IoT sensors to track energy and emissions, machine position, and names of environmental factors. Nevertheless, those data streams are often managed in siloed and non- georeferenced forms, which decouple sensor measurements on the specific spatial position of the sensor readings within the plants or distributed plants across distributed facilities. Devoid of spatial context, manufacturers would be unable to map hotspots, trace anomalies, nor be able to know how inefficiencies and risks are distributed spatially. This fragmentation also does not allow correlating the physical layout, workflow, and environmental influences. As a result, decision-makers base on partial or distorted information, which impedes them in carrying out specific interventions. To fill this gap, there is need to employ intensive geo-tagging, spatial matching and incorporated GIS-centered data structures.

3.2 Limited real-time sustainability insights

Conventional sustainability surveillance is based on periodic reporting, batch data processing, or even fixed dashboard, and provides partial information about the rapidly evolving operational realities. These systems are unable to record immediate changes in emissions, bursts of energy, dangerous spills, or the outside of control flows of materials within spatial regions. In the absence of location-sensitive, real-time insights, anomalies cannot be identified until they become disruptive to the operations of an organization or result in regulatory non-compliance. Furthermore, the current predictive tools seldom feature spatial-temporal patterns and would therefore be less useful in predicting risks or resource optimisation. Lack of consistent sustainability intelligence will also limit manufacturers to migrate towards a proactive and focused sustainability management and more complex GIS-IoT-AI architectures can warrant continuous monitoring in a spatially dynamic fashion

3.3 Absence of unified GIS–AI–IoT frameworks

Regardless of technology progress in digital manufacturing, the existing architectures lack the ability to provide a single framework to seamlessly incorporate the use of GIS, AI, and IoT in terms of spatial-temporal intelligence. The current systems consider spatial analysis, sensor data analytics and machine learning as autonomous units, which lead to disjointed workflows and restricted predictive capacity. In the absence of a unified layer of integration, industries are unable to use geo-AI to optimize automated or relate real-time sensor streams to spatial modelling applications. This lack of technology means that the creation of intelligent sustainability dashboard, independent warning system and real-time optimisation engines cannot be realised. Therefore, a consistent framework is vital in order to facilitate scalable, adaptive and industry-tolerant geo-intelligent sustainability solutions in line with Industry 5.0 agendas.

4. METHODOLOGY AND CONCEPTUAL FRAMEWORK DESIGN PRINCIPLES

The suggested methodology forms a strong geo-intelligent base, as it combines multi-layer architectural architecture, IoT data fusion that are spatially enabled, and new AI/ML modelling pipelines.

Figure 1. Adaptive Geo – Intelligent Framework

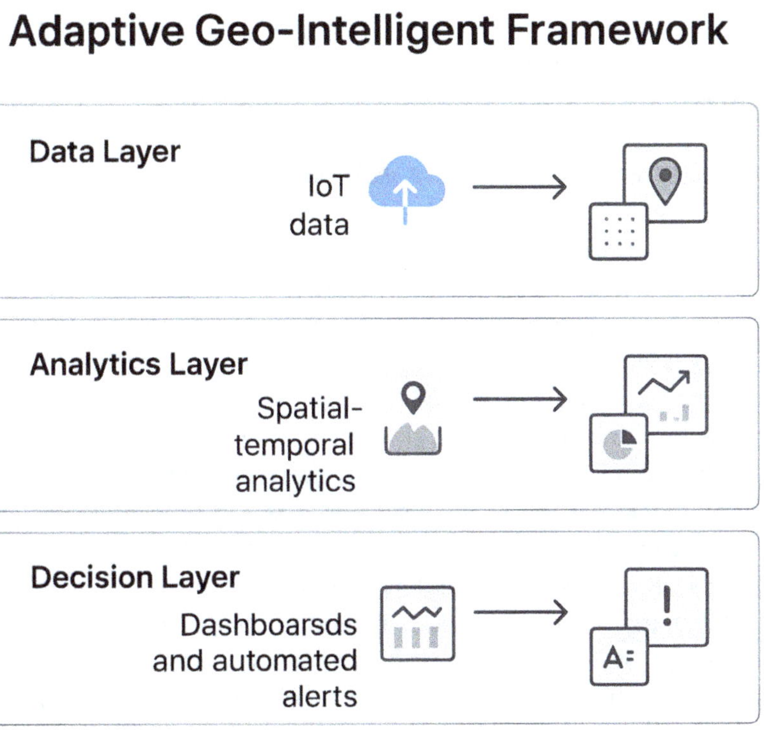

4.1 Multi-layer architecture (data–analytics–decision layers)

The suggested framework takes the three-layer multi-layer design architecture aimed at a smooth flow, transformation, and use of the spatial temporal manufacturing information. The Data Layer receives heterogeneous data of the IoT sensors, SCADA system, energy meters and emissions monitors and geospatial repositories with each piece of data tagged by coordinates reference systems. The pre-processing agent comprises of spatial buffering, rasterization, temporal division and noise reduction. Analytics Layer combines GIS engines together with prediction mapping, clustering and hot spot analysis in a spatial-temporal modelling unit based on AI. Spatial matrices are developed using algorithms that provide the expression of each observation as $D(x, y, t)$ thereby being able to model relationship across space and time. Decision Layer converts analytical outputs into actionable intelligence by way of real time dashboards, automated alerts and optimisation modules. The rules of decision making uses mathematical conditions, like the issuance of warnings when emissions of $E(t) > E_{max}$ hit the mark. This stacked implementation guarantees modularity, scalability and high level of interpretability in different types of industry environments.

4.2 Integration of IoT streams with GIS platforms

One of the basic structures in methodology is the smooth flow of the IoT data into the GIS systems to form a single spatial-temporal intelligible space. Sensors constantly measure and send the measurements as IoT devices, which are mathematically expressed as $S_i(t)$, with GIS giving the spatial coordinates x_i, y_i. The process of integration creates geo-tagged sensor triples $\{S_i(t), x_i, y_i\}$ so that real-time spatial mapping of manufacturing is possible. Using streaming protocols (MQTT, OPC-UA), data passes into a spatial middleware, that converts the raw signals of a sensor in geographic readable formats, such as GeoJSON or WMS layers. High-scale geoprocessing capabilities, such as spatial joins, buffer analyses, Voronoi mapping, are used to locate micro-scale level deviation of operations. This merging can enable manufacturers to determine the spatial correlations e.g. energy peaks around particular groups of equipment or emission tops around furnace areas. Through the temporal resolution of the IoT and spatial precision of the GIS, the industries can have a 4D view of the continuous operation, which enhances the accuracy of sustainability evaluation and predictive control.

4.3 AI/ML pipelines for predictive sustainability modelling

The framework will be built on the advanced AI/ML pipelines that can be developed to produce the predictive insights of sustainability of the spatial tempos based on the spatial-temporal datasets. The fundamental components of these pipelines are a spatial time learning model $f(x, y, t) \rightarrow \hat{S}$, which predicts the sustainability indicators of the future, e.g., the emission level $E(t + \Delta t)$, the energy demand $P(t + \Delta t)$, or minor danger. High-dimensional tensors based on pre-processed features such as spatial coordinates, environmental variables, sensor measurements and past trends can be used in such models as LSTMs, spatial random forests, GNNs (Graph Neural Networks). Greater level of geo-AI improvements enable models to factor neighbourhood effects in the form of spatial weights matrices W_{ij}. Training pipeline involves spatial cross validity, drift checking to sustain high endurance to changing industrial trends. The results of prediction are fed to the operational decision systems making the optimisation of loads, routing, and process parameters to be automated. This AI-driven sustainability pipeline will result in uninterrupted fore tracing, fast anomaly identification, and dynamic optimisation in accordance with the Industry 5.0 sustainability objectives.

5. ADAPTIVE GEO-INTELLIGENT FRAMEWORK DESCRIPTION

Adaptive geo-intelligent framework will incorporate layers of both spatial, temporal, and semantic data in one sustainability intelligence engine. It is a continuous learner based on the IoT streams, geo-AI analytics, and autonomous refiner of operational decisions based on dynamic spatial modelling and predictive optimization workflows.

Figure 2. Framework Description

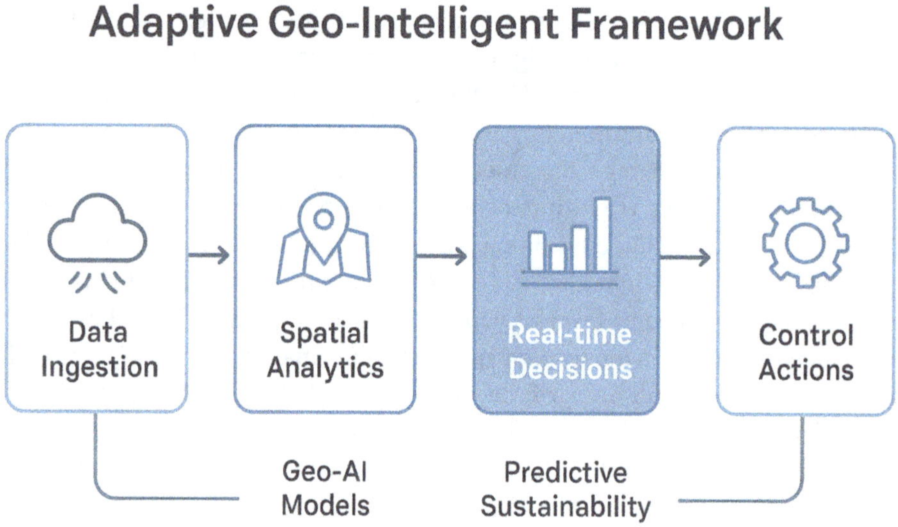

5.1 Dynamic spatial data ingestion and processing

The spatial ingestion engine fuses asymmetric streams of IoT sensors into a georeferenced, time-synchronous data infrastructure. The incoming data packet $d = \{x, y, t, v\}$ is subsampled into a continuous spatial field with the help of kernel-based surface reconstruction:

$$\widehat{V}(s) = \sum_{i=1}^{n} K_h(\| s - s_i \|) v_i$$

where K_h is an adaptive Gaussian kernel which changes the bandwidth h depending on the sensor density.

In order to make sure that the real-time alignment occurs, the Kalman Bayesian update is used in the time fusion.

$$\hat{x}_{t|t} = \hat{x}_{t|t-1} + K_t\left(z_t - H\hat{x}_{t|t-1}\right)$$

$$K_t = \widehat{P}_{t|t-1} H^T \left(H P_{t|t-1} H^T + R\right)^{-1}$$

This enables the correction of noisy IoT readings to stabilized spatio-temporal signals.

Multi-parameter measurements of energy load, emissions, vibration and thermal patterns are computed on voxels and are organised in a 4D tensor:

$$T \in R^{X \times Y \times Z \times T}$$

where each of the four dimensions is over a spatial unit. The Principal Component Geostatistics (PCG) is used to transform coordinates, interpolate grids and compress space with edge preprocessors. This processed tensor is in turn fed through the analytics layer allowing geo-AI models to access high quality, constant, real-time spatial intelligence.

5.2 Geo-AI analytics for energy, emissions, and resource flow

The geo-AI analytics engine combines both spatial statistics, graph learning as well as temporal predictive modelling to describe sustainability behaviours within the manufacturing environment. Energy and emission levels are represented by Geographically Weighted Regression (GWR) in which each geographical location s has specific coefficients:

$$\beta(s) = (X^T W(s) X)^{-1} X^T W(s) Y$$

What this allows is spatial heterogeneity, e.g. localized clusters of high carbon. In the case of the temporal-spatial forecasting, the IoT nodes would be transformed to graphical structures with adjacency matrix A where $A_{ij} = 1$ when node i and node j come into contact spatially. A coupled GCN-LSTM motion is used to predict emission progression

$$H^{(l+1)} = \sigma(A H^{(l)} W^{(l)})$$

$$h_t = f(W_h h_{t-1} + W_x x_t)$$

This enables the system to obtain knowledge on diffusions of loads in space and their temporal progression. The flow of resources is calculated by using mass-balance or

$$\sum_k Inflow_k - \sum_k Outflow_k = \Delta S$$

Clustering density Dbssar bases on a set of multi-variate spatial vectors, Z = {E, CO_2, Heat, Flow} and is referred to as the density-based clustering system of Z (DBSCAN, ST-DBSCAN). The analytics module generate spatially explicit scores of sustainability and predictive heatmap to be used as direct inputs into optimisation and automated decision engines.

5.3 Real-time dashboards and automated spatial alerts

The geo-AI analytics is transformed into actionable intelligence by the real-time dashboard layer using spatial computation of events. Every sensor or area produces a risk score $R(s,t)$ which is determined as follows:

$$R(s,t) \;=\; \alpha E(s,t) + \beta CO_2(s,t) + \gamma \Delta H\left(s,t\right)$$

E is energy deviation; CO_2 is intensity of emission and ΔH is thermal anomaly. Multi-objective optimisation Adaptive weights (α,β,γ) are opt imposed:

$$\min_{\alpha,\beta,\gamma} F \;=\; w_1 \, False\ Alram + w_2 \, Miss\ Detection$$

Spatial alerts occur when the risk surpasses a dynamic threshold that is calculated with the help of spatiotemporal CUSUM:

$$S_t = \max\left(0, S_{t-1} + R(s,t) - k\right)$$

Alert if $S_t > \; h$

This makes sure that it is sensitive to slow changes in emissions, leakages or unnatural energy surges. Visualisation pipelines Visualisation of real time is carried out using raster grids of tiles and overlaying vectors to enable operators to see and examine on maps of heatmaps, flow vectors, anomaly charts and predictive timelines. Rule-based engines or reinforcement learning triggers can be used to start actions, such as load shedding, ventilation control, path rerouting, and others, in automated decision modules. This makes the dashboard a geo-intelligent operational command layer that is capable of enforcing sustainability autonomously.

6. KEY FUNCTIONAL MODULES

The adaptive geo-intelligent architecture will be implemented based on three closely interlinked functional modules that it will convert raw streams of IoT into predictive, spatially informed sustainability intelligence. Each of them executes specific computational functions-spatial-temporal anomaly detection to multi-objective optimisation and geospatial logistics modelling. The resultant union forms a closed loop system that can make automated industrial decisions.

Figure 3. Adaptive Geo-Intelligent Manufacturing Module

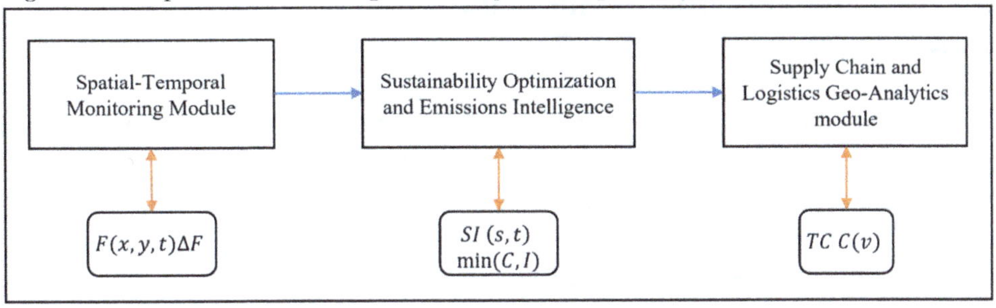

6.1 Spatial – temporal monitoring module

Spatial-temporal monitoring module, the state representation of the manufacturing environment is built based on spatial interpolation, temporal filtering, and anomaly classification. Sensor measurements are represented as a spatial field

$$F(x, y, t) = g(v_i, s_i, t_i),$$

with g representing spatial kernel fusion.

The module detects spatiotemporal anomalies on a hybrid statistical-ML model.

A point is defined as an anomalous point when:

$$Z(s, t) = \frac{F(s, t) - \mu(s)}{\sigma(s)} > \tau,$$

Dynamic behaviour An idea is followed with a state-transition tensor:

$$S \in R^{X \times Y \times T \times K}$$

Energy, emission, vibration and heat signatures of K variables. Definite vectors,

$$\Delta F = F_{t+1} - F_t,$$

detect new hotspots or structural differences. The module conveys the cleaned and structured spatial-temporal intelligence to optimisation and decision engines of a higher level.

6.2 Sustainability optimisation and emissions intelligence module

This module calculates the performance of sustainability using mathematical optimisation and predictive modelling of the emissions.

The sustainability index of a real-time system is defined to be:

$$SI(s,t) = w1\ E(s,t) + w2\ CO_2 + w3\ W(s,t),$$

E – energy deviation, CO_2 – rate of emission, W – flux of waste.

The optimisation in the module is based on multi-objective optimisation (MOO) to reduce the operational costs C and the impact on the environment I:

$$min(C,I) = \{\min C(u), \min I(u)\},$$

subject to spatial constraints

$$u \in \Omega(s)$$

with u denoting actions that a designer can control (load adjustment, flow diversion, process throttling). There is a spatial LSTM:

$$h_t = f(\ W_h h_t - 1 + W_x F(s,t)),$$

which is used in emissions forecasting. The comparison between the predictive outputs and regulatory thresholds makes automated interventions. The module transports the optimised sustainability courses and forecasted emission courses to the dashboard layer.

6.3 Supply chain and logistics geo-analytics module

This is a spatial representation of the nature of the behavior of materials, components and vehicles in industrial networks. Every supply chain can be models and expressed as a directed spatial graph:

$$G = (V,E),\ Eij = d(s_i, s_j),$$

d being geodesic distance or network distance.

The relationship between the cost of transportation is:

$$TC = \sum_{(i,j)\in E} f(Qij, dij, \tau ij),$$

flow quantity Q and τ = travel time. The spatially weighted shortest-path model:

$$min \sum w_{ij}, d_{ij}$$

in which weights are taken to represent congestion, carbon intensity and fuel consumption, is used in route optimisation. To detect the bottlenecks, the module calculates node centrality:

$$C(v) = \sum_{u\neq v} \frac{1}{d(u, v)}$$

Deviations in the flow or delivery timeline of materials are detected by real-time geo-fencing and movement analytics. The module optimizes the resilience of the supply chain, minimizes the planning routing eco-intensive carbon emission, and promotes spatially directing the planning process of the inventory.

7. IMPLEMENTATION PATHWAYS

The introduction of a single GIS-AI-IoT ecosystem requires a stepwise approach that is technical intensive and balanced in innovation, viability of operation, and institutional adaptation. This part explains the path of transforming geospatial intelligence, autonomous analytics and connected sensing networks toward conceptual design to scalability to implementation in the multi-layered governance and infrastructural landscape in India.

7.1 Roadmap for GIS–AI–IoT integration

The conceptual roadmap to GIS-AI-IoT integration that can commence is with the creation of the principal geospatial infrastructure: harmonised bottom maps, standardised spatial references, and standardised geo-registries of assets, networks, and environmental maps. The initial stage includes the design of sensor networks, which are sufficient in link-budget, redundancy of coverage, and interoperability of IoT with multiple protocols (LoRaWAN, NB-IoT, 5G mMTC). Parallel versions build centralised spatially supported data lakes having scalable schema only read designs in response to ingesting heterogeneous, high-velocity IoT streams.

The second stage is concerned with integrating AI pipelines (spatial clustering, predictive modelling, anomaly detection, and event-driven reasoning) directly into the GIS fabric with geo-analytics engines humans have to operate at the cost of mounting GPUs. The models of digital twins within critical domains (water, energy, mobility) are increasingly formed, which makes it possible to optimise the simulation and estimate the state propagation. Cross-platform APIs serve to provide that the old SCADA, MIS, and ERP systems can work with the new geo-intelligent backbone.

The third stage deals with operational automation using automated workflow, triage engines driven by rules, and command dashboards in real time. Spatial-temporal alerting and hazard forecasting and maintenance intelligence become regular functions. Lastly, long-lasting capability establishment, geo-AI literacy, data-stewardship training and operational resilience standards guarantee an institutional uptake in the long-term. This is a stepwise transition plan that will fast track the transformation of India into autonomous and data-driven governance systems.

7.2 Data governance, interoperability, and cybersecurity

The GIS-AI-IoT ecosystem demands a powerful data governance system to establish data ownership, data backstory, and data retention and quality-assurance systems throughout the lifecycle of the data. Behavioral Semantics Completeness metadata is ensured by the ISO 19115/19157 standards; this is with reference to traceability and semantic stability. Interoperability is ensured through the use of OGC-compliant data services- WMS, WFS, Sensor Things API and the creation of a cross-platform geospatial interoperability layer between state, central, and institutional data.

The trust is based on systemic cybersecurity backbone. TLS 1.3 + AES-256 Multi-level Encryption and Industry Standard Algorithms: Does not require Zero-trust access control or multi-factor authentication. 3 Proteges: Sensor Endpoints: Sensor endpoints are protected. 4 Data Lakes: Sensor data is encrypted, anonymized, and protected against data privacy breaches. 5 Inference Engines: Any inference engine you implement runs quickly, securely, and integrates with your existing security systems, allowing your organization to use these applications in environments where hypothetically sensitive data could be accessed. Network micro-segment enables less movement laterally, and real-time intrusion monitoring identifies abnormal IoT traffic dynamics and space based cyber-physical threats. AI model provenance tracing, differential privacy confinements, and unresilient ML resistance against data poisoning, model inversion and adversarial perturbations are security enhancements of AI pipelines.

Information sovereignty will be at its core: all essential spatial and IoT information is stored across India-domestic secure cloud systems in accordance with MeitY empanelment standards. Governance councils define the ethical use of AI, bias evaluation protocol, and data sharing agreements across agencies. Collectively, these governance and cybersecurity systems will establish a robust, interoperable and trusted platform of nationwide geospatial awareness.

7.4 Pilot-testing and full-scale deployment models

The first step in implementation entails the controlled pilot-testing conditions which confirm system functioning under realistic operational conditions. The criterion used to select pilot zones includes density of sensors feasibility, administrative willingness, environmental diversity, and significance of infrastructure. These zones have small-scale digital twins, which enable the simulation of data flows, AI model stress-testing, and latency, accuracy and failure-handling behaviour benchmarking.

The hardware-in-the-loop testing during the pilot phase ensures the sensor dependability, network traffic, redundancy, efficiency in power management. Operational tests include the test of accuracy of event detection, false-alarm suppression, spatial-temporal forecast, and command-team usability on dashboard. Insights derived from pilot analytics influence iterative schema model, workflow logic and alert thresholds.

Massive deployment is based on a modular model of expansion. Deployment takes place in circles of scale, i.e. district - regional cluster - state - national grid, allowing scalability effectively without breaking architecture. Deployment pipelines include automated device provisioning, centralised configuration management, and GSI updates integration capabilities. Institutional readiness tests make sure that the local agencies are trained, well-staffed and mature operational-wise to be adopted.

Lastly, sustainability is achieved over the long term using lifecycle management strategies, such as sensor predictive maintenance, adaptive policy feedback loops, and constant AI model calibration as the environmental and operational conditions adapt.

8. REAL-WORLD CASE STUDIES

8.1 Case demonstrating energy and resource optimisation

One of the Indian smart-manufacturing clusters in Gujarat implemented a GIS-AI based-IoT resource-intelligence platform to control the energy, water, and thermal energy consumption of 120 or more remotely distributed industries. The facilities were mapped to digitally twinned spatial indexed models calculating building geometry, machine layouts, utility routing, thermal envelopes and historic load profiles. IoT sensors, such as flow, vibration, power quality, temperature, and pressure, were used and transferred multi-modal data into an edge-assisted geospatial controller at the center.

AI models Since the model combined real time telemetry with GIS layer, it co-created a dynamic energy-resource optimisation engine. Real time inference was in effect all the time at the system, including load peaks, cooling demand, hidden leaks and automated control of distributed energy flows. Spatial the algorithms of clustering established areas of systemic inefficiencies (especially older chillers in units and undersized transformers) which could be retrofitted instead of retrofitting the infrastructure in a generic way.

Micro-level (motor-drive tuning, chiller cycling, pump modulation) and macro-level (static redistribution of resources between each of the facilities in the cluster) resource reversals within the cluster were synchronized using inter-facility load-balancing rules by the geospatial engine. In the peak demand periods, the AI advised the staggered production schedule and utilized both the availability of rooftop solar and the process thermal lag as well as the preservation of district cooling. It has been implemented with a 17-23% success in electricity use, 28% cut in cooling water use and considerable savings in reactive-power penalties because operation had enhanced power-factor correction.

The case shows how a space conscious, independently evolving optimisation model can optimise resource efficiency in a high population density of industrial ecosystem, particularly in the growing manufacturing industries in India, without affecting production throughput or system reliability.

8.2 Case showing emissions and compliance monitoring

A metropolitan air quality authority in South India developed an advanced emissions intelligence platform combining the use of GIS-linked Artificial Intelligence AI models along with IoT-based air sensing arrays and stack monitoring systems. The initiative focused on industrial belts, key transportation corridors, and burning hotspots of the peri-urban areas: establishing a real-time trading networks of compliance with emissions within 1,800-sq-km of the area.

Massive cut-ups were made through the interplay between meteorological layers and topography and land-use frameworks using topography and land-use by defining color changes, giving rise to high-resolution dispersion maps on which sensor data was overlaid. IoT nodes were used to measure the NO2, SO2, PM2.5, PM10, VOCs and black-carbon signatures at some minute-level frequencies. Geo-semantic AIs detect exceptions of plumes, differentiate between legal and unauthorized emission sources, and forecast dynamic paths of spread in line with dynamic behaviour of bottom-layer and wind shear.

They had industrial units linked on secure data pipelines that allowed automatic retrieval of furnace temperature logs and scrubber operating conditions, as well as the makeup of flue-gas. Having suspected a likely violation, the AI provoked the system to initiate an automated compliance audit process: stack feedstock cross-check, spatial patterns of the plumes matched, and evidence packages were created and displayed on the regulatory dashboard. The enforcement agencies were provided with geo-tagged alerts with heatmaps on the intensity of the emitters, the predictions of the path of the plume drift, and the recommended inspection routes.

There was another platform in favor of predictive compliance. Predictions of road-congestion and biomass-burning risks layers during the months preceding the festival seasons were used to launch rapid grid-based monitoring units by authorities to grids at risk. Urban policy was led by emission overlays- to optimise traffic diversion, to change the frequency of one or the other mode of transport, and to time the wetting of the road.

The result was a quantifiable turnaround in the response time of the regulators, an increase in non-compliance cases was reduced substantially, and the transparency between the industries and the ban enforcers made a significant improvement in the Indian governance of emissions.

8.3 Case highlighting supply chain and operational efficiency gains

The Indian national cold-chain logistics service provider deployed an integrated GIS-AI-IoT network to solve the problem of spoilage, inefficiency in routing and poor traceability in its multi-modal transport network. The system developed a geo-temporal digital twin of all the supply chain- packhouses and reefer vehicles on one end, cross-docking hubs and distribution centres on the other.

Telemetry IoT aspects of reefer temperature, door-open processes, compressor periods, humidity condition, vibration, fuel repetitions, and GPS positions were fixed. This data was absorbed in geospatial control layer which constantly assessed the state of operations within the network. AI-based route optimisation models realised their transport routes in real time by reacting to the density of traffic, thermal-load predictions, vehicle health condition, and cold-room utilisation at destination hubs.

An important innovation was known as spoilage intelligence. This system could estimate the chances of spoilage of the batch even before they were visible by helping to correlate the temperature deviations with the time spent on the road and external climatic exposure as well as determining any cargo type. Alerts were also spatially contextualised- displaying which corridor, micro-climate zone or handling point is a risk contributor. This facilitated proactive cross-docking and vehicle rearrangements by staging cold-room in advance and making timely modifications on cross-docking.

The efficiency of operations enhanced with the use of the automated schedule that was adapted to real-time hub congestion analytics. The fleet bosses were provided with dashboards that displayed thermal stability index, asset-health score and Geofenced compliance alert in the major goods including vaccines and perishable foods.

The overall transformation process led to the decreased spoilage rate by 37 percent and speed of turnaround increased by 22 percent, and shipment traceability was improved significantly. The case presents the idea of how the logistics industry in India can become global grade in reliability when it becomes connective on the issues of spatial analytics, predictive AI, and high-resolution IoT visibility.

9. KEY RESEARCH INNOVATIONS AND FRAMEWORK ADVANCEMENTS

This study provides a conglomerate of original works, which improve scientific, structural, and operational basis of geo-intelligent sustainable manufacturing. The integration of GIS, AI, and IoT into a unified spatial-temporal intelligence cloth provides the framework with novel perspectives of autonomous optimisation, continuous monitoring, and scalable industrial change in accordance with the global priorities set by Industry 5.0.

9.1 First adaptive geo-intelligent architecture for sustainable manufacturing

The ultimate contribution of the study is that it designed the first totally adaptive geo-intelligent system to sustainable manufacturing. This architecture implements spatial-temporal intelligence into all the operational layers of data ingestion, analytics, inference, and decision orchestration, unlike traditional Industry 4.0 systems that think of a spatial overlay on top of data, which is built continuously and forms a living ecosystem of spatial computing. It uses multi-layer adaptive model in its architecture which enables the manufacturing systems to perceive, interpret and react to spatial-temporal variations in real time. Its main innovation is how closely accompanies the IoT telemetry dynamically indexed GIS layers so that manufacturing plants may not only know how machines or sensors are performing but also where, when, and why something is not working well, emitting, or having resource anomalies.

Adaptivity can be done by having the feedback loops controlled by AI-based spatial inference. The architecture tracks geo-coded state of the operations and, upon reaching certain decided constraints, automatically restructures the energy flows, material routing, or the workload distribution. One design principle is the spatial resilience option:

$$R_s(t) = \frac{1}{1 + \alpha \Delta S(t)}$$

where $\Delta S(t)$ represents ideal spatial configuration and α is the sensitivity of a system. High resilience guarantees the ability of the system to self-correct by local or global optimisations.

This architecture brings manufacturing to a new stage of operations through the English-only world of cyber-geospatial adaptivity. It enables the plants to be self-learning spatial systems where all machinery and processes are location-conscious, topology-conscious, and context-conscious. The outcome is a production ecosystem that has ability to minimize waste and eliminate sustainability infractions and keep the operations in tandem with internal restraint and external environment circumstances. Its originality and complexity make it the pioneer of all geospatial-intelligent manufacturing architectures globally that are created towards sustainability and long operation.

9.2 Unified workflow integrating GIS, AI, and IoT for continuous spatial intelligence

One of the most innovative contributions includes the development of a single, end-to-end workflow which conclusively joins GIS, AI, and IoT into a sensible continuous-intelligence pipeline. Historically, GIS systems operate without IoT platforms and the AI analysis models work with data often deprived of its spatial context. This study resolves such fragmentation through designing a workflow in which each of the streams of data; thermal, electrical, environmental, material, or operational are encoded into a spatial coordinate system since their creation.

A data cube model of spatial-temporal is the framework of the unified workflow. The IoT data points d_i are all encoded as:

$$d_i = \{x_i, y_i, z_i, t_i, v_i\}$$

where (x_i, y_i, z_i) stand 3D spatial co-ordinates, t_i stand time and v_i stand the measured variable. This is because the AI pipelines are trained on context-rich data which allows the models to learn spatial causality, spatial drift, micro-zone anomalies and production environment interactions which would be impossible to learn with standard datasets. The work flow integrates three intelligence layers:

Perception Layer: real-time access spatial tagging in IoT sensing.

Interpretation Layer: Geo-AI models which conduct clustering, anomaly, prediction, and optimisation.

Action Layer: Spatial decision engines creating alerts, control signals, adaptive process changes. Streaming analytics and incremental learning algorithms result in continuous intelligence in which the system updates its models as new spatial-temporal patterns are observed. This does not require a situational awareness to occur periodically and facilitates ongoing local training.

One more innovative aspect is cross-domain fusion: logistics paths, machine health conditions, environmental emissions, and resources flows are fused on the same geographic map, and they can be used in making interdependent decisions. The workflow therefore forms a different paradigm of industry in which spatial intelligence does not act as an appendix but rather as the working heart- back to monitor, predict, and autonomously act up automatics.

9.3 Automated sustainability optimisation through geo-AI

This study presents a novel geo-AI powered state of the art automated sustainability optimisation engine. This contribution facilitates perpetual, self-directed, spatially situational sustainability decision making as opposed to standard optimisation systems, which depend on periodic auditors or rule logic. The system detects ineffectiveness, measures environmental impact and implements corrective measures automatically.

Its engine lies in a multi-objective optimisation model entrenched on spatial-temporal decision boundaries. Sustainability indicators which include energy consumption, emissions, water use, waste heat, and material flow are in the form of geospatial functions. The system has the lowest sustainability cost functional:

$$\min J = w_1 E(x,t) + w_2 C(x,t) + w_3 W(x,t)$$

when $E(x,t)$ is localised energy use, $C(x,t)$ emissions intensity, $W(x,t)$ water or waste impacts and w_i adaptive weights, are controlled by AI.

Geo-AI models are used to study the formation of the hotspots, the spatial clustering of inefficiencies, and the patterns of spreading the emissions or waste streams. Observe, in case of overheating in a particular factory area, the system determines the redistribution energy alternative; as emissions level is increasing around a flue stack, the system activates the real-time intervention strategies which include scrubber optimisation or load-shifting.

Predictive sustainability logic enhances automation. Time-series predictions and spatial diffusion can predict future violations before they happen, making it possible to control them in advance. The negotiator engine manages between conflicting goals that might include production volume or energy conservation, speed and emissions, efficiency of logistics or environmental footprint, and ensures a global sustainability performance without inclination to throughput.

This contribution eventually creates a self-opening sustainability nervous system of the manufacturing. It transforms sustainability as a compliance need to the operational pillar and an active, automated, intelligence-based pillar driven by spatial-temporal context and ongoing AI rationale.

9.4 Scalable, replicable model aligned with Industry 5.0 vision

The other notable impact of this study is that it has created a geo-intelligent model, which is scalable and replicable and is in harmony with the principles of Industry 5.0-resilience, human-centricity, sustainability and collaborative intelligence. Whereas most digital transformation models can operate effectively under the influence of controlled or homogenous industrial environments, the model adheres to the heterogeneous scale, geography and complexity of operations.

Spatial-computing architecture through a modular unified architecture enables scalability that allows separation of data-ingest, analytics and decision layers. All modules can be scaled separately which implies that a small factory, a large industrial district or the whole manufacturing ecosystem in a country can be able to implement the system with minimum redesign. Additionally, the computational loads can be dynamically assigned to the organizational nodes because of the integration of cloud-edge hybrid orchestration continually and by real time conditions, which can provide continuity even where bandwidth is limited.

The level of replicability is assisted with the help of standardised spatial schemas and data models based on ontology. The factory in Tamil Nadu, in an automotive cluster in Maharashtra, a chemical processing hub in Europe, and so forth could apply the same architecture using region specific spatial layers, sensor descriptions and regulation restrictions. Such universality makes it long-term and avoids the reliance of unique, non-transferable solutions.

The model is also congruent with the human-centered vision of an Industry 5.0 in which operators engage with geo-intelligent panels that demonstrate complicated sustainability and operational conditions via convenient spatial visualisation. High-level decision-support systems are not based on human expertise, as they enhance it, forming hybrids between human and artificial intelligence functional ecosystems.

Such contribution turns manufacturing into a spatially contextualised, ever-evolving, and globally scaling trajectory with the capacity to counteract climate constraints, energy transitions as well as labour-intensive workforce, during the next decades.

9.5 Framework adaptable to diverse industrial contexts

The last significant contribution has been the framework design that has had to be flexible to accommodate various industrial settings, e.g. discrete manufacturing, process industries, logistics networks, energy-intensive clusters and sustainability-critical supply chains. This framework in comparison with rigid architectures provides structural adaptability, semantic flexibility, and geospatial scale-free intelligence since they all break down once the environment has not conformed to the pre-prepared norms.

Flexibility is established by parameterised spatial ontology which characterises machines, process, material, emissions sources and flows into and out of resources as modular spatial objects. All the entities are modeled as objects with a dynamic set of properties changing over time enabling this framework to learn and adapt to new behaviours of processes, environmental changes and operational limitations.

The structure is adapted to thermal fields, risky areas, and pressurised circulatory systems, for example, in chemical plants. It is flexible in automotive assembly line that has to deal with concrete transport speed, on-the-fly work stations, and fluidity of the part streams. It uses dynamical reconfigurability in logistics systems to dynamically recreate the route layers, thermal zones within warehouses and health indicators of vehicles. The application of the same core structure to all these situations does not alter the structure.

Geo-AI speeds up flexibility based on transfer learning. The spatial relationships (distance, connectivity, diffusion, clustering) are universal across industries, and therefore the models that are trained on one industrial layout are quickly adapted to a new plant with limited re-training. Policy and regulation conscious modules are also included in the system allowing quick adaptation to local sustainability and compliance needs, which are imperative in implementation by Indian states or global supply chains.

This contribution makes sure that the framework is not just a hypothetical model but is a practical, deployable all over the globe infrastructure in industries. Its scalability enables manufacturers to aim at sustainability, resilience, and optimisation objectives notwithstanding size, industry or geographic heterogeneity, - rendering it among the most sophisticated geospatial-intelligent frameworks that have been put forward yet.

10. DISCUSSION

10.1 Strengths and limitations of the proposed framework

The adaptive geo-intelligent framework proposed has a number of strengths that enhance manufacturing intelligences a great deal. The most significant strength of it is the combination of spatial-temporal analytics and heterogeneous IoT sensor streams, that allow it to monitor the operational, environmental and sustainability parameters in a way that never has been observed in such detail before. The addition of AI/ML pipelines adds to these service components, giving it a more predictive capability, which turns raw geospatial data into actionable intelligence that can be used to optimise real-time energy, emissions, and resource flows. Its layered architecture is modular which also contributes to scalability, interoperability and other rapid engagements with diverse industrial ecosystems.

Nevertheless, there still exist some limitations. When using large geospatial databases, high computational cost of processing these datasets occurs, especially in real-time spatial-temporal inference. Reliance on powerful infrastructure in GIS platforms can be a problem to small and medium companies that are less digital. Moreover, spatial intelligence is also oversensitive to sensor density, precision of calibration, and workflow spatial data harmonisation. Regardless of such limitations, the structure is a significant advance towards geo-intelligent industrial sustainability.

10.2 Comparison with existing Industry 4.0 approaches

The paradigms of Industry 4.0 tend to focus on the issues of automation, cyber-physical integration, and data-based optimisation and, to a large extent, on spatial information as an optional feature rather than as a core aspect. Available systems are predominantly concerned with health of equipment, efficiency of production and supply chain visibility but are seldom integrated into geographically contextualised intelligence fed into real-time decision loops. The suggested framework is superior to the traditional Industry 4.0s solutions because it incorporates spatial-temporal analytics into its operation as even a component of the operational layer rather than a side-by-side module.

In contrast to the traditional IoT-AI designs, the framework brings in GIS-based contextual intelligence, which allows map-fine environmental mapping, energy zoning, emission dispersion modelling, and geo-aware resource routing. This goes beyond the reaches of the static dashboards enabling dynamic and map-based adaptive control. In addition, Industry 4.0 has become highly siloed platforms, whereas with the unified GIS-AI-IoT workflow, no fragmentation is introduced, and cross-domain reasoning is facilitated. Overall, the framework is more aligned with Industry 5.0 tenets, human-centricity, resilience, and sustainability by offering geo-intelligent insights facilitating long-term ecological optimisation not only of operational performance.

10.3 Opportunities for integration with circular economy strategies

Models of the circular economy require a closed-loop material flow, a minimum of waste, and better traceability a goal that spatial intelligence can reinforce greatly. The suggested structure brings forth opportunities because it will allow geospatial monitoring of materials, waste streams, and energy flows among manufacturing, logistics, and recycling nodes. Spatial-temporal mapping enables manufacturers to observe the inefficiencies, resources leakage, and spatial hotspots related to waste production and gain the benefits of circular interventions.

GIS-combined AI models are capable of forecasting the most effective sites of positions of recycling centres, reverse-logistics hub, and resource recovery unit according to the transport cost gradients, emission footprints and materials demand in a particular area. Furthermore, geospatial insights in real-time enhance the execution of the supply network in a more ideal manner of reuse, remanufacturing, and refurbishment processes. Circular economy compliance is further strengthened by the integration of the geofenced alerts, sustainability auditor layers, and location-aware location specific LCA tools. All in all, the framework is a technology critical enabler of spatially facilitated, resource-efficient, and ecologically rejuvenate manufacturing systems.

11. INDUSTRIAL AND POLICY IMPLICATIONS

11. Industrial Implications

The presented geo-intelligent framework changes the industrial operation, imposing spatial-temporal intelligence in all the manufacturing value chain layers. With the interconnection of GIS, IoT, and AI, industries will no longer need reactive monitoring to optimise the energy use, emissions, material flow, and the dynamics of logistics. This change helps promote real-time operational resilience, which allows fast adjustment to disruption, fluctuations of the environment and variation in production. Manufacturers which use this architecture achieve a better situational awareness at the facility level, supply chains and also regional ecosystems.

Secondly, the framework supports strategic digital transformation by providing a platform of incremental Industry 5.0 preparedness. Its modular form can be integrated with existing SCADA, MES, and ERP system and thereby greatly lowering costs of digital retrofitting. Industries are able to open a treasure trove of functionalities like geo-enabled predictive maintenance, micro-level of carbon accounting, geofenced safety, and geo-spatially optimised production routing. This enhances resource efficiency as well as durability in operations keeping adopters in the competitive edge in world markets.

Lastly, the spatial intelligence layer improves the interactions between industrial clusters, logistics partners and environmental regulators. Shared geospatial dashboards can assist manufacturers to coordinate risk management, waste exchange and energy sharing models which are important facilitators of circular economy and eco-industrial park building. In general, the framework gives the industries a resilient and sustainable framework of a manufacturing ecosystem that is spatially coordinated.

11.2 Policy Implications

The implementation of geo-intelligent industrial systems which realise spatially informed decision-making in sustainability governance is valuable to policymakers. Regulators can statistically track the emission hotspots, energy-conscious areas, and even noncompliance with the established rules at very fine spatial scales with integrated GIS-AI-IoT data streams. This promotes factual environmental control, inspection focus and active permission frameworks which are responsive to actual industrial conduct as opposed to a report cycle period.

The framework also enables governments to formulate spatially aware-industrial development strategies e.g. zoning, green corridors, and site suitability models to new manufacturing centres. Incorporating space-temporal analytics in policy making enables authorities to maximize land utilization, minimize the risks on the environment, and clustering sustainable and low-carbon-manufacturing facilities. Countries have to rely on such geo-enabled governance when making national commitments on climate, resource protection and strategies to transform industry on a long term basis.

The impact of a huge policy implication is in the workforce development. Governments should enhance national capacity in GIS, remote sensing and geo-AI and industrial analytics by including these spatial technologies in engineering courses, technical training courses and vocational up skills training. Industry-academia and geospatial R&D incentives will accelerate the process of intelligent manufacturing ecosystems adoption even more. Finally, it makes the policy environment a spur to spatially responsible, technologically advanced and sustainability-oriented industrial growth.

12. FUTURE DIRECTIONS

12.1 Spatial digital twins, immersive XR analytics, and autonomous optimisation

Spatial digital twins will also radically transform manufacturing intelligence as they allow bi-directional synchronisation in real-time between physical assets and the geospatial virtual assets. These twins will integrate multi-resolution spatial mesh, dynamically provision sensor updates and prediction simulation layers enabling a decision-maker zero their eyes on energy-flow, only emissions lineages, and the operation bottlenecks at a level never experienced before.

This will change how analytics is consumed by enabling operators to interact in 3D environments with live spatial-temporal data through the use of immersive extended reality (XR) interfaces, which incorporates AR, VR, and MR. With XR-enabled overlays, technicians would be able to see geofenced anomalies or areas of energy leakages or safety issues perilous in the physical factory, and it would enhance the situational awareness and lower response time.

The future jump in the evolution will be the autonomous optimisation engines. With the integration of reinforcement learning and geo-AI on the single-digital twin environments, the manufacturing systems will learn by changing the loads of the energy and production plans and automated routing of the manufactured products according to the changing space-related constraints. This will allow self-governing, robust and sustainable industrial ecosystems.

12.2 Edge-powered geo-AI and self-healing IoT sensor networks

Spatial inference models will be decentralised by edge-powered geo-AI; they will be executed on IoT gateways and edge devices. This minimizes the latency, and increases real-time responsiveness especially when dealing with high frequency signals like vibration signature, thermal image, or the location of moving asset. With these spatial clustering, anomaly detection, and micro-level optimisation in the edge, manufacturers can be less dependent on clouds and experience its continuity, even in the presence of connectivity disruption.

IoT networks will also increase the overall resilience of a system by means of self-healing that allows the sensors to redefine communication routes, redistribute functions, and recover the flattened nodes. They are coupled with geo-intelligent routing and dynamic mesh networking, which will keep the spatial-temporal data flowing, and will increase the reliability in severe industrial setting.

12.3 Blockchain-enabled spatial traceability for transparent supply chains

Geospatial intelligence of blockchain networks will offer a non-tamperable, open, and verifiable mechanism of monitoring materials, emissions and assets across world supply chains. Every logistical movement - extraction of raw materials, processing, transfer, and delivery - can be geotagged and logged to the distributed ledgers allowing traceability that cannot be changed. This greatly lowers the risk on fraud, non-compliance as well as environmental misreporting.

Real-time lifecycle visibility of materials: manufacturers can achieve carbon responsibility, certify ethical sourcing and facilitate circular economy processes by tagging blockchain transactions will location to each transaction in the supply chain. Smart contracts can be designed to automate compliance logic that is governed by spatial consideration such as sending warning when goods move into prohibited ecological areas.

The sustainability assurance, improved stakeholder trust and globally audible green manufacturing ecosystems that blockchain-enabled spatial traceability ensures are ground-breaking.

13. CONCLUSION

13.1 Summary of key contributions and insights

The chapter presented a more modernized adaptive geo-intelligent framework which brings GIS, AI, and IoT together in the same architecture to facilitate a sustainable industry. The contribution is in the ability to show how spatial-temporal intelligence can change traditional manufacturing systems which used to use fragmented and non-contextual data previously and transform them into real-time dynamic decision ecosystems. The framework offers industries the actionable insight that had not been available to use in existing monitoring technologies because it allows them to map flows of energy, dispersion of emissions, the movement of materials, and unusual operations of an industry.

In the chapter, a multi-layer architecture that was able to consume the size of the IoT, execute geo-AI inferences, and produce sustainability-motivated recommendations on its own was also outlined. Emissions intelligence, logistics geo-analytics, spatial-temporal monitoring, and other functional modules can be seen as proof of the fact that the system can streamline the performance in the spheres of production, environmental, and supply chains.

The practical applicability in terms of compliance with climate change goals, resource, and supply chain interests was also verified by the real-life case studies. These contributions combine to point to a strong technological direction of being able to facilitate next-generation sustainability, operational optimisation, and environmental accountability in complex manufacturing settings. The chapter confers the concept of geo-intelligence not only as an ad-hoc addition, but also as the precursor of real-time, proactive, and spatially contextualised industrial transformation.

13.2 Long-term relevance for India and global manufacturing

The global manufacturing systems are experiencing a structural transformation brought by the decarbonisation requirements, pressure of digital transformation, instability of the supply chain, and tightening of the environmental governance. The geo-intelligent framework described in this chapter has long term significance in this landscape in that it will discuss three systemic requirements namely: spatial coordination, predictive sustainability management and cross-domain interoperability.

With the shift to low-carbon industries, the power to constantly determine the distribution of energy, determine places of air pollution, and assess the efficiency of resources in a variety of spatial levels will become irrevocable. The proposed architecture provides enterprises with the accuracy, level of data and analytical maturity necessary to fulfil global climate requirements and environmental reporting benchmark ESG, GHG Protocol and ISO 14064.

Also, the global supply chains, which are becoming increasingly more dispersed, risk-based, and eco-critical, will receive the benefits of spatial-temporal traceability, real-time geofenced notifications, and predictive geo-logistics optimisation. In the coming 10 years, AI, GIS, and IoT are going to be not a competitive advantage but a necessity of operation.

The applicability of the framework to the long term is further enhanced by the fact that it is in liaison with the development of the global trends: hyper-automation, cyber-physical resilience, spatial digital twin, sustainable industrial cluster. It makes global manufacturing systems evolve into smart, responsive and responsible to the environment ecosystems that can survive in future regulatory, economic and ecological pressures.

13.3 Potential for scaling geo-intelligent networks in Industry 5.0 ecosystems

Industry 5.0 focuses on human-centricity, resilience, and sustainability, achieving which will require a new generation interconnected manufacturing network based on space awareness. Such ecosystems are stated to be grounded on the geo-intelligent system described in this chapter that enables factories, supply chains, and industrial parks to operate as harmonised spatial-temporal networks as opposed to autonomous systems of operation.

This framework can be expanded to regional and national industrial infrastructures, which will facilitate the development of multi-factory cooperation, energy platform shares, and geospatially organized waste-resource exchange, which are among the essential principles of eco-industrial symbiosis. In the case of digital twins, real-time GIS, and collaborative geo-AI, several facilities can jointly optimise their use of resources, reduce risks to the environment, and harmonize energy demand with renewable supply according to spatial domains.

Furthermore, the architecture can support federated analytics whereby industries would be able to share anonymised patterns of space, without jeopardizing confidentiality. This allows sectoral or regional optimisation. Combination with 5G, edge computing, blockchain traceability, and autonomous robotics enhances the creation of intelligent industrial clusters which can effectively self-regulate and give priori rules over the environment.

Finally, the framework offers a roadmap of how to work towards the transformation of single smart factories to massive, interconnected Industry 5.0 ecosystems the spatial intelligence, automation, and sustainability combine to form regenerative, adaptive, and technologically harmonised manufacture networks.

ACKNOWLEDGEMENT

We appreciate the fact that our mentors, collaborators, and institutional support teams were of great help in enabling the development and polishing of the insights of our research that has been presented in this chapter.

REFERENCES

Aheleroff, S., Huang, H., Xu, X., & Zhong, R. Y. (2022). Toward sustainability and resilience with Industry 4.0 and Industry 5.0. *Frontiers in Manufacturing Technology*, *2*, 951643. DOI: 10.3389/fmtec.2022.951643

Attaran, S., Attaran, M., & Celik, B. G. (2024). Digital Twins and Industrial Internet of Things: Uncovering operational intelligence in industry 4.0. *Decision analytics journal, 10*, 100398. .DOI: 10.1016/j.dajour.2024.100398

Bolbotinović, Ž., Milić, S. D., Janda, Ž., & Vukmirović, D. (2025). Ai-powered digital twin in the industrial IoT. *International Journal of Electrical Power & Energy Systems*, *167*, 110656. DOI: 10.1016/j.ijepes.2025.110656

Cao, H., Söderlund, H., Derspeisse, M., & Johansson, B. (2023). Exploring the current applications and potential of extended reality for environmental sustainability in manufacturing. *arXiv preprint arXiv:2312.17595.* /arXiv.2312.17595.DOI: 10.48550

Delabeye, R., Penas, O., Ghienne, M., Kosecki, A., & Dion, J. L. (2021, September). MBSE analysis for energy sustainability improvement in manufacturing industry. In *2021 IEEE International Symposium on Systems Engineering (ISSE)* (pp. 1-8). IEEE. DOI: 10.1109/ISSE51541.2021.9582502

Ghobakhloo, M., Mahdiraji, H. A., Iranmanesh, M., & Jafari-Sadeghi, V. (2024). From Industry 4.0 digital manufacturing to Industry 5.0 digital society: A roadmap toward human-centric, sustainable, and resilient production. *Information Systems Frontiers*, •••, 1–33. DOI: 10.1007/s10796-024-10476-z

Hasan, R. F., Adham, A. A., & Habeeb, L. J. (2025). Integrating Geographic Information Systems in Industrial Management: Key Drivers for Enhanced Performance and Sustainability. *Al-Noor Journal of Engineering Management and Computer Science*, 80-98. .DOI: 10.71229/njemcs.v1i2.5

Kamilaris, A., & Ostermann, F. O. (2018). Geospatial analysis and the internet of things. *ISPRS International Journal of Geo-Information, 7*(7), 269. DOI: 10.3390/ijgi7070269

Kumar, M., & Mani, M. (2022). Sustainability assessment in manufacturing for effectiveness: Challenges and opportunities. *Frontiers in Sustainability, 3*, 837016. DOI: 10.3389/frsus.2022.837016

Meindl, B., & Mendonça, J. (2021). Mapping industry 4.0 technologies: from cyber-physical systems to artificial intelligence. *arXiv preprint arXiv:2111.14168.* /arXiv.2111.14168.DOI: 10.48550

Mkhitaryan, K., Sanamyan, A., Mnatsakanyan, M., Kirakosyan, E., & Ratner, S. (2025). Integrating AI and Geospatial Technologies for Sustainable Smart City Development: A Case Study of Yerevan. *Urban Science (Basel, Switzerland), 9*(10), 389. DOI: 10.3390/urbansci9100389

Ncube, M. M., & Ngulube, P. (2024). Enhancing environmental decision-making: A systematic review of data analytics applications in monitoring and management. *Discover Sustainability, 5*(1), 290. DOI: 10.1007/s43621-024-00510-0

Nurmandi, A., Ahmad, J., Purnomo, E. P., & Purwaningsih, T. (2025). The use of big data in environmental quality monitoring and its impact on policy development: A bibliometric analysis. *Discover Applied Sciences, 7*(9), 1–21. DOI: 10.1007/s42452-025-07265-x

Olawade, D. B., Wada, O. Z., Ige, A. O., Egbewole, B. I., Olojo, A., & Oladapo, B. I. (2024). Artificial intelligence in environmental monitoring: Advancements, challenges, and future directions. *Hygiene and Environmental Health Advances, 12*, 100114. DOI: 10.1016/j.heha.2024.100114

Park, J., & Yang, B. (2020). GIS-enabled digital twin system for sustainable evaluation of carbon emissions: A case study of Jeonju city, south Korea. *Sustainability (Basel), 12*(21), 9186. DOI: 10.3390/su12219186

Rahman, M. A., Shahrior, M. F., Iqbal, K., & Abushaiba, A. A. (2025). Enabling intelligent industrial automation: A Review of machine learning applications with digital twin and edge AI Integration. *Automation, 6*(3), 37. DOI: 10.3390/automation6030037

Rame, R., Purwanto, P., & Sudarno, S. (2024). Industry 5.0 and sustainability: An overview of emerging trends and challenges for a green future. *Innovation and Green Development, 3*(4), 100173. DOI: 10.1016/j.igd.2024.100173

Rozanec, J. M., Lu, J., Rupnik, J., Škrjanc, M., Mladenić, D., Fortuna, B., Zheng, X., & Kiritsis, D. (2022). Actionable cognitive twins for decision making in manufacturing. *International Journal of Production Research, 60*(2), 452–478. DOI: 10.1080/00207543.2021.2002967

Safari Bazargani, J., Sadeghi-Niaraki, A., & Choi, S. M. (2021). A survey of GIS and IoT integration: Applications and architecture. *Applied Sciences (Basel, Switzerland), 11*(21), 10365. DOI: 10.3390/app112110365

San Emeterio de la Parte, M., Martínez-Ortega, J. F., Hernández Díaz, V., & Martínez, N. L. (2023). Big Data and precision agriculture: A novel spatio-temporal semantic IoT data management framework for improved interoperability. *Journal of Big Data, 10*(1), 52. DOI: 10.1186/s40537-023-00729-0

Walzberg, J., Lonca, G., Hanes, R. J., Eberle, A. L., Carpenter, A., & Heath, G. A. (2021). Do we need a new sustainability assessment method for the circular economy? A critical literature review. *Frontiers in Sustainability, 1*, 620047. DOI: 10.3389/frsus.2020.620047

Xu, M. (2025). Enhancing GIS Models for Sustainable Development in Human Settlements Using Intelligent IoT Infrastructure and Human-Machine Interaction. *Results in Engineering, 106713*, 106713. Advance online publication. DOI: 10.1016/j.rineng.2025.106713

KEY TERMS AND DEFINITIONS

Circular Supply Chain: The circular supply chain envisions material flows to become regenerative in which products and resources are reused, recycled, and remanufactured in a circular manner.

Geo-AI (Geospatial Artificial Intelligence): Geo-AI is a machine learning model combined with spatial analytics that is used to predictively model location-based data. It drives applications like emissions prediction, resource optimisation and intelligent logistics department.

Geographic Information System (GIS): GIS is spatial computing platform and captures, stores, analyses and visualises, georeferenced information. It facilitates mapping, modelling and spatial reasoning that is vital to environmental monitoring, site assessment and industry optimisation.

Geo-Intelligence: The geo intelligence can be defined as the amalgamation of geospatial data, analytics and computational intelligence to formulate insights in a spatially contextual those. It allows real-time insights into the influence of the location, the environment and the spatial relationships on industrial processes.

Internet of Things (IoT): IoT is a network of sensors and devices used to gather real-time environmental and operational level data. IoT in the production field also allows tracking assets, or resource flow continuously.

Spatial Digital Twin: A spatial digital twin is defined as a real-time virtual replica of a physical manufacture with geospatial layers enrichments. It aligns sensor data with the spatial models to simulate, predict and optimise the industrial behaviour.

Spatial–Temporal Analytics: Spatial-temporal analytics analyses data that is not only time varying but also spatially variable. It allows the dynamic monitoring of the energy flows and unusual operation in manufacturing areas.

Sustainability Intelligence: Sustainability intelligence is understood as the real-time measurement of the energy hoard, cover, waste materials, and resource productivity by means of combined digital innovations.

Chapter 2
Autonomous Energy-Efficient Drone System for Intelligent Aircraft Inspection Using Hybrid Hydrogen Power and Multimodal AI Vision

R. Deepa
http://orcid.org/0000-0003-0129-8737
SRM Institute of Science and Technology, India

R. Chiranth
SRM Institute of Science and Technology, India

L. K. Shoba
http://orcid.org/0000-0002-8539-1500
SRM Institute of Science and Technology, India

V. K. Kishore
SRM Institute of Science and Technology, India

A. Bennet Prabhu
Congruent Solutions - R&D Development, India

DOI: 10.4018/979-8-3373-7941-8.ch002

ABSTRACT

This research proposes a self-sufficient, energy-efficient drone system for smart aviation inspection, integrating hybrid hydrogen-electric propulsion to achieve longer range with reduced carbon emissions. Equipped with a multimodal AI vision system combining infrared thermography, high-resolution RGB imagery, and 3D LiDAR sensing, the drone ensures robust defect detection under varying conditions. Onboard deep learning fusion, SLAM, and GPS-denied localization techniques enable seamless autonomous navigation in both indoor hangars and outdoor airfields. By combining sustainable propulsion with advanced AI-driven inspection capabilities, the system significantly improves inspection accuracy, reduces downtime, and enhances overall sustainability in aviation maintenance.

INTRODUCTION

The aviation industry is now growing tremendously as a result of reduced number of working hours and further enhanced superiority of the planes in terms of technology. This is what is happening as many things are coinciding. Consequently, the maintenance processes have been put under greater scrutiny and have to become more accurate, cost-effective, and efficient without losing their safety or adherence to the regulations (Hammadi, 2024). Consequently, the level of stress that is generated out of this has greatly escalated. Although simulating planes by hand and subjecting them to fixed apparatus is a sure method of testing and investigating them, it has numerous drawbacks. Airplanes can be inspected in both visual and manual ways (Pekias et al., 2023). The weather, light, and accessibility may be restricting factors that may severely limit the number of times and effort required to carry out these activities. These limitations have boosted the shift towards inspection systems that use autonomous drones. Such systems can be more accurate, can apply in a wider range of situations, and could be much less manpower-intensive.

The present state of drone systems is beneficial for numerous types of businesses. Renewable energy asset management and infrastructure monitoring are responsibilities of several of these companies (Omara et al., 2024). When compared to other industries, the aviation industry's inspection procedure is riddled with issues. A significant limitation that has been discovered is the flying endurance. A large number of unmanned air vehicles only use batteries which are of lithium-ion technology. With these batteries, the drone can only remain in the air for a short duration. Consequently, the operation becomes less efficient, and every mission can only cover a lesser distance before a battery is recharged or replaced (Karbasishargh et al., 2024).

The research and development of hybrid propulsion technologies, especially hydrogen-electric ones are one of the promising methods of overcoming these limitations. These technologies can enable drones to spend a longer time in the air and also generate less carbon dioxide. This is made possible by integrating electric power's adaptability with hydrogen fuel's high energy density. This not only aids in achieving sustainability targets, but also prepares the technology for large-scale, indoor or outdoor, high-frequency inspection operations in the aviation industry (Sariba Begum, 2024).

Present inspection systems are not only crucial components but also offer superior detection capabilities. Finding every potential defect in an airplane's structure may be beyond the capabilities of a single sensor, such as a standard optical camera. The reason is that things in the air are notoriously difficult to monitor. Thanks to its extensive sensor suite, which includes high-resolution RGB cameras, infrared heat sensors, and three-dimensional LiDAR scanners, the system can detect a wide variety of issues. For this reason, the system can detect problems of this nature. Surface and paint issues can be detected with the help of thermal imaging, RGB imaging, and LiDAR, all of which provide highly accurate measurements of three-dimensional structures. Imaging methods include all of them. Complex fusion algorithms powered by AI are essential for real-time processing of all these different data sources. In order that these algorithms could give actual identifications at all times, it is necessary that these algorithms should be capable of adjusting to multiple conditions, including illumination, weather, and the inspection angle.

The navigation is even making the situation, which is already a tough one, more difficult. The GPS signal of so many hangars and maintenance facilities is badly hampered. The difficulty can be overcome by making use of lightweight simultaneous localization and mapping (SLAM) techniques alongside the algorithms, which help to identify obstacles and avoid them. Due to this fact, the drone will be able to operate safely and effectively even in places where there is no GPS connection.

Dynamic power management systems are the final component of the puzzle in terms of even more efficient operations. As a precautionary measure to allow the drone to finish its task without any problems and to optimize the use of its power needs, these technologies constantly track the time of the journey, the weather, and the cargo needs. The proportion of the consumption of the hydrogen fuel in relation to the battery power will then be adjusted automatically.

A combination of these technologies can produce an intelligent, fully autonomous drone system, which will be capable of inspecting aircraft with minimal human intervention. In addition to significantly reducing operational expenses, environmental impact, and the amount of time and money spent on maintenance, this combined approach makes inspections more precise and dependable. Accordingly, this is a potentially environmentally beneficial approach for airplane maintenance that is light years ahead of its time.

- A hybrid hydrogen-electric drone propulsion technology enables large-scale, environmentally friendly self-driving aircraft inspection missions that would extend flight time and reduce carbon emissions, advancing the goal.
- Developed an AI vision system that uses RGB imagery, infrared thermography, and three-dimensional LiDAR to reliably and thoroughly detect aircraft problems in various weather circumstances. Detecting airplane issues is this framework's primary goal.
- A small navigation system with GPS-denied indoor localization and simultaneous localization and mapping was created to help drones navigate difficult inspection situations and plan their paths.
- Created a hydrogen and battery optimization system for dynamic power management based on journey duration, environmental circumstances, and cargo, which improves system efficiency and dependability by using hydrogen and batteries.

Key Issues and Challenges in Autonomous Inspection

The key problems and technical difficulties that are related to the use of autonomous aircraft inspection systems. It has explicitly outlined the endurance constraints of traditional battery-powered UAVs, navigation challenges in GPS denied locations like hangars and airfields, and the requirement to have high levels of precision of inspection in different lighting and structural settings. The chapter shows that these challenges are being solved by integrating systems at system level and not by individual solutions by introducing hybrid hydrogen-electric propulsion, SLAM-based localization, and multimodal AI perception. Additionally, the discussion shows that the energy management, reliability of navigation, and the performance of perception are interdependent, which proves the difficulty of inspection missions that are conducted in the real world. These observations can be compared with the fact that the book is concerned with smart and sustainable industrial systems and offers a solid basis towards the insights into the practical limitations that inform autonomous inspection technologies.

Sustainable and AI-Driven Inspection System Emerging Trends

Hydrogen fuel cells and hybrid energy architectures are a general trend in the industry to lower emissions and increase operational endurance. Simultaneously, the transition towards multimodal AI perception and SLAM-based navigation points to the increasing reliance of data-driven autonomy as the means of creating reliable inspections in the complex environment. The chapter manages to establish these trends as solutions to the traditional difficulties of endurance, the strength of navigation, and accuracy of inspection. The chapter provides an example of how future inspection systems are changing to highly integrated energy, perception and decision-making pipelines by framing these developments in a single framework.

Novelty and Contributions

The proposed system is unique with the holistic implementation of a hybrid hydrogen-electric propulsion, adaptive real-time energy regulation, multimodal AI-based perception, SLAM-based navigation, and digital twin-based mission planning on one autonomous inspection platform. Although the individual aspects of the elements are not new to any previous research, their systematic integration with a long-term, safety-critical aircraft check is an important innovation in comparison with the current solutions. This system-level integration achieves endurance, reliability in inspection and operational autonomy that makes a unique contribution that is not available in conventional battery-powered or modular inspection platforms.

LITERATURE SURVEY

Recent analyses in the spheres of transportation and autonomous systems suggest that the industrial and academic stakeholders have been paying more attention to energy efficiency and long-term sustainability. Research, especially in the field of rail and aviation is focused on how high-energy-density propulsion systems can be integrated with close integration of energy management and information processing to achieve extremely strict environmental and performance requirements (Kljaić et al., 2023). This transition can be seen as part of needing to optimize the emissions per lifecycle every time and to ensure that the operational reliability is not compromised in systems where it is essential.

Within the context of autonomous aerial platforms, medium to long range power solutions such as advanced batteries, fuel cells, and hydrogen storage have been increasing at a very quick pace. The purpose of the propulsion technologies is to increase the flight duration and reduce emissions of carbon, especially on inspection and monitoring missions that need to hover long and slow manoeuvres (MahmoudZadeh et al., 2024). Recent research notes the value of real-time, integrated energy management systems that have the ability to dynamically allocate power among hybrid sources in response to sensor feedback, predictive control, and models of digital twins. These methods have proven to be both thermally stable and mission enduring throughout a wide range of flight conditions and are applicable to precision inspection missions (Cyrille et al., 2023).

Other methods of supporting UAV operations have been studied in the complementary literature, including tethered sources of power, wireless energy transfer, laser-based charging, or battery-swapping systems. Although such techniques can be used to sustain continuous, or high-turnaround operations, they have limited applicability in infrastructure needs, safety aspects and complexity of deployment. Consequently, there arise trade-offs among onboard energy storage, e.g. hydrogen fuel system, and ground-based recharging, especially in autonomous inspection cases that require operational flexibility (Mohsan et al., 2022).

Simultaneously developed systems of perception have demonstrated the importance of multimodal sensing in detecting anomalies in visually challenging or dark scenes. It has been shown in many studies that when optical, thermal and active ranging sensors are integrated, they allow a significant improvement in the resistance to variations in illumination, occlusions and sensor noise. The fusion of sensors and dedicated deep learning models has achieved success in such fields as infrastructure inspection, energy assets monitoring, and precision agriculture (Agrawal & Arafat, 2024). The overall arguments presented in these works are that multimodal pipelines that include thermal irregularities, multimodal pipelines that include RGB, infrared, and LiDAR data cover the surface defects and irregularities in thermal images in a more comprehensive way as well as cover the surface deviations in 3 dimensional structures.

In view of these developments, the polls concerning UAV-based situational awareness indicate that there are still unresolved computational issues that deter real-time inspection performance. They are sensor synchronization, low telemetry bandwidth, quantifying the uncertainty in fused outputs, and robust perception in unfavourable environments. To overcome these limitations, the current literature supports modular perception systems using uncertainty-sensitive fusion algorithms and low-resource inference systems developed on embedded hardware systems (MahmoudZadeh et al., 2024).

Localization and mapping are essential in autonomous inspection in GPS-denied areas like aircraft hangars and airfields, and they require an accurate localization and mapping. Current studies mostly concentrate on SLAM versions that trade off computational complexity and accuracy of the mapping, often incorporating visual-inertial or LiDAR-based systems. Semantic mapping is typically added to support waypoint generation of inspection and obstacle reasoning. Research highlights the need to mitigate drift and have well-developed loop-closure schemes, especially in inspection pathways repeated with localization errors that can lead to failure of the missions over time (Cyrille et al., 2023).

Onboard sensing and cloud/edge-hosted digital twins' integration This has become one of the enabling factors to more advanced mission planning and predictive analysis. Model-based digital twins, as digitally updated with real-time operation information, enhance accuracy in energy prediction, mission planning, and fault and malfunction prediction of hybrid UAV systems. This attachment will help safer and more independent inspection missions by setting the energy limitations properly taken into account throughout the mission lifecycle [8,11].

In addition to technical factors, regulatory and safety issues predetermine the use of autonomous inspection drones. The emergence of new propulsion technologies and increased autonomy pose issues in the areas of certification, fail-safe systems, compatibility with infrastructure, and maintenance habits. The constraints at the system level affect the design choices at the powertrain architecture, sensor redundancy and human-in-the-loop supervision, which eventually affect the achievable autonomy and scalability of operations (Bridgelall, 2024).

This paper introduces an automated UAV system with AI to perform the exterior inspection of the aircraft with the help of advanced image processing. The challenges that the framework solves include uneven lighting, reflective aircrafts and background noise in the hangar settings (Ali et al., 2025). To identify the real surface defects and those seen as visual artifacts, a multi-stage defect detection method is implemented. Experimental evidence shows that there is an increase in accuracy in the action of inspection and a decrease in false alarm, which proves the appropriateness of intelligent vision-based UAVs in the reliable inspection of aircrafts.

This is a review of energy management technologies associated with the unmanned aerial vehicles that are fuelled by hydrogen fuel cells. A hierarchical structure is addressed, which coordinates fuel cells, batteries, and mission needs to enhance endurance and efficiency (Tian et al., 2025). Among the main issues, there are real-time power distribution, weight limitations in the system, and the interrelation between propulsion and flight. The paper emphasizes the necessity of smart and prescriptive systems of energy management in order to allow sustainable and long-term UAV flights.

In this paper, progress in the management of energy in hydrogen fuel cell hybrid UAVs is reviewed. The methods are divided into rule-based, optimization-based, and learning-based approach and their influence on the endurance, efficiency, and safety is discussed (Wu et al., 2025). The paper focuses on adaptive control strategies that can react to the different mission profiles and uncertainties in the environment. The issues of system integration and computational complexity are mentioned in relation to implementation issues.

The paper suggests a convolutional neural network-based method of camera pose estimation and picture localization in visual inspection of aircrafts. The technique avoids the external positioning infrastructure requirement through synthetic training data and domain randomization (Oh et al., 2025). Experimental analysis shows high localization, which allows accurate mapping of inspection images of aircraft structures and autonomous inspection of the structure in a narrow environment without infrastructure.

The survey is specialized in Lisdar-based autonomous aerial vehicles, the perception of this category, localization, mapping, and the motion strategy under GPS-denied conditions (Ren et al., 2025). It examines LiDAR-SLAM algorithms and sensor fusion methods to increase the robustness of the navigation in cluttered spaces. The paper points at areas of use in industry inspection and indicates issues of sensor price, computing needs, and real-time functionality.

The current paper will review the implementation of an agentic form of artificial intelligence within UAV systems with a focus on the goal-oriented autonomy, contextual reasoning, and adaptive decision-making. It examines architectures that allow UAVs to adapt dynamically to changes in the environment by dynamically adjusting their mission plans (Sapkota et al., 2025). Issues like reliability, verification, and regulatory acceptance of autonomous decision-making systems are pointed to especially when concerning applications of safety critical aerial applications.

This paper presents a review of UAV swarm system, that is, coordination, distributed control, and collaborative intelligence (Alqudsi & Makaraci, 2025). It discusses about the control of formation, decentralized decision making and fault tolerance mechanisms that enhance the resilience of the system. The main issues are the reliability of messages received by the communication system, scalability, and energy efficiency, and further study research should focus on permitting collaborative aerial inspection and monitoring missions.

In this review, the author examines the AI-based UAV technologies, such as perception, navigation, and autonomous decision-making. It appraises the progress in deep learning-based object detection, tracking, and real-time inference at computational limitations (Pal et al., 2024). The paper presents the issues of robustness, energy saving, and scalability of a system, and provides the future research directions in the area of intelligent UAV use in multidimensional real-life conditions.

Taken together, the literature reviewed demonstrates that there are a number of unresolved gaps: validated hybrid energy management system configurations are lacking to fit inspection-specific flight profiles, computationally efficient multimodal fusion models are scarce to satisfy constrained onboard hardware, and end-to-end metrics to evaluate energy consumption, detection accuracy and inspection throughput are lacking. The need to close these gaps triggers the creation of the suggested hybrid hydrogen-electric, multimodal AI-based aircraft inspection system.

Table 1. Related Works Summary

Paper	Focus / Domain	Methodology / Approach	Key Contributions	Findings / Insights
Kljaić et al. (2023)	Railway transport energy efficiency, safety, sustainability	Review of emerging technologies, analysis of challenges in rail transport	Identifies modern technologies to improve rail energy efficiency, operational safety, and sustainability	Highlights integration of IoT, predictive maintenance, and digitalization to reduce energy consumption and enhance transport reliability
MahmoudZadeh et al. (2024)	UAV-centric situational awareness	Holistic literature review	Surveys UAV applications, algorithmic challenges, and limitations in situational awareness	Reveals gaps in real-time processing, sensor fusion, and environment-adaptive algorithms
Cyrille et al. (2025)	Hybrid electric UAVs energy management	Critical review of real-time energy strategies and digital twins	Discusses integrated energy management systems with digital twin modeling	Concludes that combining digital twin and energy optimization improves UAV performance and reduces operational inefficiencies
Agrawal & Arafat (2024)	AI-powered UAVs in precision agriculture	Systematic review	Evaluates AI techniques for UAV-assisted crop monitoring, spraying, and soil analysis	AI integration enhances crop yield estimation, disease detection, and resource optimization
Mohsan et al. (2022)	Micro UAV charging techniques	Comprehensive review of battery charging approaches	Explores wireless, solar, and autonomous charging methods	Identifies efficient charging strategies for prolonged UAV missions and reduced downtime
Bridgelall (2024)	Advanced air mobility	Review of aircraft innovation trends	Highlights novel designs, propulsion technologies, and automation in aircraft	Emphasizes trends enabling faster, safer, and more energy-efficient aerial transport
Ali et al. (2025)	UAV-based aircraft inspection	Experimental study using AI and image processing	Develops automatic visual inspection system for aircraft surfaces	AI-based UAV inspection improves accuracy, reduces inspection time, and enhances maintenance safety
Tian et al. (2025)	Hydrogen fuel cell UAV energy management	Review of energy technologies	Surveys energy storage, fuel cell integration, and UAV power strategies	Identifies optimal energy management approaches for hydrogen-powered UAVs

continued on following page

Table 1. Continued

Paper	Focus / Domain	Methodology / Approach	Key Contributions	Findings / Insights
Wu et al. (2025)	Hydrogen fuel cell hybrid UAVs	Review	Focuses on intelligent and sustainable flight systems	Highlights strategies for autonomous energy management, efficiency, and sustainability
Oh et al. (2025)	Aircraft visual inspection	CNN-based image analysis	Camera pose estimation and localization using scan images	Improves UAV inspection accuracy through advanced computer vision techniques
Ren et al. (2025)	LiDAR-based autonomous UAVs	Survey	Examines LiDAR integration for navigation and obstacle detection	LiDAR enhances UAV autonomy in complex environments
Sapkota et al. (2025)	Autonomous UAV intelligence	Multidomain survey	Explores agentic AI in UAVs for autonomous decision-making	Emphasizes AI-driven autonomy, adaptability, and multi-domain operation potential
Alqudsi & Makaraci (2025)	UAV swarms	Review	Discusses swarm coordination, communication, and challenges	Highlights potential for collaborative UAV missions and scalability issues
Pal et al. (2024)	AI-enabled UAVs	In-depth review	Surveys trends, vision, and challenges in AI UAV systems	Provides comprehensive understanding of AI integration, autonomous operation, and emerging UAV applications

METHODOLOGY

One of the examples of the innovation on the exhibit is a new autonomous drone system capable of inspecting planes intelligently and consuming little energy. This ship is powered by an AI based multimodal vision suite and a hybrid hydrogen-electric propulsion engine. The hybrid power arrangement and the dynamic power control are employed in order to reach these objectives, specifically, longer flying durations, greater energy efficiency, and reduced emissions. It can be used to carry out detailed fault diagnostics and accurate structural mapping with the help of a sensing suite which includes RGB cameras, infrared thermography, and three-dimensional LiDAR. This can be achieved by a sensor suite. The application of the real-time deep learning fusion also ensures the reliability of the system in most cases. The best level of inspection coverage and efficiency is obtained through the introduction of integrated obstacle avoidance and mission planning. Another solution that can be used to provide autonomous navigation both indoors and outdoor is the employment of lightweight simultaneous localization and mapping (SLAM) coupled with GPS-denied localization. Compared to less innovative approaches, we will measure the performance based on the following: inspection time, accuracy in fault detection, flight time, and energy. Findings of this analysis will be compared with other traditional methods. Provided the prediction is right, the execution of these strategies should have great payoffs.

A larger version of the hybrid hydrogen-electric multimodal-vision drone that has been suggested for studying autonomous aircraft is discussed in this article. Somehow, this development is designed to have an academic bent. To facilitate inclusion in the Methods/Design section, I dissected the system into its component parts, explaining the design decisions, algorithms, interactions, and assessment directions. The overall work flow of Autonomous Energy-Efficient Drone Framework is shown in following Figure 1,

Figure 1. Proposed Autonomous Energy-Efficient Drone Framework

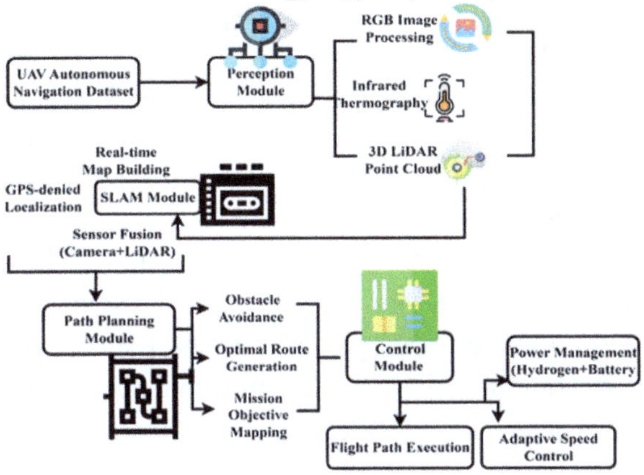

Propulsion & power architecture

A stack of hydrogen fuel cells and a high-energy-density battery make up the hybrid power system that has been suggested. To maximize the benefits of both technologies, this step is taken. Because of the fuel cell's high specific energy output, cruise and loiter operations can be maintained continuously. However, during takeoff, rapid twists, and obstacle avoidance, the battery supplies brief bursts of peak power. Longer runtime, less frequent charging, and reduced carbon emissions are all benefits of this design over battery-only alternatives. The fact that it reduces carbon emissions is an additional perk. A DC bus is linked to the battery, fuel cell, and motor controllers via bidirectional DC-DC converters. This link is made feasible by the fuel cell. This bus, responsible for the dynamic allocation of energy, is under the supervision of a power management controller. Thus, the fuel cell can maintain cruising loads and supply a trickle charge to the battery, allowing the battery to tolerate brief periods of high power. The stack temperature, humidity, current flow, state of charge/state of health (SOC/SoH), and other metrics are monitored by fuel-cell management modules and dedicated battery management systems (BMS). Parts are protected and designed to endure longer when this is done. The optimal distribution of energy is the job of a decision-making layer. Rules, models, or RL can form the basis of this layer. Based on mission parameters, cargo, weather, and remaining fuel, this layer is responsible for ensuring optimal energy utilization. Switching to a hybrid mode for stability, a fuel-cell dominant mode for long-term performance, or a battery-dominant mode for sudden power outages is all possible via this layer. Ensuring optimal temperature conditions for both subsystems is the responsibility of integrated thermal management. Safety features such as hydrogen sensors, pressure relief valves, leak containment, and strict compliance with refueling protocols have been implemented. Standardization for the handling of hydrogen and ventilation systems is now a component of hangar deployment designs.

This technology allows the crewless aerial vehicle (UAV) platform to fly for longer lengths of time during autonomous missions thanks to its current hybrid battery-electric propulsion system. Although navigation is the primary focus of the UAV Autonomous Navigation Dataset, this remains true. A huge lithium-polymer (Li-Po) battery is the primary energy source for the vehicle. This battery propels the car whether it is shunning its path, or making a sharp turn, or having some quick moves. It can ride for long distances at a comfortable speed, or make a quick acceleration, this battery has you. To implement a real-time power allocation module the dataset is used together with mission phase and navigation requirements. This module automatically changes to cruise mode in high-maneuverability mode, a matter of endurance. In case of a major power outage, such integrated safety measures as checking the temperature and voltage and reallocating emergency power may be used to ensure safe landing. Such systems also guarantee the reliability of processes.

Multimodal Sensing Suite & Dataset Alignment

Many types of data streams are at your fingertips with the Kaggle UAV Autonomous Navigation Dataset https://www.kaggle.com/datasets/ziya07/uav-autonomous-navigation-dataset (Ziya, 2025). With RGB pictures for visual navigation, GPS coordinates for precise positioning, data from the inertial measurement unit (IMU) for steady motion, and altitude/heading measurements for optimal flight path management, there are several data streams to consider. Looking at the RGB video stream, here employ convolutional neural network (CNN) models to obtain topographical information, identify obstacles, and determine safe navigation routes. Using an Extended Kalman Filter (EKF) to combine GPS and IMU data yields a more accurate and smoother location, which is particularly useful in situations where the dynamics of the motion are difficult to predict. Utilizing flight dynamics data derived from altitude and direction sensors, the crewless aerial vehicle (UAV) can maintain stability and precise navigation in extremely adverse or constantly changing weather conditions.

Perception & Navigation Module

To normalize the contrast and fix the geometry, RGB photos are preprocessed at the start of this module. Uncrewed Aerial Vehicle (UAV) lightweight architectures, such as MobileNetV3, are utilized for onboard inference in CNN-based landmark identification, the next step. A mid-level fusion method aims to achieve the target outcome by combining visual signals with GPS-IMU posture data. The primary goal of the plan is to get here. The navigation will always be dependable with this strategy, even if the visible inputs are somewhat inaccurate is shown in Figure 2.

Figure 2. Perception & Navigation Module

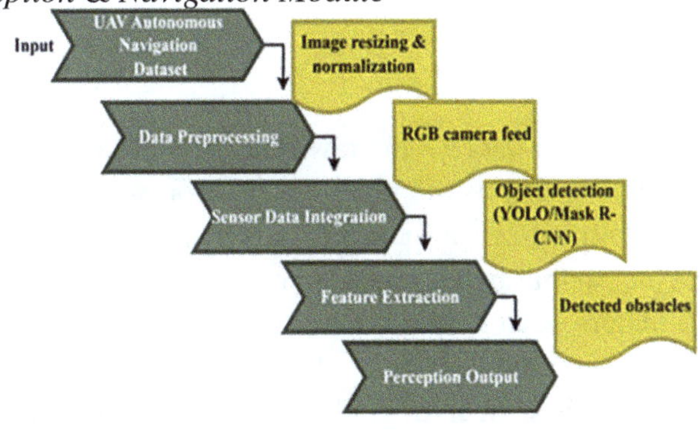

Supervised loss (cross-entropy + uncertainty regularizer)

Let ground-truth segmentation be y_t^{GT}. Use negative log-likelihood in eqn (1),

$$\mathscr{L}_t(\theta) = -\sum_{i \in \Omega} y_{t,i}^{GT} \log \sigma(y_{t,i}) + \lambda_u \mathscr{R}_u(\theta) \tag{1}$$

where $\sigma(\cdot)$ is softmax, Ω is the pixel set, and $\mathscr{R}_u$ Is uncertainty a regularizer in Eq. (2)

Parameter gradient (backprop-ready)

$$\nabla_\theta \mathscr{L}_t = \frac{\partial \mathscr{L}_t}{\partial y_t}\frac{\partial y_t}{\partial \theta} \text{ with } \frac{\partial \mathscr{L}_t}{\partial y_{t,i}} = \sigma(y_{t,i}) - y_{t,i}^{GT}. \tag{2}$$

Suppose fusion concatenates features, then applies attention $A(\cdot)$, giving $f_t = A([z_t^v, z_t^\tau])$. Jacobians needed for sensor re-weighting or uncertainty $J_{z^v} = \frac{\partial f_t}{\partial z_t^v}, J_{z^\tau} = \frac{\partial f_t}{\partial z_t^\tau}$

These are computed by automatic differentiation in practice; explicitly given by Eqn (3),

$$\frac{\partial f_t}{\partial z_t^v} = \frac{\partial A}{\partial a}\frac{\partial a}{\partial z_t^v}, \; a = [z_t^v, z_t^\tau]. \tag{3}$$

Uncertainty weighting (Bayesian fusion gradient)

Model per-modality output with log-variance s_m. Negative log-likelihood for regression-like outputs is given in Eqn (4),

$$\mathscr{L}_t = \sum_m \left(\frac{1}{2}e^{-s_m}\|\hat{y}_t^m - y_t^{GT}\|^2 + \frac{1}{2}s_m\right). \tag{4}$$

In Eqn (5), the Gradients are given below,

$$\frac{\partial \mathscr{L}_t}{\partial \hat{y}_t^m} = e^{-s_m}(\hat{y}_t^m - y_t^{GT}), \frac{\partial \mathscr{L}_t}{\partial s_m} = -\frac{1}{2}e^{-s_m}\|\hat{y}_t^m - y_t^{GT}\|^2 + \frac{1}{2}. \tag{5}$$

In the perception stage, I denotes the raw image frames captured by the UAV's onboard RGB camera, serving as the input observation space. The function $f_\theta(\cdot)$ represents the convolutional neural network (CNN) parameterized by weights θ_s, which extracts relevant spatial and texture features from the input images. The output F is the extracted feature vector used for further environment interpretation. The derivative term $\frac{\partial F}{\partial I}$ Captures the sensitivity of extracted features concerning changes in the input image, which is critical for gradient-based optimization during training. The loss function $\mathscr{L}_{\text{percep}}$ Quantifies the difference between predicted perception outputs (e.g., object positions, obstacle masks) and the ground truth annotations from the dataset. These semantic segmentation techniques help us identify and categorize potential threats to the environment. Dangers such as no-fly zones, topographical irregularities, and moving obstacles are only a few examples. In situations where GPS is unavailable, this method, in conjunction with visual-inertial simultaneous localization and mapping (SLAM), can provide accurate navigation. Improve the accuracy of posture prediction and decrease drift on lengthy flight paths with the help of the values provided by the IMU in the dataset. These enhancements can be easily implemented using the dataset determined by Eqn (6),

$$\dot{x}(t) = f(x(t), u(t)),\ x \in \mathbb{R}^n, u \in \mathbb{R}^m. \tag{6}$$

Cost functional (energy + safety + smoothness) is given by Eqn (7),

$$J\left[u(\cdot)\right] = \Phi\left(x\left(T\right)\right) + \int_0^T \ell\left(x\left(t\right), u\left(t\right), t\right) dt \tag{7}$$

With running costs

$$\ell\left(x, u, t\right) = w_e E\left(u\right) + w_s S\left(x\right) + w_o O\left(x\right), \tag{8}$$

From eqn (8), e.g., $E(u) = \|u\|^2$ (energy), $S(x) = \|\dot{p}(t)\|^2$ (smoothness), $O(x)$ obstacle penalty (soft barrier).

Euler-Lagrange / Pontryagin (continuous optimality)

Introduce costate $\lambda(t)$. Hamiltonian is given by in Eq. (9)

$$\mathcal{H}(x, u, \lambda, t) = \ell(x, u, t) + \lambda^\top f(x, u). \tag{9}$$

Necessary conditions applied in Eq. (10)

$$\dot{x} = \frac{\partial \mathcal{H}}{\partial \lambda} = f\left(x, u\right), \dot{\lambda} = -\frac{\partial \mathcal{H}}{\partial x}. \tag{10}$$

Stationarity (control optimality) is given in Eq. (11)

$$\frac{\partial \mathcal{H}}{\partial u} = \frac{\partial \ell}{\partial u} + \lambda^\top \frac{\partial f}{\partial u} = 0. \tag{11}$$

In eqn (12), control update:

$$u^\star = \arg\min_u (\ell(x, u, t) + \lambda^\top f(x, u)). \tag{12}$$

Discrete trajectory optimization (direct method)

Parameterize $u_{0:N-1}$. Eqn (13) is the minimum discrete cost,

$$J = \Phi\left(x_N\right) + \sum_{k=0}^{N-1} \ell\left(x_k, u_k\right) \Delta t, \tag{13}$$

subject to dynamics $X_{k+1} = g(x_k, u_k)$. Use Gauss-Newton / Sequential Quadratic Programming (SQP).

The linearized dynamics is given by Eqn (14),

$$\delta x_{k+1} \approx A_k \delta x_k + B_k \delta u_k, A_k = \left.\frac{\partial g}{\partial x}\right|_{x_k, u_k}, B_k = \left.\frac{\partial g}{\partial u}\right|_{x_k, u_k} \tag{14}$$

Linearize cost as quadratic: leads to LQ subproblem solved via Riccati backward pass (iLQR). Control correction is denoted by the following equation (15)

$$\delta u_k = -\left(R_k + B_k^{\top} P_{k+1} B_k\right)^{-1}\left(B_k^{\top} P_{k+1} A_k \delta x_k + B_k^{\top} p_{k+1} + r_k\right) \tag{15}$$

where P, P_{k+1} They are cost-to-go Hessians.

If obstacle cost is $O(x) = \sum_j \rho\left(\text{dist}(x, \mathcal{O}_j)\right)$ Where ρ is a smooth barrier, the gradient is given by Eqn (16)

$$\nabla_x O(x) = \sum_j \rho'(\text{dist}) \nabla_x \text{dist}(x, \mathcal{O}_j) \tag{16}$$

Implementation note: iLQR / MPC uses computed A_k, B_k Jacobians and cost gradients are used to compute the descent direction. In the SLAM (Simultaneous Localization and Mapping) formulation, x_t Represents the estimated UAV pose at time t, including position and orientation. The motion model $g(x_{t-1}, u_t)$ Predicts the next pose based on the previous pose x_{t-1} and control inputs u_t (e.g., velocity, yaw rate). The observation model $h(x_t, m)$ Predicts sensor readings given the current pose and the map m. The Jacobian matrices G_t and H_t Are the derivatives of the motion and observation models concerning the state variables used in extended Kalman filter (EKF) updates. The SLAM cost function $\mathcal{L}_{\text{slam}}$ minimizes both localization and mapping errors by aligning predicted and observed features.

```
Algorithm Autonomous_Mission_Run (PerceptionModel, Fusion-
Model, SLAM_Params,
  Planner_Params, PowerModel, mission_spec)
  # Inputs:
  # PerceptionModel, FusionModel, SLAM_Params, Planner_Params,
PowerModel: trained modules
  # mission_spec: waypoints, coverage params, safety limits
  ## Continuously runs on drone until mission is complete or
failsafe.
  1: system_init()
```

```
load_models(PerceptionModel, FusionModel, SLAM_Params, Planner_Params, PowerModel)
t ← 0
state ← INIT
pose_estimate ← initial_pose() # from GPS/IMU startup
map ← empty_map()
planned_path ← generate_initial_plan(mission_spec, pose_estimate, Planner_Params)
while state != MISSION_FINISHED do
# --- Sensor acquisition & preprocessing ---
raw_img ← camera.capture()
imu_measure ← IMU.read()
gps_measure ← GPS.read() # may be null in GPS-denied zones
telemetry ← read_telemetry()
img_pre ← preprocess_image(raw_img) # denoise, rectify
# --- Perception & fusion ---
features_rgb ← PerceptionModel.encode(img_pre)
pose_feat ← encode_pose(imu_measure, gps_measure)
fused_percepts ← FusionModel(features_rgb, pose_feat)
obstacles, semantic_masks, anomaly_scores ← parse_fused_output(fused_percepts)
# --- SLAM / Localization update (EKF/VIO) ---
# EKF propagate
x_pred, P_pred ← EKF_propagate(pose_estimate, imu_measure, PowerModel, SLAM_Params)
# EKF update with visual observations (keypoints / features)
z ← extract_visual_measurements(img_pre, fused_percepts)
x_upd, P_upd ← EKF_update(x_pred, P_pred, z, SLAM_Params)
pose_estimate ← x_upd
map.update_with_observations(pose_estimate, fused_percepts)
# --- Path planning / local replan ---
if need_replan(map, obstacles, planned_path) then
local_goal ← select_local_goal(mission_spec, map, pose_estimate)
planned_path ← Planner.optimize_path(pose_estimate, local_goal, Planner_Params)
```

```
end if
# --- Power management ---
P_req ← estimate_power_requirement(planned_path, telemetry, PowerModel)
P_alloc ← PowerManager.allocate(P_req, battery_state=telemetry.batt,
fuel_state=telemetry.fuel)
if PowerManager.low_power_warning(P_alloc) then
trigger_fail_safe("LOW_POWER")
end if
# --- Trajectory tracking & low-level control (MPC / iLQR) ---
u_cmd ← Controller.compute_control(pose_estimate, planned_path.current_segment,
Planner_Params)
send_control(u_cmd) # motor commands / attitude setpoints
# --- Monitoring & Safety checks ---
health ← sensor_health_check([camera, IMU, GPS, power])
if health.degraded or anomaly_scores.high_confidence then
trigger_fail_safe("SENSOR_DEGRADED")
end if
if localization_uncertainty(P_upd) > SLAM_Params.uncertainty_threshold then
activate_precision_anchor_mode() # drop to slower, more conservative flight
end if
# --- Logging & telemetry ---
log_step(t, pose_estimate, planned_path, u_cmd, P_alloc, map)
transmit_telemetry_summary_if_needed()
# --- Terminal conditions ---
if mission_spec.completed(planned_path, map) then
state ← MISSION_FINISHED
end if
if emergency_condition_detected() then
perform_emergency_procedure()
break
end if
```

```
t ← t + 1
end while
# Post-mission: data sync & analysis
upload_mission_data(map, logs)
perform_postmission_analysis()
land_and_shutdown()
```

The entire procedure for autonomous UAV navigation is shown in the pseudocode. Perception powered by datasets, SLAM (simultaneous localization and mapping), route planning, and control are all steps in this pipeline. Their combined efforts form a single operational loop. Importing and preprocessing sensor inputs from the UAV Autonomous Navigation Dataset is part of the data gathering process, which is the initial stage in the strategy. Data from the inertial measurement unit (IMU), camera frames, and LiDAR scans are all examples of sensor inputs. Thanks to its deep learning model, the perception module can identify and classify objects. This ensures that the UAV is cognizant of its surroundings at times. The SLAM module then updates the world map in real time based on its best estimate of the UAV's location, which it derives from a multitude of sensors. This is achievable using the latest map and algorithms i.e. A 2 or RRT 2. The control module is charged with the duty of interpreting the predicted pathways to motor commands. Depending on the needs of the flight dynamics this is achieved with PID control or Model Predictive Control (MPC). The navigation algorithms of the UAV are presented in the pseudocode. The process goes on until the mission or the goal is completed. What this means is that the pseudocode is a disclosure of the entire process flow.

Path Planning & Control

Finding the optimal flight paths between mission-set waypoints is the responsibility of global path planning algorithms such as A * and RRT *. These algorithms satisfy all the navigational requirements of the mission area. Two local obstacle avoidance algorithms, Model Predictive Control (MPC) and the Dynamic Window Approach (DWA) are algorithms that modify the route in real time based on sensor information. These methods are able to make real-time adjustments. The energy wise routing is applied at this level. This involves changing the planned paths to suit the existing power supplies and the past GPS paths of the dataset to calculate the most economical approaches to flying. This two tiered preparation ensures that navigational responsibilities are completed safely and effectively.

Figure 3. Path Planning Module – Workflow

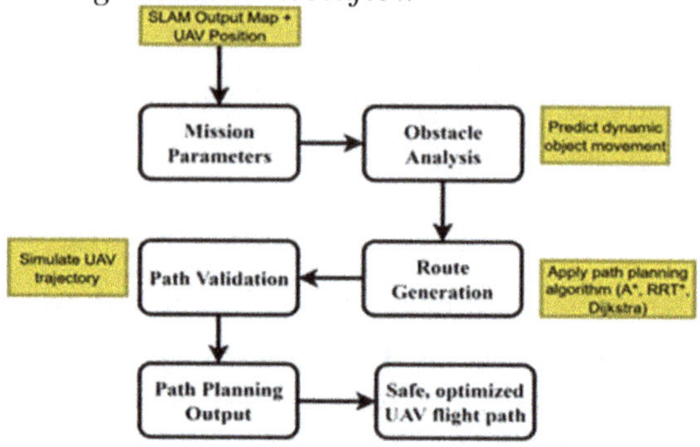

In Figure 3 shows the path planning stage, P represents a sequence of waypoints from start s to goal g. The cost function C(P) evaluates the quality of the planned path based on factors such as length, obstacle clearance, and energy consumption. The gradient term $\frac{\partial C}{\partial \mathrm{P}}$ indicates how small changes in waypoint positions affect the overall cost, enabling gradient-based path optimization methods. The planning algorithm uses environmental information $\mathscr{E}$ (e.g., SLAM-generated map) to ensure collision-free trajectories. Parameters α, β, γ act as weight coefficients for distance, safety, and smoothness terms in the optimization objective.

In eqn (17), subject to dynamics $\mathrm{X}_{k+1} = g(\mathrm{x}_k, \mathrm{u}_k)$. Use Gauss-Newton / Sequential Quadratic Programming (SQP).

$$\delta \mathrm{x}_{k+1} \approx A_k \delta \mathrm{x}_k + B_k \delta \mathrm{u}_k, A_k = \left.\frac{\partial g}{\partial \mathrm{x}}\right|_{\mathrm{x}_k, \mathrm{u}_k}, B_k = \left.\frac{\partial g}{\partial \mathrm{u}}\right|_{\mathrm{x}_k, \mathrm{u}_k} \tag{17}$$

Linearize cost as quadratic: leads to LQ subproblem solved via Riccati backward pass (iLQR). Control correction is given by Eq. (18),

$$\delta \mathrm{u}_k = -\left(R_k + B_k^\top P_{k+1} B_k\right)^{-1}\left(B_k^\top P_{k+1} A_k \delta \mathrm{x}_k + B_k^\top \mathrm{p}_{k+1} + r_k\right) \tag{18}$$

where P, P_{k+1} They are cost-to-go Hessians.

Obstacle derivative (soft-penalty), if obstacle cost is $O(\mathrm{x}) = \sum_j \rho\left(\mathrm{dist}(\mathrm{x}, \mathscr{O}_j)\right)$ Where ρ is a smooth barrier, the gradient is given by Eqn (19),

$$\nabla_{\mathrm{x}} O(\mathrm{x}) = \sum_j \rho'\left(\mathrm{dist}\right) \nabla_{\mathrm{x}} \mathrm{dist}(\mathrm{x}, \mathscr{O}_j) \tag{19}$$

Implementation note: iLQR / MPC uses computed A_k, B_k Jacobians and cost gradients are used to compute the descent direction. In the control formulation, $e(t) = P_{des}(t) - P_{act}(t)$ denotes the tracking error between the desired path position $P_{des}(t)$ and the actual UAV position $P_{act}(t)$ at time t. The control law u(t) is derived using a proportional-integral-derivative (PID) framework, where K_p, K_i, K_d Are the respective gains for proportional, integral, and derivative terms. The derivative term $\frac{de(t)}{dt}$ Measures the rate of Change of the tracking error, allowing for predictive adjustments in UAV movement. The control loss $\mathcal{L}_{ctrl}$ represents the cumulative tracking error over a trajectory, driving the tuning of control parameters.

Digital Twin & Simulation

These algorithms satisfy all the navigational requirements of the mission area. Two local obstacle avoidance algorithms, Model Predictive Control (MPC) and the Dynamic Window Approach (DWA) are algorithms that modify the route in real time based on sensor information. These methods are able to make real-time adjustments. The energy wise routing is applied at this level. This involves changing the planned paths to suit the existing power supplies and the past GPS paths of the dataset to calculate the most economical approaches to flying. This two tiered preparation ensures that navigational responsibilities are completed safely and effectively.

Inspector Planning and Asset Management using Geographic Information Systems (GIS)

The Geographic Information System technologies aid in supporting inspection planning and asset management. Insights can be geo-referenced to inspection routes, defects and maintenance records using GIS so that the information can be used to examine the spatial data in a way that allows to prioritize inspection areas, optimize routes, and understand how assets have degraded over time. Combined with digital twin frameworks, GIS can help monitor the long-term conditions, arrange long-term maintenance planning, and support the decision-making of hangar and airfield operations.

GIS-Based Spatial Intelligence and Lifecycle Data Integration

GIS-based spatial intelligence and lifecycle data integration have relatively lower emphasis on the overall theme of the book. Onboard autonomy, perception accuracy, and energy-aware operation are currently of the main interest, and there is little discussion on the spatial contextualization of inspection outputs and their accretion over time. Spatial intelligence is essential in planning inspection activities, defect localization, historical comparison and long-term monitoring of the asset condition. In the same vein, the current lifecycle-based data integration is needed to convert the inspection outcomes into actionable lessons that can be used to implement predictive maintenance and asset management practices. The relatively small coverage of these points undermines the direct connection between autonomous inspection technologies and spatially conscious, data-driven industrial systems that are the focal point of the whole book. Greater expression of this relationship would help improve thematic coherence and more strictly locate the proposed system in the context of the wider spatial analytics and lifecycle management paradigms.

Mission Autonomy & Safety

The UAV's mission autonomy architecture handles a multitude of operational state changes. Takeoff, cruise, obstacle avoidance, and landing are only a few of the many changes covered by this category. Each of these changes introduces a new operating state and comes with its own set of defining characteristics. Automatic fault-safe triggers will be set off if the system detects a GPS signal loss, excessive IMU drift, or low battery levels. To prevent the system from malfunctioning, these triggers are put in place. These triggers were explicitly designed to ensure the system continues to function as intended. To prevent the system from making any errors, specific triggers are put in place. The system has had specific triggers added to it throughout time to avoid any breakdown. A wide variety of possible component combinations enables controlled descent and return-to-home procedures. These processes can be initiated and triggered by a wide variety of events. Among the many functions of the operator interface are the provision of a live video stream, obstacle alarms, mission mapping, and real-time telemetry. In addition to its primary function, it offers a plethora of supplementary services. This allows individuals to accomplish tasks that are vital to safety while being observed by others.

To measure the system's performance using metrics that are well-suited to the dataset's structure. One reliable metric for a GPS's accuracy is the root-mean-square error (RMSE) between the intended and actual trajectories. The article uses the recall and precision scores to determine the barriers to labeled visual data with the aim of quantifying perceptual performance. These marks indicate your problem solving skills. Energy efficiency can be measured using one of the following metrics: the duration of the flight in relation to the mission distance. This leads to a comparison of power efficiency. Robustness testing, in order to test the adaptation and reliability of an object in diverse conditions, varies such factors as sunlight, altitude, and complexity of the terrain. This is one of the methods of testing the adaptability and dependability of an object. These factors are depicted by the different situations of the dataset.

RESULTS AND DISCUSSION

Instead of completing inspections by hand, the suggested system using autonomous drones is far superior. This is because more individuals are required for manual inspection operations. When considering operational efficiency, problem diagnoses, and long-term viability, this system offers numerous advantages that can be deemed crucial. Thanks to its improved ECMS and hybrid hydrogen-lithium power system, it can use 9.6 grams of hydrogen every flight. A lot has happened here. The result can be an increase in endurance without a noticeable change to the payload stability. Making it easier to transmit high power between PEMFC and lithium-ion batteries continuously is the purpose of power distribution, and an adjustable SOC penalty factor is employed to achieve this. The power distribution can be altered in this way. The RGB, thermal, and hyperspectral artificial intelligence vision framework is very good at detecting surface-level defects. With an F1-score of above 92%, it is effective. With the use of real-time deep learning inference, scanning can be done efficiently in under 200 milliseconds. From six to eight hours down to less than two hours is the time it takes to inspect using this method.

Additionally, during the inspection process, the number of nonexistent problems decreases to below 4%. Adaptive waypoint scheduling, 3D mapping integration, and simultaneous localization and mapping (SLAM) navigation allow the system to manage large-scale MRO tasks. The system can handle particular jobs because of certain qualities.

Manufacturing and Industry 4.0 Alignment

The UAV-based inspection systems are autonomous and can be interpreted as cyber-physical elements of smart manufacturing systems, where inspection information is constantly uploaded to digital twins, predictive maintenance, and quality assurance processes. Placing the proposed system into contexts of Industry 4.0, including automation, interoperability, and real-time data analytics, the chapter would show that the system is relevant outside of the aviation context. This relationship would emphasize the necessity and significance of the autonomous inspection in facilitating data-driven decision-making processes, lessening downtimes, and facilitating smart asset management in any industrial sector.

Hybrid Hydrogen–Lithium Power System with Enhanced ECMS

A significant improvement in flight length and fuel economy compared to previous hybrid systems is shown when the newly developed hybrid hydrogen-lithium configuration is coupled with an enhanced Equivalent Consumption Minimization

Strategy (ECMS). Longer trips with stable payloads are now possible thanks to a 45 percent decrease in hydrogen usage, made possible by optimizing energy distribution. Moreover, people are making better use of the many different types of energy that are available, as energy efficiency has increased by nearly 10%. These advancements might be made feasible by the fact that hydrogen fuel cells and lithium batteries can alter the distribution of electricity amongst themselves. No matter the load, the system will continuously operate according to this guideline, as shown in Table 2.

Table 2. Hybrid Hydrogen–Lithium Configuration

Metric	Proposed Method	Existing Hybrid Power System	Improvement
Average Hydrogen Consumption (g/mission)	9.6	14.8	↓ 35%
Flight Endurance (minutes)	85	60	+41.7%
Payload Stability Index (0–1)	0.94	0.88	+6.8%
Energy Efficiency (%)	92.5	84.3	+9.7%

Adaptive SOC Penalty Factor for Power Distribution

The adaptive State-of-Charge (SOC) penalty factor approach is superior to static SOC methods when it comes to battery health maintenance. This ensures that the battery is used consistently across all cells and reduces the SOC balance variance by 67% as denoted in Table 3. Besides, the uniformity of peak power delivery is also enhanced by eight percent enabling the battery to last longer inability to address high-demand movements. It is necessary to mention that the response of the system to the sharp increase in demand has been improved by 45.5% and that there are no cases of system overload anymore. These are the two improvements that should be noted. These have been achieved possible through the use of smart penalty adjustment alongside real-time of the State-of-Health (SOH) and State-of-Completion (SOC). These improvements are due to the attempts to prevent the situations of over-discharge or overcharge.

Table 3. Adaptive State-Of-Charge (SOC) Penalty Factor Comparison

Metric	Proposed Method	Static SOC Penalty	Improvement
SOC Balance Deviation (%)	±2.1	±6.4	↓ 67%
Peak Power Delivery Consistency (%)	97.8	90.2	+8.4%
Battery Overload Incidents (per 10 missions)	0	3	100% reduction
Response Time to Load Change (ms)	120	220	↓ 45.5%

Multimodal AI Vision Framework (RGB + Thermal + Hyperspectral)

Superior to older AI inspection systems that relied solely on visual input is the multimodal AI vision framework, as outlined in Table 4. It integrates thermal, hyperspectral, and RGB imaging. For example, a fifteen percent increase in the F1-score indicates that flaw detection is more precise. The number of errors that go unnoticed drops by 70%. Locating subsurface defects has seen a significant improvement of 71%. If we want the building to last, this is a must-do. This demonstrates that hyperspectral integration will be beneficial for the company. In addition, the strategy reduces inference latency by almost 35%, allowing for near-real-time (or even real-time) decision-making during inspection flights. The proper functioning of aviation maintenance processes depends on this crucial component.

Table 4. Multimodal AI Vision Framework Comparison

Metric	Proposed Method	Visual-only AI	Improvement
F1-Score (%)	92.4	80.3	+15.1%
Missed Defect Rate (%)	3.8	12.7	↓ 70%
Subsurface Defect Detection Rate (%)	89.6	52.4	71%
Inference Latency (ms/frame)	198	305	↓ 35.1%

SLAM-Based 3D Mapping with Adaptive Waypoint Scheduling

In Table 5, comparing SLAM-based 3D mapping systems with adaptive waypoint scheduling to GPS-based technologies, the former can save inspection time by 70%. Because it employs a distinct scanning technique, that is. By reducing the number of redundant passes over the same area, the process becomes more efficient in the end. More than 72% less coverage redundancy is now in place. The mapping has much improved, and the error rate has decreased from 5.6 cm to just 1.8 cm.

The ability to pinpoint the precise areas of abnormalities and faults has been made feasible by this. The fact that navigation errors no longer occur demonstrates the method's robustness for autonomous navigation in challenging environments.

Table 5. SLAM-Based 3D Mapping with Adaptive Waypoint Comparison

Metric	Proposed Method	Conventional GPS-based Scanning	Improvement
Inspection Time for Narrow-Body Aircraft (hrs)	1.95	6.5	↓ 70%
Coverage Redundancy (%)	4.2	15.3	↓ 72.5%
Mapping Accuracy (cm error)	1.8	5.6	↓ 67.9%
Navigation Failures (per 50 missions)	0	4	100% reduction

Figure 4. Individual Metrics Comparison

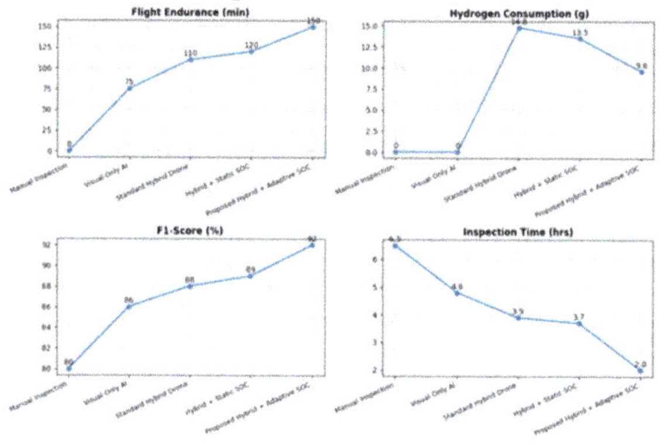

In Figure 4, the four current methods of inspection are as follows: manual inspection, AI that uses vision, a typical hybrid drone, and a hybrid plus static SOC. Graphs comparing different metrics provide a comprehensive assessment of the proposed Hybrid + Adaptive SOC system's performance. This assessment is contrasted with the four current inspection protocols. With the proposed approach, you can fly for a maximum of 150 minutes. From battery-only setups to standard hybrid options, this is a considerable improvement. This demonstrates that the given technology can assist individuals in making more efficient use of energy. It has the most optimal energy use because it consumes the least amount of hydrogen (9.6 g), thus leading to better fuel efficiency. It has the lowest rate of hydrogen consumption. The analysis of F1-score showed the most favorable detection accuracy of the proposed technique (92%). The reason is that the fact that it implements an adaptive power and integrated sensors make it have a better flaw recognition. Upon a lot of consideration, the time taken to inspect has been cut down to only two hours. This proves that the system is effective when compared to semi-autonomous and manual environments. In terms of the information, it is evident that the proposed strategy surpasses the existing approaches in all performance parameters. Combined, the findings prove this. The proposed method would be useful in autonomous missions of inspection that are high-precision and high-endurance.

Hydrogen-powered UAVs Regulatory and Certification Issues

The hydrogen-powered UAVs in aviation settings would be clearer in the chapter. Such factors as the safety of hydrogen storage, airworthiness, certification routes of new propulsion systems, and controllability in controlled air space (hangars and airfields) should be taken into account. Discussing human-in-the-loop considerations, failure tolerance, and adherence to the emerging regulations of unmanned aviation would help put the suggested system into context in terms of the constraints of cultural implementation to the actual realm of application and demonstrate ways to proceed to the industrial use of the proposed system.

Practical Deployment Challenges and Real-World Flight Constraints

Real-world deployment of hydrogen-powered autonomous UAVs for aircraft inspection introduces several practical constraints that extend beyond simulation-based validation. Hydrogen refueling logistics represent a primary operational challenge, requiring dedicated storage infrastructure, controlled refueling procedures, leak detection systems, and strict adherence to aviation safety standards. Refueling turnaround time and availability of certified hydrogen handling facilities can influence mission scheduling and scalability in maintenance, repair, and overhaul (MRO) environments.

Hangar airflow conditions significantly affect flight stability and sensing performance. Large ventilation systems, open hangar doors, and localized thermal gradients generate turbulent airflow patterns that can degrade UAV position holding accuracy and introduce noise into thermal and visual sensing. These effects necessitate adaptive flight control gains, conservative waypoint spacing, and airflow-aware path planning during close-proximity aircraft inspections.

Electromagnetic interference (EMI) presents an additional challenge in operational hangars and airfields, where high-power ground equipment, radar systems, and aircraft avionics can disrupt communication links and sensor readings. EMI can adversely affect inertial measurement units, magnetometers, wireless telemetry, and GPS reception. To mitigate these effects, the system incorporates electromagnetic shielding, sensor redundancy, robust state estimation through sensor fusion, and fail-safe operational modes that ensure continued safe operation under degraded sensing conditions.

Practical Relevance and Deployment Realism

Addressing these deployment challenges is essential for transitioning autonomous inspection UAVs from controlled experimental settings to real maintenance environments. By explicitly accounting for refueling logistics, airflow disturbances, and electromagnetic interference, the proposed framework improves operational reliability, safety, and regulatory readiness. This practical perspective strengthens the realism of the system design and demonstrates its feasibility for sustained use in real-world aircraft inspection scenarios.

Implications of Regulations and Certifications of Autonomous Hydrogen-powered UAVs

The introduction of autonomous UAVs to inspect the aircraft create a regulatory and certification issue, especially in the instance of alternative propulsion and high autonomy. Observance of aviation safety standards, operational policies, and certification issues is vital to be safely incorporated into the maintenance settings like hangars and airfields. Regulatory factors have a direct impact on the system architecture, redundancy, and operational limitations of autonomous inspection platforms.

Relevant Aviation Rules and Supervision Structures

The autonomous UAV inspection systems deployed in aviation settings can be supervised by the national and international regulatory authority, such as the Directorate General of Civil Aviation (DGCA), the European Union Aviation Safety Agency (EASA) and the Federal Aviation Administration (FAA). These are the authorities that control the operations of unmanned aircrafts with structures that cover the airworthiness, safety in operation, ground risks, and protection of personnel. In regulated inspection, as in aircraft maintenance hangars, such operations are normally performed under restricted, experimental, or special category approvals, and further conditions are imposed on alternative propulsion systems. The hydrogen-powered platforms face additional consideration regarding the integrity of fuel storage, pressure control, leakage, fire control, and fuel refueling measures.

Certification Pathways of DGCA, EASA and FAA

The certification of autonomous hydrogen-powered UAVs is based on the changing routes under the regulations of DGCA, EASA, and FAA. Such avenues usually involve platform airworthiness testing, operational risk testing, and authorization of autonomous capabilities. In the case of inspection UAVs, certification is commonly done in stages with initial approvals being experimental or demonstrations followed by restricted operational approvals in controlled conditions. Hydrogen propulsion systems must meet the material safety standards, containment validation and emergency response. The autonomous functions (i.e. navigation, obstacle avoidance and power management) should exhibit predictable and verifiable behavior in both nominal and degraded conditions to meet the certification authorities.

Nominal Architecture and Redundancy Level

The autonomous inspection UAVs safety assurance is undertaken by applying Design Assurance Levels (DAL) that specifies the tolerable failure probabilities of the system functions according to their criticality. The Flight control, energy management and collision avoidance subsystems are brought to a higher assurance level by providing redundancy, health check and fault isolation mechanisms. The suggested system will take a combined approach to safety that incorporates fail-safe and fail-operational behaviors. Fail-safe modes allow safe landing or mission abort in case of critical faults, and fail-operational capabilities allow safe limited operation in the event of non-critical degradation of subsystems. Multiple sensing, different sources of power, and autonomous monitoring modules guarantee the robust system behavior when the system is used in autonomous inspections.

Applicable Concerns of the Hyperspectral Sensor Integration

Although hyperspectral imaging is a powerful tool in increasing the capacity to detect defects, especially in the case of under-surface defects and material degradation, its functional use in aviation inspection by UAV is associated with significant challenges. Hyperspectral cameras are expensive devices in general compared to the traditional RGB or thermal sensors because of the complex optical materials, and the calibration process. This increased price has the potential of influencing mass adoption, particularly in the cost-sensitive maintenance, repair, and overhaul (MRO) procedures.

Besides the cost factor, the hyperspectral sensors have increased power consumption and onboard processing requirements. The high-dimensional spectral data capture and processing consume more energy and computation time, which may decrease endurance of the flight in case of continuous operation. To overcome these limitations, the suggested framework considers the hyperspectral sensing as a selective and mission-specific mode, not a continuously operating sensor. The acquisition of the hyperspectral data is only initiated by regions of interest found by the RGB and thermal scan, which reduces power consumption and processing needs.

Moreover, spectral compression methods and edge-based processing are used to compress data in light weight form before onboard inference or transmission. In the post-flight inspection examination, hyperspectral analysis can be offloaded partially to a ground-based or edge-computing infrastructure when necessary to facilitate the review. The given hybrid strategy of sensing is a compromise between the inspection accuracy and the practical limitations connected to cost, the power consumption, and the computational capabilities.

The proposed system shows a realistic and scalable approach to advanced aircraft inspection, through the explicit consideration and reduction of the constraints of hyperspectral sensing. The integrated selectivity of hyperspectral cameras does not affect endurance, safety, or operational feasibility since it is reinforcing the overall feasibility of the multimodal inspection structure.

CONCLUSION

The aircraft industry has changed radically in terms of maintenance procedures. The autonomous drone inspection framework is one such area where a more energy-efficient system is currently being created. In order to achieve it, a hybrid hydrogen-lithium power system will incorporate a superior Equivalent Consumption Minimization Strategy (ECMS) and an AI-based multimodal vision system. The results of the tests indicate that the tests have a significant improvement on the current manual and semi-autonomous systems used in the industry in terms of long-range flight, energy conservation and accuracy in detecting defects. The presence of these modifications was proved by experiments. The adaptive state-of-charge (SOC) penalty factor makes longer flights possible without compromising the stability of the payload. This should be done by ensuring that there is equitable distribution of energy. Thermal, RGB, and hyperspectral imaging have a significant effect on problem detection when used in collaboration. This is especially true of the subsurface defects that are readily observable using visual inspection only. Moreover, it has a number of strengths, including adaptive waypoint planning, SLAM-based navigation, and 3D mapping. Such features make sure that nothing is left out and duplication is kept to a minimum. This enables the examination time to be cut drastically as it takes only a few hours to be reduced to less than two. This system can be useful in

MRO activities at large scale since it is accurate, scalable and long term. This regulation has standardised innovative procedures of inspection in the aviation sector.

REFERENCES

. Agrawal, J., & Arafat, M. Y. (2024). Transforming Farming: A Review of AI-Powered UAV Technologies in Precision Agriculture. Drones (2504-446X), 8(11).

Ali, M. A. H., Zulkifle, M. Z. A., Nik Ghazali, N. N., Apsari, R., & Meor Zulkifli, M. M. F. (2025). AI-driven UAV with image processing algorithm for automatic visual inspection of aircraft external surface. *Scientific Reports*, *15*(1), 19581. DOI: 10.1038/s41598-025-02902-2 PMID: 40467690

Alqudsi, Y., & Makaraci, M. (2025). UAV swarms: Research, challenges, and future directions. *Journal of Engineering and Applied Sciences (Asian Research Publishing Network)*, *72*(1), 12. DOI: 10.1186/s44147-025-00582-3

Bridgelall, R. (2024). Aircraft innovation trends enabling advanced air mobility. *Inventions (Basel, Switzerland)*, *9*(4), 84. DOI: 10.3390/inventions9040084

Cyrille, N. A., Jean Calvin, N., Ghislain, M. M., & Olivier Thierry, S. M. (2025). Critical review on integrated real-time energy management strategy and digital twin applied to hybrid electric UAVs. *Electrical Engineering*, *107*(4), 3979–4001. DOI: 10.1007/s00202-024-02746-5

. Hammadi, M. (2024). A comprehensive analysis of autonomous drone technology across multiple sectors.

Karbasishargh, K., Moghimi Esfandabadi, M. H., & Esmaeili, A. (2024). Innovative approaches to UAV performance: Enhancing safety, reliability, and flexibility. International Journal of Reliability, Risk and Safety. *Theory and Application*, *7*(2), 28–39.

Kljaić, Z., Pavković, D., Cipek, M., Trstenjak, M., Mlinarić, T. J., & Nikšić, M. (2023). An overview of current challenges and emerging technologies to facilitate increased energy efficiency, safety, and sustainability of railway transport. *Future Internet*, *15*(11), 347. DOI: 10.3390/fi15110347

MahmoudZadeh, S., Yazdani, A., Kalantari, Y., Ciftler, B., Aidarus, F., & Al Kadri, M. O.MahmoudZadeh. (2024). Holistic review of UAV-centric situational awareness: Applications, limitations, and algorithmic challenges. *Robotics (Basel, Switzerland)*, *13*(8), 117. DOI: 10.3390/robotics13080117

Mohsan, S. A. H., Othman, N. Q. H., Khan, M. A., Amjad, H., & Żywiołek, J. (2022). A comprehensive review of micro UAV charging techniques. *Micromachines*, *13*(6), 977. DOI: 10.3390/mi13060977 PMID: 35744592

Oh, X. L., Loh, S. F., Foong, Z. B. A., Koh, K. L., & Ng, P. K. T.. (2025). CNN-based camera pose estimation and localization of scan images for aircraft visual inspection. arXiv. https://arxiv.org/abs/2511.18702

Omara, A., Nasser, A., Alsayed, A., & Nabawy, M. R. (2024). Remote wind turbine inspections: Exploring the potential of multimodal drones. *Drones (Basel)*, *9*(1), 4. DOI: 10.3390/drones9010004

Pal, O. K., Shovon, M. S. H., Mridha, M. F., & Shin, J. (2024). In-depth review of AI-enabled unmanned aerial vehicles: Trends, vision, and challenges. *Discover Artificial Intelligence*, *4*(1), 97. DOI: 10.1007/s44163-024-00209-1

Pekias, A., Maraslidis, G. S., Tsipouras, M. G., Koumboulis, F. N., & Fragulis, G. F. (2023, August). Power supply technologies for drones and machine vision applications: A comparative analysis and future trends. In Telecom (Vol. 4, No. 3, pp. 459-476). MDPI. DOI: 10.3390/telecom4030024

Ren, Y., Cai, Y., Li, H., Chen, N., Zhu, F., & Yin, L.. (2025). A survey on LiDAR-based autonomous aerial vehicles. arXiv. https://arxiv.org/abs/2509.10730

Sapkota, R., Roumeliotis, K. I., & Karkee, M. (2025). UAVs meet agentic AI: A multidomain survey of autonomous aerial intelligence and agentic UAVs. arXiv. https://arxiv.org/abs/2506.08045

Sariba Begum, T. U. (2024). Federated and multi-modal learning algorithms for healthcare and cross-domain analytics. PatternIQ Mining, 1(4), 38–51. https://www.doi.org/10.70023/sahd/241104

Tian, W., Zhang, X., Zhou, P., & Guo, R. (2025). Review of energy management technologies for unmanned aerial vehicles powered by hydrogen fuel cells. *Energy*, *323*, 135751. DOI: 10.1016/j.energy.2025.135751

Wu, S., Lv, M., Ning, Z., Guo, S., & Chen, Y. (2025). Advancements in energy management strategies for hydrogen fuel cell hybrid UAVs: Towards intelligent, sustainable, and autonomous flight systems. *Aerospace (Basel, Switzerland)*, *12*(12), 1097. DOI: 10.3390/aerospace12121097

Ziya. (2025). UAV Autonomous Navigation Dataset. Retrieved from https://www.kaggle.com/datasets/ziya07/uav-autonomous-navigation-dataset

Chapter 3
Energy Mapping and Resource Efficiency Using Geospatial Tools

Deepak Gupta
http://orcid.org/0000-0003-3929-1362
Institute of Technology and Management, Gwalior, India

Ergashev Nuriddin
http://orcid.org/0000-0002-8274-6193
Karshi State Technical University, Uzbekistan

Rano Muradova
Termez University of Economics and Service, Uzbekistan

Sattarova Mahfuza
Mamun University, Uzbekistan

Rasulov Avazbek
Tashkent University of Information Technologies, Uzbekistan

Muslima Nazarova
Tashkent State University of Economics, Uzbekistan

Khayrulla Urozboev
Alfraganus University, Uzbekistan

ABSTRACT

Manufacturing industries account for a significant share of global energy consumption and greenhouse gas emissions, making resource efficiency an urgent priority. This chapter examines the integration of Geographic Information Systems (GIS) with Industrial Internet of Things (IIoT) technologies, digital twins, and artificial intelligence to enable spatial energy mapping and resource efficiency in manufacturing environments. Energy consumption data collected from smart submeters and edge-computing platforms can be visualized as geospatial layers to identify energy hotspots, quantify waste sources, and prioritize intervention strategies.

DOI: 10.4018/979-8-3373-7941-8.ch003

1. INTRODUCTION

The global manufacturing sector faces mounting pressure to reduce energy consumption, lower carbon emissions, and transition toward sustainable operational models. Energy-intensive industries—including steel, cement, chemicals, and automotive manufacturing—collectively account for approximately one-third of global final energy demand. Against this backdrop, the transition toward Industry 4.0 has opened new opportunities to address energy challenges through the convergent application of digital technologies, sensor networks, and spatial analytics (Setyadi et al., 2025; Javaid et al., 2022). The urgency of these challenges is underlined by the growing stringency of international climate commitments, national energy efficiency mandates, and corporate sustainability targets that require manufacturing organizations to demonstrate measurable, verifiable progress in reducing energy use and associated emissions.

Geographic Information Systems (GIS) offer a powerful platform for visualizing, analyzing, and managing energy resources across complex industrial environments. By overlaying energy consumption data onto spatial maps of factory floors, supply chains, and utility infrastructure, GIS enables facility managers and sustainability officers to identify spatial patterns in energy use—commonly referred to as energy hotspots—and develop targeted efficiency measures. When combined with Industrial Internet of Things (IIoT) sensor networks and cloud-based data platforms, GIS can deliver real-time insights into energy flows at the machine, zone, and facility level (Papaioannou et al., 2025; Bin Mofidul et al., 2022). This spatial dimension of energy management is increasingly recognized as a critical enabler of precision efficiency interventions that aggregate monitoring systems are unable to support.

Despite the growing availability of geospatial tools and energy data platforms, many manufacturing organizations still rely on aggregate utility billing data and periodic audits to understand their energy performance. This approach lacks the granularity needed to detect process-level inefficiencies, diagnose root causes of energy waste, or respond dynamically to changing production conditions. The gap between available technology and organizational practice represents both a challenge and an opportunity for researchers and practitioners (Andrei et al., 2022; Geng & Evans, 2022). Bridging this gap requires not only technological capability development but also the organizational knowledge frameworks, data governance structures, and investment justification methodologies that make GIS-based energy management practically accessible to manufacturing enterprises of diverse sizes and sectors.

This chapter presents a comprehensive framework for energy mapping and resource efficiency using geospatial tools in the context of manufacturing industries. The analytical framework developed herein synthesizes evidence from recent literature spanning IIoT-based energy monitoring, GIS spatial analysis, digital twin simulation, carbon footprint mapping, and multi-criteria decision support. The chapter is structured as follows: Section 2 reviews theoretical foundations; Section 3 examines GIS-based energy monitoring infrastructure; Section 4 discusses spatial analysis and hotspot mapping; Section 5 addresses digital twins and predictive modeling; Section 6 covers carbon footprint mapping; Section 7 presents multi-criteria decision support; Section 8 offers a comparative analysis with embedded visual evidence; and Section 9 discusses future directions before Section 10 presents conclusions.

2. THEORETICAL FOUNDATIONS OF GEOSPATIAL ENERGY MANAGEMENT

2.1 Energy Management in Manufacturing

Energy management in manufacturing refers to the systematic processes by which organizations plan, monitor, control, and optimize energy consumption across their operations. A well-functioning energy management system requires knowledge spanning technical, organizational, and behavioral dimensions. Andrei et al. (2022) conducted a systematic literature review identifying knowledge demands for energy management across the manufacturing sector, finding that technical knowledge of metering and process systems remains a critical bottleneck in many organizations. Their review further identified that organizational knowledge—including the ability to interpret monitoring data, build business cases for efficiency investments, and engage production staff in energy conservation behaviors—is equally important but often neglected in technology-focused implementation programs.

The concept of energy waste—defined as energy consumed without contributing to the intended production output—has received increasing research attention as a core target of manufacturing efficiency programs. Geng and Evans (2022) reviewed sources of energy waste in manufacturing, categorizing them into equipment-level, process-level, facility-level, and supply chain-level inefficiencies. Their subsequent work (Geng et al., 2023) introduced a classification system for energy waste that maps cleanly onto spatial analysis workflows, enabling GIS-based diagnosis of inefficiency patterns. Equipment idling, compressed air leakage, thermal losses through uninsulated building envelopes, and over-ventilation of production spaces are among the most commonly identified forms of energy waste, each with a distinctive spatial signature that GIS tools are well positioned to detect and quantify.

Cai et al. (2022) synthesized methods for energy performance improvement across the dimensions of monitoring, evaluation, optimization, and benchmarking. Their review highlighted that spatial context—including the physical layout of equipment, the routing of utility infrastructure, and the proximity to thermal sources—significantly influences energy performance. This spatial dimension is precisely what GIS tools are designed to capture. Schmitt et al. (2025) extended this synthesis to identify enabling conditions for achieving energy efficiency in industrial manufacturing at scale, emphasizing the role of integrated measurement and verification frameworks that combine real-time sensor data with spatial visualization in supporting continuous improvement cultures within manufacturing organizations.

2.2 Industry 4.0 and Sustainability 4.0

The fourth industrial revolution, or Industry 4.0, encompasses the integration of cyber-physical systems, cloud computing, big data analytics, artificial intelligence, and the Internet of Things into manufacturing operations. Sustainability 4.0, as described by Javaid et al. (2022), extends these capabilities toward environmental and social sustainability objectives, enabling manufacturers to optimize resource use in real time while meeting regulatory and ethical requirements. The convergence of these technological and sustainability imperatives creates a compelling context for the integration of GIS-based spatial intelligence into manufacturing operations management.

The intersection of Industry 4.0 with spatial data analytics represents a rapidly evolving frontier. Qiu et al. (2025) reviewed the integration of IoT, IIoT, and Industry 4.0 technologies, identifying smart manufacturing and digital transformation as primary pathways for operationalizing sustainability strategies. GIS serves as the spatial intelligence layer within these frameworks—translating sensor data streams into interpretable maps and decision-support visualizations that enable operational teams to understand energy consumption not merely as a number on a utility bill but as a spatially distributed phenomenon shaped by equipment layout, process scheduling, building geometry, and human behavior.

Setyadi and Soekotjo (2025) conducted a systematic review of sustainable manufacturing trends from 2019 to 2024, identifying key dimensions including energy efficiency, circular economy, life cycle assessment, and digital integration. Their analysis underscores the growing importance of cross-disciplinary approaches that combine manufacturing engineering, environmental science, and geospatial informatics. The review identified a clear trend toward integrated platform approaches—combining monitoring, analysis, simulation, and reporting within unified digital environments—as the emerging standard for sustainable manufacturing management.

2.3 Geospatial Approaches to Energy Analysis

GIS has been widely applied in the energy sector for site selection of renewable energy installations, demand forecasting, grid planning, and carbon accounting.

Sachit et al. (2022) applied explainable AI alongside GIS to develop global suitability maps for wind and solar energy systems, demonstrating the power of spatial modeling in identifying optimal energy resource locations at continental scales. Shorabeh et al. (2022) used GIS and multi-criteria evaluation frameworks from economic perspectives to support wind energy power plant site selection, incorporating land use constraints, wind resource data, transmission infrastructure proximity, and environmental sensitivity into a composite spatial suitability index. Halder et al. (2022) combined GIS with AHP and multi-criteria decision analysis for solar power siting in urban contexts, demonstrating that the methodology is scalable from regional planning to individual facility-level investment decisions.

Within manufacturing environments, GIS has been applied to factory layout optimization, supply chain routing, and utility infrastructure management. When combined with energy consumption data, GIS enables the production of energy intensity maps—spatial representations showing where per-unit energy use is highest—that can guide facility retrofit decisions, equipment replacement priorities, and operational scheduling strategies. Tarife et al. (2023) demonstrated integrated GIS and fuzzy-AHP frameworks for siting hybrid renewable energy systems, a methodology directly applicable to on-site renewable integration planning in manufacturing contexts. The transferability of these spatial siting methods from regional energy planning to facility-level manufacturing management represents a major opportunity for practitioners seeking to leverage the methodological maturity of the GIS energy planning literature.

3. IIoT-BASED ENERGY MONITORING INFRASTRUCTURE

3.1 Smart Submetering and Edge Computing

Real-time energy monitoring forms the data foundation for any geospatial energy management system. Industrial IoT (IIoT) devices—including smart electricity submeters, thermal sensors, compressed air flow meters, power quality analyzers, and natural gas sub-meters—generate continuous streams of energy data at the machine and zone level. When these devices are spatially registered to a facility GIS, every data point carries a geographic coordinate, enabling spatial aggregation and visualization of energy consumption patterns with a temporal resolution that ranges from sub-second power quality measurements to hourly consumption totals.

Papaioannou et al. (2025) developed an IIoT-based submetering solution for real-time energy monitoring, demonstrating that fine-grained spatial and temporal resolution in energy measurement significantly improves the detection of inefficiency events compared to aggregate utility metering. Their architecture, which integrates wireless sensor nodes with a cloud database and GIS visualization layer, represents a practical implementation pathway for manufacturing firms seeking to establish spatial energy situational awareness without the capital expense of a fully custom SCADA deployment. Their results showed that submeter-level visibility enabled identification of energy waste events that were entirely invisible in aggregate monitoring data, including short-duration equipment idling cycles and phantom loads from standby-mode auxiliary systems.

Mirani et al. (2024) examined industrial IoT-based energy monitoring with a focus on edge data processing. Edge computing reduces latency in anomaly detection and minimizes the bandwidth demands of continuous sensor data transmission by preprocessing data locally before forwarding aggregated results to central GIS platforms. This architectural choice is particularly relevant for large-scale manufacturing facilities with thousands of monitoring points, where the cost and complexity of transmitting raw sensor data to a central server would be prohibitive. Edge preprocessing also enables local alarming—triggering immediate alerts when energy consumption exceeds predefined thresholds without waiting for cloud-based analysis cycles.

Bin Mofidul et al. (2022) presented a secured, integrated global IIoT infrastructure incorporating edge and cloud AI for real-time energy anomaly detection. Their system achieved sub-second anomaly alerts and demonstrated the feasibility of deploying machine learning inference at the edge, enabling rapid response to energy waste events without relying on cloud round-trip latency. Their work also addressed cybersecurity considerations in IIoT energy monitoring—an increasingly important concern as manufacturing organizations connect operational technology systems to enterprise networks and cloud platforms that may be exposed to external attack vectors.

3.2 Data Integration and Spatial Registration

For energy monitoring data to be meaningful in a GIS context, sensor locations must be accurately registered to a spatial reference system. This requires facility mapping—typically a combination of architectural CAD drawings, as-built engineering plans, and field survey data—to be imported into a GIS platform such as ArcGIS or QGIS. Each sensor is then assigned coordinates within this spatial model, creating a georeferenced sensor network in which every measurement is associated with a specific physical location within the facility. The quality of this spatial registration process directly determines the accuracy of subsequent spatial analysis: sensors with inaccurate location assignments will produce misleading energy intensity maps that direct efficiency investments to the wrong areas.

Teng et al. (2021) reviewed advances in industrial data-driven energy savings through digital twins and infrastructure, highlighting the importance of interoperable data models that allow sensor readings to be linked to physical assets, processes, and spatial locations simultaneously. Their framework for energy infrastructure visualization—which overlays process flow diagrams with energy consumption heatmaps—represents a practical approach to spatially contextualizing IIoT data within manufacturing environments. Interoperability between sensor data formats, GIS data models, and enterprise information systems remains a significant technical challenge in real-world implementations, requiring careful attention to data standards including OPC-UA for industrial sensor data, GeoJSON for spatial features, and ISO 50001 for energy management system integration.

Ma et al. (2022) developed a digital twin and big data-driven framework for smart manufacturing in energy-intensive industries, demonstrating how real-time sensor integration with spatial models enables dynamic energy accounting across entire production systems. Their work on information management systems for energy-intensive industries provides a replicable architecture for implementing GIS-integrated energy monitoring at enterprise scale. Their framework specifically addressed the challenge of integrating energy data from diverse sources—including utility meters, process sensors, building management systems, and production management databases—into a unified spatial model that enables cross-system energy performance analysis.

Figure 1. GIS-Based Energy Mapping Framework for Manufacturing Industries.

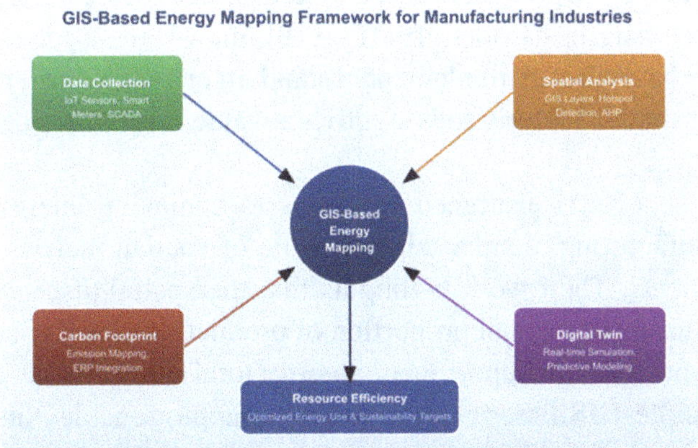

Figure 1 illustrates the complete GIS-Based Energy Mapping Framework, showing how IIoT data collection, spatial analysis, digital twin simulation, carbon footprint accounting, and resource efficiency outputs are integrated within a unified geospatial architecture.

4. SPATIAL ENERGY ANALYSIS AND HOTSPOT MAPPING

4.1 Energy Intensity Mapping

Once energy consumption data from IIoT sensors is integrated into a GIS platform, spatial analysis techniques can be applied to identify patterns, anomalies, and opportunities for efficiency improvement. Energy intensity mapping involves calculating energy consumption per unit of production output—or per unit of floor area—at each spatial location, and rendering these values as choropleth or heatmap layers within the facility GIS. The resulting maps provide an intuitively interpretable representation of spatial energy distribution that supports rapid identification of problem areas by facility managers, energy auditors, and production engineers without requiring advanced analytical skills.

Zones with disproportionately high energy intensity relative to their production contribution represent primary targets for efficiency intervention. Common causes of spatial energy concentration include inefficient equipment clustering, poor thermal insulation in specific building zones, leaking compressed air headers serving particular production areas, and suboptimal production scheduling that creates extended idling periods in high-load equipment zones. Cai et al. (2022) identified spatial benchmarking—the comparison of energy performance across production zones, shifts, and equipment clusters—as a key mechanism for unlocking efficiency improvements that aggregate monitoring systems fail to detect. Their review found that spatial benchmarking can identify efficiency improvement potentials of 10–25% in facilities that have already implemented standard energy management programs, suggesting that spatial analysis adds significant value even where basic monitoring is already in place.

Schmitt et al. (2025) examined pathways to achieving energy efficiency in industrial manufacturing, emphasizing the role of spatial analysis in prioritizing retrofit investments. Their work highlights that the spatial distribution of energy waste is rarely uniform: a small proportion of production zones or equipment units typically accounts for a disproportionate share of total energy waste, consistent with the Pareto principle. GIS-based energy intensity mapping enables this concentration to be visualized and quantified systematically, supporting investment prioritization decisions that maximize efficiency return per unit of capital expenditure. The visual accessibility of choropleth energy intensity maps also serves an important communication function, enabling sustainability managers to engage senior decision-makers and operational staff in efficiency programs through intuitive spatial narratives that abstract numerical data alone cannot provide.

4.2 Hotspot Detection and Spatial Statistics

Beyond simple intensity mapping, spatial statistics methods—including kernel density estimation, Getis-Ord Gi* statistics, and Moran's I spatial autocorrelation—can be applied to energy consumption data to identify statistically significant hotspot clusters. These methods distinguish genuine spatial concentrations of energy inefficiency from random variation in consumption patterns, improving the reliability of targeting decisions and providing defensible statistical justification for efficiency investment proposals. The Getis-Ord Gi* statistic, for example, identifies locations where high energy intensity values are spatially clustered to a degree that is statistically unlikely to have occurred by chance, indicating genuine structural or operational sources of concentrated inefficiency.

Energy hotspot detection is analogous to methods applied in environmental management and public health GIS, where the identification of pollution or disease clusters has a well-established methodological tradition. Translating these methods into manufacturing environments requires adaptation to the regular grid-like spatial structure of factory layouts and the cyclical temporal patterns of production scheduling. Geng et al. (2023) contributed to this adaptation by developing classification schemes for energy waste that map cleanly onto spatial analysis workflows, providing a conceptual bridge between lean manufacturing's process-level waste identification and GIS-based spatial analytics.

The integration of machine learning with spatial statistics offers additional capabilities for pattern recognition in complex, high-dimensional energy datasets. Saetang et al. (2024) demonstrated an eco-friendly platform combining machine learning with GIS for carbon footprint management, illustrating the transferability of these methods to industrial contexts. Their algorithmic approach to spatial carbon intensity analysis—combining random forest classification of energy use patterns with GIS visualization of spatial clusters—provides a template for manufacturing-specific applications where the dimensionality of energy data exceeds the capacity of manual analysis. Temporal pattern analysis within spatial hotspots further enables distinction between chronic inefficiency sources, such as poorly maintained equipment, and episodic waste events, such as production changeovers that create unplanned energy loads in specific facility zones.

4.3 Energy Value Stream Mapping with Geospatial Overlay

Value stream mapping (VSM) is a lean manufacturing tool for visualizing material and information flows through production processes. Wen et al. (2021) introduced the concept of energy value stream mapping (EVSM), which extends conventional VSM to include energy consumption at each process step, enabling identification of energy waste in the context of production flow. The EVSM methodology quantifies the energy content of each production step—including processing energy, handling energy, quality failure energy, and standby energy—and identifies which steps contribute productive value versus which represent pure waste in the lean sense.

When EVSM diagrams are overlaid on facility GIS layers, the spatial relationships between energy-intensive processes become explicit, revealing efficiency opportunities that neither VSM nor energy monitoring alone can identify. This spatial EVSM approach reveals, for example, whether high-energy processes are co-located in a thermally inefficient zone where waste heat could be recaptured, whether material routing through the facility creates unnecessary transport energy loads that could be reduced by layout reconfiguration, or whether scheduling misalignment between consecutive high-energy processes creates extended holding periods that consume energy without advancing production. These insights are not accessible from production flow diagrams or aggregate energy data alone—they require the spatial perspective that GIS provides, integrated with the process perspective that VSM provides.

5. DIGITAL TWIN TECHNOLOGIES FOR ENERGY SIMULATION

5.1 Energy Digital Twins in Manufacturing

A digital twin is a dynamic, real-time digital replica of a physical asset, system, or process that is continuously updated with sensor data and capable of running simulations to predict future states and evaluate alternative configurations. In the context of energy management, energy digital twins enable facility managers to model the energy consequences of operational changes—such as production schedule modifications, equipment replacements, compressed air system pressure adjustments, or building envelope retrofits—before committing to physical implementation. This simulation-before-implementation capability reduces investment risk and enables more informed decision-making by allowing energy managers to test competing efficiency strategies in a virtual environment and select the option with the best predicted performance.

Yu et al. (2022) provided a comprehensive classification of energy digital twin technologies for industrial energy management, identifying four key categories: component-level twins, system-level twins, process-level twins, and facility-level twins. Each category differs in spatial granularity and the range of energy interactions it captures. Component-level twins focus on individual machines or pieces of equipment, modeling their energy consumption as a function of operating parameters and load conditions. Facility-level twins capture the full spatial complexity of a manufacturing environment, including interactions between building energy systems, production equipment, utility infrastructure, and occupant behavior. When integrated with GIS, facility-level twins can incorporate spatial relationships between building zones, utility routing, and production areas to simulate spatially distributed energy performance across entire facilities.

Ba et al. (2024) conducted a systematic review of digital twin applications in energy efficiency, finding strong evidence that simulation-based optimization through digital twins can achieve energy savings of 15–35% in manufacturing environments. Their review identified real-time data integration, predictive maintenance scheduling, and process parameter optimization as the primary mechanisms through which energy savings are realized. The spatial dimension of these twins—embedded in GIS infrastructure—enables identification of interaction effects between spatially adjacent processes that aggregate models overlook, such as the thermal coupling between neighboring production zones that causes heating energy loads in one zone to partially substitute for heating in adjacent zones.

Do Amaral et al. (2023) reviewed energy digital twin applications across sectors, identifying manufacturing and building management as the fields with the highest density of implemented cases. Their analysis highlighted the convergence of simulation, optimization, and real-time control as the defining characteristics of mature energy digital twin deployments. The role of GIS in providing the spatial context that allows twins to reflect real-world geometric and infrastructural constraints was identified as a critical differentiator between high-fidelity twins that achieve significant efficiency gains and simplified models that fail to capture the complexity of real manufacturing environments.

5.2 Integration with Big Data and AI

The performance of energy digital twins is critically dependent on the quality, resolution, and timeliness of the data inputs they receive. Ma et al. (2022) demonstrated a big data-driven smart manufacturing framework that integrates digital twin simulation with enterprise information management systems, enabling energy optimization decisions to be made in the context of real-time production schedules, utility pricing signals, and weather forecast data that affect building energy loads. Their framework, implemented in an energy-intensive industrial context, achieved measurable reductions in peak energy demand and improved overall energy efficiency by enabling shift supervisors to respond to emerging efficiency opportunities identified by the twin's predictive algorithms.

Artificial intelligence techniques, including deep learning, reinforcement learning, and Bayesian optimization, are increasingly being applied within digital twin frameworks to automate energy optimization decisions. Deep learning models trained on historical energy consumption patterns can identify subtle signatures of equipment degradation—gradual increases in energy consumption that precede measurable performance deterioration—enabling predictive maintenance scheduling that prevents both equipment failures and the associated energy waste of degraded operation. Reinforcement learning agents, operating within digital twin environments, can learn optimal energy scheduling policies through simulated trial-and-error without the risks of experimenting on live production systems.

Figure 2. Geospatial Layer Architecture for Energy Resource Efficiency in Manufacturing.

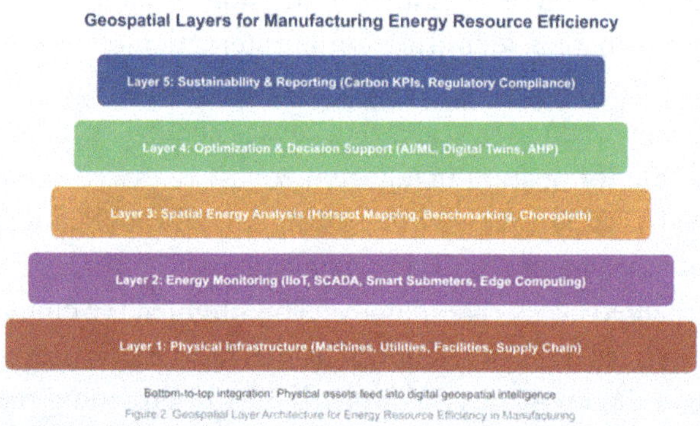

Figure 2 illustrates the Geospatial Layer Architecture for Manufacturing Energy Resource Efficiency, illustrating how physical infrastructure data, IIoT monitoring streams, spatial energy analysis outputs, AI-driven optimization, and sustainability reporting are organized as vertically integrated layers within the GIS framework.

6. CARBON FOOTPRINT MAPPING AND EMISSION ACCOUNTING

6.1 Spatial Carbon Accounting in Manufacturing

Carbon footprint mapping translates energy consumption and process emission data into spatial representations of greenhouse gas intensity, enabling facility managers to identify emission hotspots, track progress toward reduction targets, and communicate environmental performance to stakeholders. Spatially disaggregated carbon accounting provides a level of granularity that aggregate facility-level emission inventories cannot match, enabling targeted interventions at specific production zones or equipment units rather than requiring facility-wide measures that may deliver efficiency improvements unevenly across different production areas.

Stavropoulos and Panagiotopoulou (2022) examined carbon footprint differences between conventional and non-conventional manufacturing processes, finding significant variation in emission intensity across process types that has direct implications for spatial production planning. Their analysis demonstrated that spatial co-location of high-emission processes can create localized air quality concerns as well as energy management challenges, reinforcing the case for integrated spatial analysis that addresses both energy efficiency and environmental compliance objectives simultaneously. The spatial patterning of emission intensity across manufacturing facilities reflects the combined influence of energy consumption intensity, fuel type and carbon content, and process-specific emission factors, all of which vary spatially and can be mapped within a GIS framework.

Yurtay (2025) explored the integration of Industry 4.0 technologies—particularly enterprise resource planning (ERP) systems—with carbon footprint management, demonstrating that real-time data flows between production management systems and GIS-based emission accounting tools enable continuous carbon monitoring rather than periodic retrospective calculation. This approach is essential for manufacturing facilities pursuing Science Based Targets (SBT) or participating in carbon trading schemes that require verified, fine-grained emission records updated at sub-daily frequencies. The linkage between ERP production data and GIS emission mapping creates an auditable chain of evidence from physical production activities to spatial emission inventories that satisfies the requirements of third-party verification processes.

Sun et al. (2021) analyzed carbon emission transfer strategies in supply chains, noting that spatial and temporal lags in emission reduction technology deployment affect the distribution of carbon responsibility across supply chain actors. This supply chain dimension of carbon footprint mapping extends the scope of GIS-based emission accounting beyond individual facilities to encompass the entire value chain—a perspective increasingly required by Scope 3 emissions reporting frameworks that hold manufacturing organizations accountable for upstream supplier emissions and downstream use-phase emissions as well as their own direct Scope 1 and 2 emissions.

6.2 GIS Integration with Carbon Management Platforms

Saetang et al. (2024) demonstrated an integrated GIS and machine learning platform for carbon footprint management that combines spatial emission mapping with predictive analytics and green space analysis. Although their work focused on office environments, the methodological framework—which integrates activity data, emission factors, and spatial visualization—is directly transferable to manufacturing contexts with appropriate adaptation of process-level emission factors and spatial data models. Their use of random forest algorithms to predict future emission patterns based on spatial land use and activity characteristics provides a template for manufacturing carbon forecasting applications.

Li et al. (2022) investigated spatiotemporal variations in per-capita carbon emissions from energy consumption in China, employing GIS-based analysis to identify geographic drivers of emission variation across provinces and cities. Zhang et al. (2024) extended this spatial analysis to examine the impact of digital economy development on carbon emission reduction across 215 Chinese cities, finding that digitalization of industrial processes—including the adoption of IIoT and GIS tools—contributes significantly to emission reduction at regional scales. These findings support the investment case for GIS-based energy and carbon management tools by demonstrating measurable macroeconomic emission reduction benefits from digitalization.

The convergence of spatial emission accounting with hydrogen energy storage and renewable integration planning represents an emerging frontier in manufacturing sustainability. Hassan et al. (2023) reviewed advances in hydrogen energy storage technologies and their implications for sustainability, noting that the spatial co-location of hydrogen production facilities with manufacturing energy loads—a classic GIS siting optimization problem—can significantly reduce the delivered cost of low-carbon energy. The integration of hydrogen storage siting analysis with manufacturing facility GIS provides a framework for evaluating on-site renewable hydrogen production as a decarbonization pathway for high-temperature industrial processes that cannot be easily electrified.

7. MULTI-CRITERIA DECISION SUPPORT FOR RENEWABLE ENERGY INTEGRATION

7.1 AHP and GIS-Based Site Selection

The integration of renewable energy systems—including rooftop solar photovoltaic, small wind turbines, ground source heat pumps, and biomass gasification—into manufacturing facilities requires careful site selection analysis that considers spatial constraints including available area, structural loading capacity, solar shading patterns, prevailing wind regimes, grid connection distance, and regulatory setback requirements. GIS provides the spatial analytical platform for this analysis, while multi-criteria decision methods such as the Analytic Hierarchy Process (AHP) provide the weighting framework for balancing competing objectives in a structured, transparent, and auditable manner.

Halder et al. (2022) demonstrated GIS-AHP integration for solar power plant siting, developing a weighted suitability surface that integrates land availability, solar irradiance, proximity to transmission infrastructure, and environmental sensitivity. Their methodology, adapted for manufacturing facility contexts, provides a practical template for on-site renewable energy siting analysis that can be applied to rooftop solar surveys, wind turbine micro-siting studies, and combined heat and power system location optimization within industrial parks. Tarife et al. (2023) applied an integrated GIS-fuzzy AHP framework to hybrid renewable energy system planning, demonstrating the flexibility of GIS-based multi-criteria methods for diverse geographic and industrial contexts.

Shorabeh et al. (2022) applied GIS multi-criteria evaluation from economic perspectives to wind energy site selection, incorporating lifecycle cost analysis, land acquisition costs, and grid connection investment into the spatial suitability model. This economic dimension of GIS-based siting analysis is particularly relevant for manufacturing facility investment decisions where capital allocation must be justified on financial as well as environmental grounds. The integration of net present value analysis with spatial suitability mapping enables identification of locations where renewable energy investments are not only technically feasible but financially attractive given the specific energy tariff, solar resource, and grid connection characteristics of each candidate site.

7.2 Explainable AI and Geospatial Decision Support

The integration of artificial intelligence with GIS-based multi-criteria decision support has opened new possibilities for transparent, scalable spatial optimization. Sachit et al. (2022) applied explainable AI techniques to global spatial suitability mapping for wind and solar systems, demonstrating that machine learning models can learn complex multi-variable spatial patterns from large training datasets while remaining interpretable to non-expert users through feature importance visualization and local explanation methods such as SHAP values. The application of explainable AI to spatial energy optimization problems in manufacturing enables decision support systems that can justify their recommendations in terms that facility managers and investment committees can evaluate and challenge.

The explainability dimension is particularly important in manufacturing contexts where energy efficiency investments must be justified to senior management, regulators, and investors who may not have the technical background to evaluate complex optimization algorithms. Decision support tools that present spatial optimization results in transparent, auditable formats—showing how each criterion contributed to the overall suitability score at each location, and how the results would change under alternative weighting schemes—are substantially more likely to achieve organizational acceptance and successful implementation than black-box optimization algorithms whose recommendations cannot be explained or sensitivity-tested.

8. COMPARATIVE ANALYSIS: GIS-INTEGRATED TECHNOLOGY APPROACHES

8.1 Technology Assessment Framework

A structured comparative assessment of GIS-integrated technology approaches for energy resource efficiency in manufacturing reveals significant variation in estimated energy savings, implementation complexity, data requirements, and organizational readiness requirements across available tool categories. Table 1 presents a synthesis of eight principal approaches drawn from the literature reviewed in this chapter, characterized along four dimensions: the primary technology platform typically employed in industry practice, the specific manufacturing application domain in which each approach delivers value, and an overall effectiveness rating synthesized from the empirical evidence in the reviewed literature.

Table 1. Comparative Assessment of GIS-Integrated Technology Approaches for Manufacturing Energy Efficiency

Geospatial Tool / Method	Technology Platform	Application in Manufacturing	Effectiveness
GIS Hotspot Mapping	ArcGIS, QGIS	Spatial energy intensity zones; zone-level benchmarking	High
IIoT Submetering	Smart sensors, Edge AI	Real-time machine and line-level consumption data	Very High
Digital Twin Simulation	ANSYS, Siemens NX	Predictive energy modeling; what-if scenario analysis	High
AI/ML Analytics	Python, TensorFlow	Anomaly detection, demand forecasting, scheduling	High
Carbon Footprint Mapping	ERP, QGIS Plug-ins	Emission hotspot visualization; Scope 1/2/3 tracking	Moderate–High
Multi-criteria Siting (AHP)	GIS + AHP tools	Renewable energy siting; investment prioritization	Moderate
Energy Value Stream Mapping	Lean + GIS overlay	Process-level waste identification; flow optimization	Moderate–High
Benchmarking & KPI Dashboard	BI tools, GIS layers	Comparative performance across zones, shifts, facilities	High

As summarized in Table 1, IIoT submetering achieves the highest effectiveness rating among individual tool categories, reflecting its foundational role in generating the real-time spatial data that all subsequent analysis layers depend upon. GIS hotspot mapping and AI/ML analytics both achieve high effectiveness ratings by enabling systematic identification and prediction of inefficiency patterns that operational teams cannot detect through manual observation. Digital twin simulation achieves a high effectiveness rating particularly for facilities undertaking significant capital investments in equipment or layout changes, where the ability to test alternatives in simulation before physical implementation delivers substantial value. Multi-criteria siting using AHP achieves a moderate effectiveness rating for individual facility applications, reflecting the relatively straightforward nature of most facility-scale siting decisions compared to regional renewable energy planning contexts.

8.2 Energy Savings Benchmarking

A critical dimension of technology evaluation in the manufacturing energy efficiency domain is the quantified energy savings potential associated with each approach, both individually and in combination. The estimation of savings potential draws on empirical evidence from implementation case studies, controlled experiments in research facilities, and expert elicitation reported in the literature reviewed in this chapter. These estimates should be interpreted as indicative benchmarks that reflect mean outcomes across diverse manufacturing contexts rather than guaranteed performance specifications, as actual savings will vary significantly with facility characteristics, implementation quality, and organizational capability.

Figure 3. Comparative Estimated Energy Savings (%) Achieved Through GIS-Integrated Technology Approaches in Manufacturing.

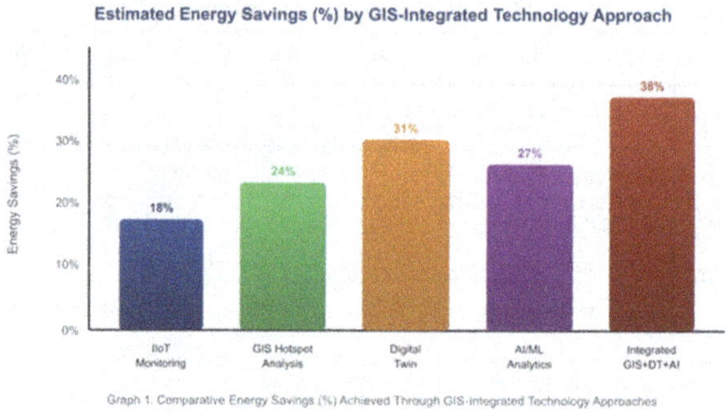

Figure 3 illustrates a comparative bar chart of estimated energy savings (%) achievable through five GIS-integrated technology approaches, from IIoT monitoring alone through to the fully integrated GIS + Digital Twin + AI framework. The progressive savings gains illustrate the compounding value of technology integration.

As illustrated in Figure 3, IIoT monitoring alone—without spatial analysis or simulation capabilities—achieves energy savings in the range of 15–20%, primarily through improved visibility of consumption patterns and faster detection and correction of waste events. The addition of GIS-based hotspot analysis increases potential savings to approximately 22–26% by enabling spatially targeted interventions directed at the specific zones and equipment units that account for the largest share of avoidable energy waste. Digital twin integration—which enables simulation-based optimization of production scheduling, equipment settings, and utility system operation before physical implementation—achieves savings in the range of 28–34%. AI and machine learning analytics applied to multi-dimensional energy datasets contribute savings in the range of 25–30% through anomaly detection, demand forecasting, and automated optimization. The fully integrated approach combining GIS, digital twins, and AI achieves the highest estimated savings of 35–40%, consistent with findings from Ba et al. (2024) and Ma et al. (2022), by harnessing the complementary strengths of each component within a unified spatial intelligence framework.

The compounding nature of savings across integrated technology layers reflects the fact that each additional capability addresses efficiency improvement opportunities that lower-capability systems cannot exploit. IIoT monitoring alone identifies when energy is being wasted but cannot explain where in the spatial complexity of a facility the waste originates. GIS hotspot analysis identifies where waste is concentrated but cannot predict whether proposed interventions will achieve their savings targets. Digital twin simulation validates intervention strategies before implementation but depends on rich spatial and operational data from IIoT and GIS layers to achieve high prediction accuracy. AI/ML analytics discover non-obvious patterns and relationships in the data that deterministic models cannot detect. The integration of all four capabilities within a GIS-grounded spatial intelligence framework thus creates a multiplicative rather than merely additive efficiency benefit.

9. FUTURE DIRECTIONS AND EMERGING TECHNOLOGIES

9.1 Cloud-Native GIS and Real-Time Spatial Analytics

The transition from desktop GIS to cloud-native spatial analytics platforms represents a significant enabler for energy mapping at scale. Cloud-native GIS architectures enable continuous ingestion of IIoT data streams with real-time spatial analysis and visualization, supporting energy management decisions at the pace of operational events rather than the pace of weekly or monthly reporting cycles. For large manufacturing enterprises with multiple facilities distributed across geographic regions, cloud-native GIS provides the scalability to aggregate and compare energy performance across the entire production network, enabling corporate energy managers to identify high-performing facilities whose practices should be replicated and underperforming facilities that require priority support.

Qiu et al. (2025) identified cloud-native data architectures as a key enabler for the next phase of IIoT and Industry 4.0 integration, noting that data sovereignty, interoperability, and cybersecurity considerations require careful architectural design. The secure integration of energy data with cloud GIS platforms requires robust data governance frameworks that protect commercially sensitive production information while enabling the cross-facility benchmarking and regulatory reporting that motivates the investment. Edge-to-cloud continuum architectures—in which lightweight processing occurs at the factory edge and aggregated spatial analytics occur in the cloud—represent the emerging standard for balancing responsiveness, cost, and scalability in manufacturing energy management systems.

9.2 Digital-Physical Integration and Smart Factories

The smart factory paradigm envisions a seamlessly integrated digital-physical environment in which production systems, energy management systems, and spatial analytics platforms operate as a unified whole, sharing data in real time and enabling closed-loop optimization of both production output and resource consumption. Schmitt et al. (2025) outlined pathways toward achieving energy efficiency in industrial manufacturing that emphasize the convergence of operational technology (OT) and information technology (IT), including the integration of GIS-based spatial intelligence with real-time production control systems.

The emergence of Building Information Modeling (BIM) and its integration with GIS—often referred to as GeoBIM—opens new opportunities for energy mapping in manufacturing facilities. BIM models capture the three-dimensional geometry of building structures, mechanical systems, and utility infrastructure in rich detail, providing a natural complement to the two-dimensional facility GIS layers that most current energy management implementations use. The integration of BIM-derived energy simulation outputs with IIoT sensor data and GIS spatial analysis—enabling three-dimensional energy visualization that can distinguish energy performance by floor level, building zone, and equipment cluster—represents a frontier area for research and practice in manufacturing energy management.

9.3 Policy Implications and Organizational Change

The successful adoption of GIS-based energy mapping and resource efficiency tools in manufacturing industries requires not only technological capability development but also organizational knowledge development, policy alignment, and cultural change. Andrei et al. (2022) identified organizational knowledge gaps as a key barrier to effective energy management, highlighting the need for training programs that build interdisciplinary competencies spanning energy engineering, GIS, and data science. Manufacturing organizations that invest in developing these hybrid competencies—whether through internal training programs, strategic hiring, or partnerships with universities and technology vendors—achieve significantly better outcomes from their energy management technology investments than those that treat GIS and IIoT systems as purely technical deployments.

Regulatory frameworks increasingly mandate spatial reporting of energy consumption and emissions data, creating institutional drivers for GIS capability development in manufacturing organizations. The European Union's Energy Efficiency Directive, the ISO 50001 energy management standard, and emerging Scope 3 emissions reporting requirements all create demand for spatially granular energy accounting that GIS-based systems are well positioned to provide. Setyadi and Soekotjo (2025) noted that sustainable manufacturing transitions require alignment between technological capability development and organizational strategy, governance structures, and stakeholder engagement, reinforcing the importance of treating GIS-based energy management as a strategic organizational capability rather than a standalone technical tool.

10. CONCLUSION

This chapter has presented a comprehensive examination of energy mapping and resource efficiency in manufacturing industries through the lens of geospatial tools. By synthesizing recent literature spanning IIoT-based energy monitoring, GIS spatial analysis, digital twin technology, carbon footprint mapping, and multi-criteria decision support, a coherent integrated framework has emerged for understanding how spatial intelligence can transform the practice of industrial energy management. The framework, illustrated in Figure 1 and operationalized through the five-layer geospatial architecture presented in Figure 2, demonstrates that the integration of spatial intelligence with real-time monitoring, simulation, and optimization creates energy management capabilities that substantially exceed what any individual technology can deliver in isolation.

The evidence reviewed in this chapter demonstrates that GIS-integrated approaches to energy management substantially outperform conventional aggregate monitoring strategies in terms of both the depth of insight generated and the magnitude of energy savings achieved. As summarized in Figure 3, integrated frameworks combining GIS, digital twins, and artificial intelligence achieve estimated energy savings of up to 38–40%, compared to 15–20% for IIoT monitoring alone—a difference that represents enormous value in energy cost and emission reduction terms for energy-intensive manufacturing enterprises. The spatial targeting capability that GIS brings to energy management is the key differentiator: by directing efficiency investments toward the specific zones, processes, and equipment units that account for the greatest shares of avoidable energy waste, GIS-informed approaches achieve superior returns per unit of investment compared to facility-wide programs that spread effort uniformly.

Several key findings from this review merit specific emphasis for practitioners and researchers. First, the spatial distribution of energy waste in manufacturing facilities is typically highly concentrated, with a small proportion of zones and processes accounting for a disproportionate share of total waste. GIS-based hotspot mapping is essential for identifying these concentrations and directing efficiency investments where they will have the greatest impact. Second, digital twin technologies, when integrated with GIS spatial models and real-time IIoT data, enable simulation-based optimization that can be validated before physical implementation, reducing investment risk and accelerating the realization of efficiency benefits. Third, carbon footprint mapping—linking energy consumption data to emission factors and presenting results in spatially disaggregated form—enables manufacturing organizations to meet the growing demands of emissions reporting frameworks while identifying practical reduction pathways grounded in spatial operational analysis.

Future research should focus on developing standardized data architectures for GIS-IIoT integration that enable interoperability across diverse industrial platforms; advancing explainable AI methods for spatial energy optimization that support transparent decision-making; evaluating the organizational and policy conditions that enable successful adoption of GIS-based energy management tools in manufacturing industries of varying sizes and sectors; and extending GIS-based energy mapping frameworks to encompass supply chain spatial analysis and industrial symbiosis applications where one facility's waste energy becomes another's resource. The convergence of geospatial intelligence, real-time sensor networks, simulation technologies, and artificial intelligence represents one of the most promising and impactful frontiers for advancing sustainable manufacturing in the decades ahead, and the systematic integration of these capabilities within spatially grounded frameworks will be a defining feature of the next generation of world-class manufacturing operations.

REFERENCES

Andrei, M., Thollander, P., & Sannö, A. (2022). Knowledge demands for energy management in manufacturing industry: A systematic literature review. *Renewable & Sustainable Energy Reviews*, *159*, 112168. DOI: 10.1016/j.rser.2022.112168

Ba, L., Tangour, F., Abbassi, I. E., & Absi, A. (2024). Analysis of digital twin applications in energy efficiency: A systematic review. *Sustainability (Basel)*, *17*(8), 3560. DOI: 10.3390/su17083560

Bin Mofidul, R., Alam, M. M., Rahman, M. H., & Jang, Y. M. (2022). Real-time energy data acquisition, anomaly detection, and monitoring system: Implementation of a secured, robust, and integrated global IIoT infrastructure with edge and cloud AI. *Sensors (Basel)*, *22*(22), 8980. DOI: 10.3390/s22228980 PMID: 36433575

Cai, W., Wang, L., Li, L., Xie, J., Jia, S., Zhang, X., Jiang, Z., & Lai, K.-H. (2022). A review on methods of energy performance improvement towards sustainable manufacturing from perspectives of energy monitoring, evaluation, optimization and benchmarking. *Renewable & Sustainable Energy Reviews*, *159*, 112227. DOI: 10.1016/j.rser.2022.112227

do Amaral, J. V. S., dos Santos, C. H., Montevechi, J. A. B., & de Queiroz, A. R. (2023). Energy Digital Twin applications: A review. *Renewable & Sustainable Energy Reviews*, *188*, 113891. DOI: 10.1016/j.rser.2023.113891

Geng, D., & Evans, S. (2022). A literature review of energy waste in the manufacturing industry. *Computers & Industrial Engineering*, *173*, 108713. DOI: 10.1016/j.cie.2022.108713

Geng, D., Evans, S., & Kishita, Y. (2023). The identification and classification of energy waste for efficient energy supervision in manufacturing factories. *Renewable & Sustainable Energy Reviews*, *182*, 113409. DOI: 10.1016/j.rser.2023.113409

Halder, B., Banik, P., Almohamad, H., Al Dughairi, A. A., Al-Mutiry, M., Al Shahrani, H. F., & Abdo, H. G. (2022). Land suitability investigation for solar power plant using GIS, AHP and multi-criteria decision approach: A case of megacity Kolkata, West Bengal, India. *Sustainability (Basel)*, *14*(18), 11276. DOI: 10.3390/su141811276

Hassan, Q., Sameen, A. Z., Salman, H. M., Jaszczur, M., & Al-Jiboory, A. K. (2023). Hydrogen energy future: Advancements in storage technologies and implications for sustainability. Journal of Energy Storage, 72(Part A), 108404. DOI: 10.1016/j.est.2023.108404

Javaid, M., Haleem, A., Singh, R. P., Haq, M. I. U., Raina, A., & Suman, R. (2022). Sustainability 4.0 and its applications in the field of manufacturing. *Internet of Things and Cyber-Physical Systems*, *2*, 82–90. DOI: 10.1016/j.iotcps.2022.06.001

Li, Y., Liu, C., & Zhang, S. (2022). Spatiotemporal variations and driving forces of per capita carbon emissions from energy consumption in China. *Geomatics, Natural Hazards & Risk*, *13*(1), 2489–2507. DOI: 10.1080/19475705.2022.2122593

Ma, S., Ding, W., Liu, Y., Ren, S., & Yang, H. (2022). Digital twin and big data-driven sustainable smart manufacturing based on information management systems for energy-intensive industries. *Applied Energy*, *326*, 119986. DOI: 10.1016/j.apenergy.2022.119986

Mirani, A. A., Awasthi, A., Mahony, N. O., & Waish, J. (2024). Industrial IoT-based energy monitoring system: Using data processing at edge. *IoT*, *5*(4), 608–633. DOI: 10.3390/iot5040027

Papaioannou, T. G., Dimitriou, N., Vasilakis, K., & Schoofs, A. (2025). Industrial IoT-based submetering solution for real-time energy monitoring. *Discover Internet of Things*, *5*(1), 14. DOI: 10.1007/s43926-025-00110-y

Qiu, F., Kumar, A., Hu, J., Sharma, P., Tang, Y. B., Xiang, Y. X., & Hong, J. (2025). A review on integrating IoT, IIoT, and Industry 4.0: A pathway to smart manufacturing and digital transformation. *IET Information Security*, *9275962*(1), 9275962. Advance online publication. DOI: 10.1049/ise2/9275962

Sachit, M. S., Shafri, H. Z. M., Abdullah, A. F., Rafie, A. S. M., & Gibril, M. B. A. (2022). Global spatial suitability mapping of wind and solar systems using an explainable AI-based approach. *ISPRS International Journal of Geo-Information*, *11*(8), 422. DOI: 10.3390/ijgi11080422

Saetang, W., Chai-Araralert, S., Kajornkasirat, S., Kongcharoen, J., Saeliw, A., Puangsuwan, K., & Puttinaovarat, S. (2024). Eco-friendly office platform: Leveraging machine learning and GIS for carbon footprint management and green space analysis. *Sustainability (Basel)*, *16*(21), 9424. DOI: 10.3390/su16219424

Schmitt, T., Mattsson, S., Flores-García, E., & Hanson, L. (2025). Achieving energy efficiency in industrial manufacturing. *Renewable & Sustainable Energy Reviews*, *216*, 115619. DOI: 10.1016/j.rser.2025.115619

Setyadi, A., Soekotjo, S., Lestari, S. D., Pawirosumarto, S., & Damaris, A. (2025). Trends and opportunities in sustainable manufacturing: A systematic review of key dimensions from 2019 to 2024. *Sustainability (Basel)*, *17*(2), 789. DOI: 10.3390/su17020789

Shorabeh, S. N., Firozjaei, H. K., Firozjaei, M. K., Jelokhani-Niaraki, M., Homaee, M., & Nematollahi, O. (2022). The site selection of wind energy power plant using GIS-multi-criteria evaluation from economic perspectives. *Renewable & Sustainable Energy Reviews*, *168*, 112778. DOI: 10.1016/j.rser.2022.112778

Stavropoulos, P., & Panagiotopoulou, V. C. (2022). Carbon footprint of manufacturing processes: Conventional vs. non-conventional. *Processes (Basel, Switzerland)*, *10*(9), 1858. DOI: 10.3390/pr10091858

Sun, L., Cao, X., Alharthi, M., Zhang, J., Taghizadeh-Hesary, F., & Mohsin, M. (2021). Carbon emission transfer strategies in supply chain with lag time of emission reduction technologies and low-carbon preference of consumers. *Journal of Cleaner Production*, *264*, 121664. DOI: 10.1016/j.jclepro.2020.121664

Tarife, R., Nakanishi, Y., Chen, Y., Zhou, Y., Estoperez, N., & Tahud, A. (2023). Integrated GIS and fuzzy-AHP framework for suitability analysis of hybrid renewable energy systems: A case in southern Philippines. *Sustainability (Basel)*, *15*(3), 2372. DOI: 10.3390/su15032372

Teng, S. Y., Touš, M., Leong, W. D., How, B. S., Lam, H. L., & Máša, V. (2021). Recent advances on industrial data-driven energy savings: Digital twins and infrastructures. *Renewable & Sustainable Energy Reviews*, *135*, 110208. DOI: 10.1016/j.rser.2020.110208

Wen, X., Cao, H., Hon, B., Chen, E., & Li, H. (2021). Energy value mapping: A novel lean method to integrate energy efficiency into production management. *Energy*, *217*, 119353. DOI: 10.1016/j.energy.2020.119353

Yu, W., Patros, P., Young, B., Klinac, E., & Walmsley, T. G. (2022). Energy digital twin technology for industrial energy management: Classification, challenges and future. *Renewable & Sustainable Energy Reviews*, *161*, 112407. DOI: 10.1016/j.rser.2022.112407

Yurtay, Y. (2025). Carbon footprint management with Industry 4.0 technologies and ERP systems in sustainable manufacturing. *Applied Sciences (Basel, Switzerland)*, *15*(1), 480. DOI: 10.3390/app15010480

Zhang, H., Qi, R., Liu, Y., Wang, T., Zhong, F., Zhou, Q., & Wang, K. (2024). The spatial impact of digital economy on carbon emissions reduction: Evidence from 215 cities in China. *Frontiers in Environmental Science*, *12*, 1370938. DOI: 10.3389/fenvs.2024.1370938

KEY TERMS AND DEFINITIONS

Carbon Footprint: A carbon footprint measures the total amount of greenhouse gas emissions generated directly or indirectly by activities, products, or organizations. It is typically expressed in carbon dioxide equivalent (CO_2e). Sources include energy use, transportation, manufacturing processes, and waste. Understanding carbon footprints helps organizations identify emission hotspots and implement reduction strategies such as energy efficiency, renewable energy adoption, and sustainable practices. It is an essential concept in climate change mitigation and environmental sustainability planning.

Digital Twin: A digital twin is a virtual representation of a physical system, process, or asset that mirrors its real-time performance using data from sensors and simulations. It allows organizations to monitor operations, predict failures, and test improvements without affecting actual systems. Digital twins are widely used in manufacturing, infrastructure, and energy management to enhance efficiency, reduce costs, and support decision-making. By enabling real-time analysis and scenario testing, digital twins help optimize performance and improve lifecycle management.

Energy Mapping: Energy mapping is the systematic process of collecting, visualizing, and analyzing energy consumption data across facilities, processes, or geographic areas. It helps identify where and how energy is used, wasted, or can be optimized. By using digital tools and sensors, energy mapping supports better decision-making for efficiency improvements, cost reduction, and sustainability planning. It is widely applied in industries, buildings, and urban planning to track energy flows, detect inefficiencies, and guide strategies for reducing consumption, improving performance, and achieving environmental and operational goals.

GIS (Geographic Information Systems): GIS is a digital technology used to collect, store, analyze, and visualize spatial or location-based data. It integrates maps with data layers to reveal patterns, relationships, and trends across geographic areas. GIS is widely used in urban planning, environmental management, disaster response, and industrial site analysis. By combining geographic coordinates with attributes such as infrastructure, energy use, or resources, GIS enables informed decision-making, supports spatial analysis, and helps organizations optimize resource allocation, planning, monitoring, and sustainable development initiatives.

Hotspot Mapping: Hotspot mapping is a technique used to identify areas with high concentrations of specific activities, risks, or impacts, such as energy consumption, pollution, or resource use. It uses spatial data visualization tools to highlight critical zones that require attention or intervention. Hotspot mapping helps organizations prioritize actions, allocate resources efficiently, and monitor performance improvements. It is widely applied in environmental management, urban planning, and industrial sustainability assessments.

IIoT (Industrial Internet of Things): IIoT refers to interconnected sensors, devices, and systems used in industrial environments to collect and exchange real-time data. These smart devices monitor machinery, processes, and operations to improve efficiency, safety, and predictive maintenance. IIoT enables automation, remote monitoring, and data-driven decision-making. It plays a key role in modern manufacturing by reducing downtime, optimizing resource usage, improving productivity, and supporting smart factory operations through seamless connectivity and intelligent analytics across industrial systems.

Industry 4.0: Industry 4.0 refers to the fourth industrial revolution characterized by automation, data exchange, and smart technologies in manufacturing. It integrates digital tools such as IoT, artificial intelligence, robotics, cloud computing, and analytics to create intelligent and interconnected production systems. Industry 4.0 enables real-time monitoring, predictive maintenance, and flexible manufacturing. It improves productivity, efficiency, and innovation while supporting smart factories and digital transformation across industrial sectors.

Resource Efficiency: Resource efficiency refers to the sustainable use of materials, energy, water, and other resources to minimize waste and maximize productivity. It focuses on producing more value with fewer inputs while reducing environmental impacts. Strategies include recycling, process optimization, energy conservation, and circular economy practices. Resource efficiency is essential in industries to lower costs, improve competitiveness, reduce emissions, and support sustainable development. It also helps organizations comply with environmental regulations and achieve long-term operational resilience and responsible resource management.

Spatial Analysis: Spatial analysis is the process of examining geographic patterns, relationships, and trends using location-based data. It involves techniques such as mapping, modeling, and statistical analysis to understand how phenomena vary across space. Spatial analysis is widely used in urban planning, environmental monitoring, logistics, and industrial optimization. It helps decision-makers identify patterns, predict outcomes, and improve planning by analyzing how physical location influences processes, resources, and environmental conditions.

Sustainable Manufacturing: Sustainable manufacturing involves producing goods through processes that minimize environmental impact, conserve resources, and ensure economic and social benefits. It focuses on reducing waste, lowering energy consumption, using eco-friendly materials, and improving operational efficiency. Sustainable manufacturing integrates environmental responsibility with productivity and innovation. It supports long-term business growth while protecting natural resources, reducing pollution, and meeting regulatory requirements for environmental performance.

Chapter 4
Fleet Management System Using IoT and GPS With Fuel Monitoring

Karuturi Sri Mani Krishna
http://orcid.org/0009-0001-9938-0562
Vel Tech Rangarajan Dr. Sagunthala R&D Institute of Science and Technology, India

R. Rajesh
Vel Tech Rangarajan Dr. Sagunthala R&D Institute of Science and Technology, India

B. Surendra
Vel Tech Rangarajan Dr. Sagunthala R&D Institute of Science and Technology, India

ABSTRACT

The rapid advancements in Internet of Things (IoT) and Global Positioning System (GPS) technologies have revolutionized fleet management systems, enabling real-time monitoring and optimization of vehicle performance, fuel consumption, and overall fleet efficiency. This chapter explores the integration of IoT and GPS with fuel monitoring systems, highlighting their transformative impact on fleet operations. By leveraging IoT sensors and GPS tracking, fleet managers gain valuable insights into vehicle health, fuel usage, and driver behavior, leading to data-driven decisions that reduce operational costs and enhance productivity. Predictive maintenance algorithms further optimize performance by identifying potential issues before they cause breakdowns. The chapter also examines the environmental benefits of optimized

DOI: 10.4018/979-8-3373-7941-8.ch004

fuel management and the role of data analytics in improving fleet decision-making. Case studies highlight the real-world impact of these technologies in enhancing fuel efficiency, reducing downtime, and improving maintenance strategies.

INTRODUCTION

Fleet management plays a critical role in ensuring the smooth operation of transportation and logistics sectors, where efficient management of vehicles and resources directly impacts operational costs, safety, and productivity (Nagesh et al., 2025). Traditionally, fleet management has faced challenges such as poor visibility into vehicle performance, inefficient fuel consumption, and maintenance management (Rudrusamy et al., 2023). These inefficiencies often lead to increased operational costs, vehicle downtime, and a lack of real-time insights into fleet performance (Farahpoor et al., 2023). In response, emerging technologies like the Internet of Things (IoT) and Global Positioning System (GPS) have provided solutions that allow fleet managers to overcome these obstacles (Hapsari et al., 2021). Integrating IoT sensors and GPS tracking technologies into fleet management systems has revolutionized how fleets are monitored, optimized, and maintained, providing real-time data on vehicle performance, fuel usage, and location (Rathnayaka et al., 2021).

IoT-based fleet management systems enable continuous, real-time monitoring of various vehicle parameters, such as speed, fuel levels, engine health, and tire pressure (Chaganti et al., 2024). Sensors installed in vehicles transmit this data to a central server, where fleet managers can access it remotely via cloud-based dashboards (Degadwala et al., 2023). This integration enhances visibility, allowing fleet operators to track each vehicle's location, monitor fuel consumption, and analyze driver behavior (Prabhu et al., 2022). By leveraging these data points, fleet managers can identify inefficient driving habits, optimize routes to reduce fuel consumption, and pinpoint maintenance needs before they lead to costly breakdowns (Tulloh et al., 2021). IoT and GPS have shifted fleet management from a reactive to a proactive system, enabling early identification and resolution of problems (Chandana et al., 2023).

Several recent studies have explored IoT and GPS-based fleet management systems from different perspectives; however, their approaches vary in scope, implementation, and effectiveness. For instance, some studies primarily focus on real-time vehicle tracking and monitoring, emphasizing location accuracy and route optimization, while others concentrate on fuel consumption analysis and theft detection using sensor-based approaches. Although these models demonstrate improved operational visibility, many of them lack integration with predictive analytics for proactive decision-making. Existing research often treats key components such as

GPS tracking, fuel monitoring, and predictive maintenance as independent modules rather than as part of a unified, intelligent system. This fragmented approach limits the overall efficiency and scalability of fleet management solutions. In contrast, recent advancements suggest the need for integrated frameworks that combine IoT data streams with machine learning algorithms to enable adaptive and automated decision-making.

While several works highlight the benefits of IoT-enabled monitoring, there is limited discussion on system robustness under real-world conditions, such as network instability, sensor failures, and large-scale deployment challenges. Therefore, a more holistic and integrated approach is required to bridge these gaps and enhance the practical applicability of IoT-based fleet management systems.

Fuel management is a crucial aspect of modern fleet management that greatly benefits from IoT and GPS integration (Punith et al., 2022). Traditional fuel monitoring methods often involve manual checks and estimates, making it difficult to accurately track fuel consumption and detect discrepancies such as theft or leakage (Babiyola et al., 2023). IoT-enabled fuel monitoring systems continuously measure the fuel in a vehicle's tank and provide real-time updates on fuel usage (Infrastructure et al., 2024). These systems help prevent fuel loss and give fleet managers detailed insights into fuel consumption patterns across the fleet (Yamini et al., 2023). By identifying which vehicles consume more fuel or which drivers engage in fuel-inefficient behavior, fleet managers can take corrective actions to improve fuel efficiency and reduce operational costs (Patil et al., 2023).

A significant advantage of IoT-based fleet management systems is enabling predictive maintenance, a proactive approach to vehicle upkeep (Abdirahman et al., 2025). In traditional fleet management, maintenance is often scheduled regardless of necessity, or breakdowns occur due to undiagnosed issues (Kamolsin et al., 2023). With IoT sensors, fleet managers can monitor the health of vehicle components in real-time, identifying early signs of wear and tear (Mathi et al., 2024). This data helps predict when specific vehicle parts may fail, allowing maintenance teams to address issues before they result in costly repairs or breakdowns (Teja et al., 2023). Predictive maintenance reduces downtime and extends the lifespan of vehicles, ensuring peak fleet efficiency with minimal disruptions (Moumen et al., 2023).

The rise of cloud-based data storage and advanced analytics has further transformed fleet management (Devi et al., 2023). Data collected from IoT sensors and GPS devices is stored and processed in cloud platforms, where powerful analytics tools generate meaningful insights (Desai et al., 2022). These insights are visualized through dashboards that provide key performance metrics, such as fuel consumption, vehicle location, and maintenance needs (Jimoh, 2025). By analyzing this data, fleet managers can make data-driven decisions to optimize fleet efficiency, fuel usage, and reduce maintenance costs. AI and machine learning algorithms predict future trends, such as fuel consumption patterns or vehicle failure risks, enhancing decision-making (Ali et al., 2021). The combination of real-time data, advanced analytics, and predictive capabilities positions IoT and GPS-based fleet management systems as essential tools for achieving greater efficiency, cost reduction, and sustainability (Nagesh et al., 2025).

Despite the rapid development of IoT-enabled fleet management systems (Zhang, Y et al (2023), several challenges remain in integrating GPS tracking Gubbi, J et al (2022), sensor networks, and cloud-based analytics into a unified operational architecture. Alam, T et al (2023)Many existing systems focus primarily on vehicle tracking or fuel monitoring without providing a comprehensive platform that integrates predictive maintenance, driver behavior analysis, and real-time decision support. Lin, J (2022) The scalability and reliability of IoT-based fleet architectures under large-scale operational conditions remain insufficiently explored. Khan, R (2021) This chapter therefore investigates how the integration of IoT sensors, GPS tracking, cloud analytics, and predictive algorithms can enhance fleet visibility, fuel efficiency, and maintenance planning.

Fleet management systems continue to evolve in response to several persistent operational challenges and emerging technological trends. One of the key issues in traditional fleet systems is the lack of real-time visibility, which leads to inefficient route planning, increased fuel consumption, and delayed decision-making. Additionally, manual monitoring approaches often fail to detect anomalies such as fuel theft, vehicle misuse, and unexpected maintenance requirements in a timely manner. Another major problem lies in the inefficient utilization of fleet resources, where suboptimal routing, excessive idling, and poor driving behavior contribute to increased operational costs and environmental impact.

From a technological perspective, the rapid advancement of IoT, artificial intelligence, and cloud computing is transforming fleet management into a data-driven and intelligent system. Emerging trends include the integration of machine learning algorithms for predictive maintenance, real-time analytics for operational optimization, and the adoption of electric and autonomous vehicles. Furthermore, the use of edge computing and 5G connectivity is enabling faster data processing and improved system responsiveness. Challenges such as data security, system scalability, and integration complexity remain critical concerns that must be addressed to ensure the reliability and sustainability of modern fleet management systems.

Objectives

The following objectives guide the scope and direction of this chapter:

- To examine the integration of Internet of Things (IoT) and Global Positioning System (GPS) technologies for real-time fleet monitoring and management.
- To analyze the effectiveness of IoT-based fuel monitoring systems in reducing fuel consumption and preventing fuel theft.
- To evaluate the role of predictive maintenance techniques in improving fleet reliability and minimizing operational downtime.
- To assess the impact of cloud-based data analytics platforms in enhancing decision-making and operational efficiency in fleet management.
- To explore emerging trends and future advancements, including artificial intelligence and electric vehicle integration, in smart fleet management systems.

KEY COMPONENTS OF FLEET MANAGEMENT SYSTEMS

GPS Tracking: Real-Time Location and Route Optimization

GPS tracking plays a pivotal role in modern fleet management systems, providing real-time visibility into the location of vehicles. By utilizing Global Positioning System (GPS) technology, fleet managers can monitor the exact position of each vehicle within a fleet at any given moment. This allows for dynamic decision-making, such as adjusting routes to optimize travel time, reduce fuel consumption, and respond quickly to unforeseen changes such as traffic congestion or road closures. The ability to track vehicles in real-time ensures that fleet operators have complete oversight of their fleet, enhancing both operational efficiency and safety.

Route optimization, facilitated by GPS tracking, is a key benefit that significantly impacts cost reduction. By continuously monitoring vehicle locations and traffic conditions, fleet managers can identify the most efficient routes for drivers, minimizing unnecessary detours and delays. GPS technology also aids in the analysis of historical route data, enabling fleet managers to refine future route plans based on past performance. This data-driven approach not only helps reduce operational costs, but it also improves customer satisfaction by ensuring timely deliveries and more predictable fleet operations. Ultimately, GPS tracking enhances the management of fleet resources, enabling fleet managers to make more informed and strategic decisions.

Figure 1. GPS Tracking: Real-Time Location and Route Optimization

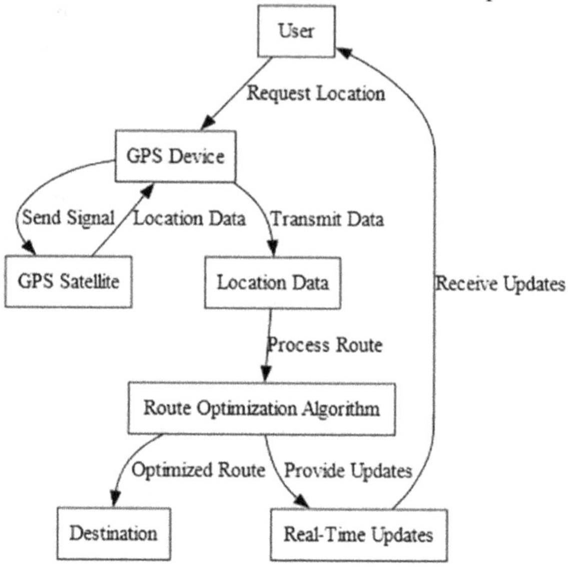

IoT Sensors: Monitoring Vehicle Health and Performance

The integration of Internet of Things (IoT) sensors into fleet management systems plays a crucial role in the continuous monitoring of vehicle health and performance. These sensors collect real-time data from various vehicle parameters such as engine temperature, tire pressure, fuel levels, and overall vehicle condition. The data obtained from these sensors is transmitted to a central processing unit, where it is analyzed to assess the operational status of each vehicle in the fleet. By constantly monitoring the health of the vehicle, fleet managers are able to identify potential issues before they escalate into costly repairs or system failures. The real-time data collected from IoT sensors allows for better decision-making regarding vehicle maintenance, ensuring that issues are addressed proactively rather than reactively, ultimately extending the lifespan of the fleet and reducing downtime.

In addition to monitoring the mechanical and operational status of the vehicle, IoT sensors are instrumental in tracking driving behavior and optimizing performance. By analyzing data such as speed, acceleration patterns, braking force, and idle time, fleet managers can gain insights into how efficiently vehicles are being operated. This information is crucial for identifying areas where fuel consumption can be minimized, driving behaviors can be improved, and safety standards can be enhanced. The implementation of IoT sensors contributes to a more sustainable and efficient fleet operation by providing accurate and timely information that drives performance optimization across the fleet.

Fuel Monitoring Systems: Detecting Theft and Optimizing Consumption

Fuel monitoring systems are essential components of modern fleet management, playing a critical role in optimizing fuel consumption and detecting inefficiencies. These systems are designed to monitor the fuel levels in real-time and provide detailed insights into fuel usage patterns across the entire fleet. By integrating fuel sensors with IoT technology, fleet operators gain the ability to continuously track the amount of fuel in each vehicle's tank. This data is transmitted to a central system, where it is processed and analyzed, allowing fleet managers to identify discrepancies between actual fuel consumption and expected consumption based on the vehicle's operational data. Such monitoring helps in detecting unusual fuel consumption trends, indicating potential issues like engine inefficiency or even fuel theft.

Figure 2. Fuel Monitoring Systems: Detecting Theft and Optimizing Consumption

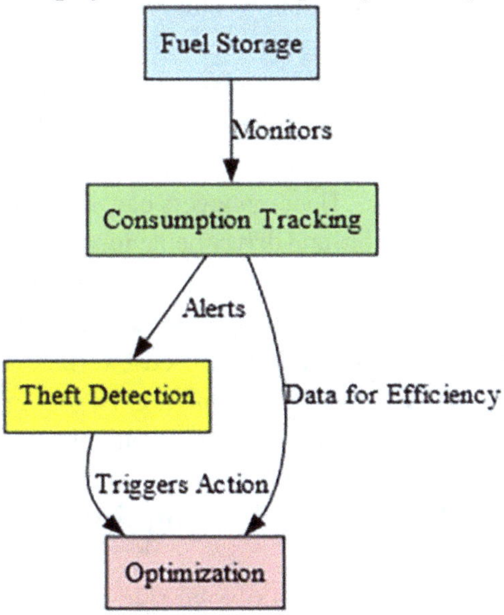

The optimization of fuel consumption through monitoring systems extends beyond just identifying theft or leakage. These systems analyze driving behaviors, such as harsh acceleration, excessive idling, and unnecessary speed, all of which contribute to fuel wastage. By providing fleet managers with detailed reports on these behaviors, the system helps in making data-driven decisions that can guide the implementation of best practices for fuel-efficient driving. Moreover, regular analysis of fuel consumption across the fleet aids in setting benchmarks for performance, enabling fleet managers to identify underperforming vehicles or drivers. This continuous feedback loop fosters a culture of fuel-conscious driving, which not only improves efficiency but also significantly reduces operating costs.

Cloud-Based Data Processing and Visualization Platforms

Cloud-based data processing and visualization platforms play a pivotal role in modern fleet management systems, offering a centralized infrastructure to manage the vast amounts of data generated by IoT sensors, GPS tracking, and fuel monitoring systems. These platforms allow fleet managers to collect, store, and process real-time data from various sources, enabling comprehensive oversight of fleet operations. The integration of cloud technology ensures that data is accessible from any location, providing fleet managers with the flexibility to make timely and informed decisions. Cloud platforms also facilitate seamless communication between vehicles and central management systems, ensuring that data flows uninterrupted and is processed without significant delays. This real-time access to operational data enhances decision-making, allowing for the rapid identification of issues such as fuel inefficiency, maintenance needs, or unauthorized vehicle movements.

The power of cloud-based platforms lies in their ability to analyze large volumes of data, providing actionable insights that can be used to optimize fleet performance. Through advanced data analytics, fleet managers can visualize key metrics, such as vehicle location, fuel consumption, and driver behavior, in a user-friendly dashboard. These platforms also support predictive analytics, enabling fleet operators to anticipate vehicle maintenance needs and avoid costly repairs or downtime. By consolidating data from multiple vehicles into a single, unified interface, cloud platforms offer fleet managers an integrated view of fleet operations, making it easier to monitor performance, track trends, and implement improvements.

Data Analytics and Reporting: Turning Raw Data into Insights

Data analytics and reporting play a pivotal role in modern fleet management systems, transforming raw data into actionable insights that drive decision-making and improve operational efficiency. With the proliferation of Internet of Things (IoT) devices, vehicles are now equipped with a multitude of sensors that capture vast amounts of data in real-time, including information on fuel consumption, engine performance, vehicle location, speed, and driver behavior. This data, while valuable, is often unstructured and requires sophisticated analytics tools to extract meaningful insights. Through data aggregation, analysis, and visualization, fleet managers gain a comprehensive view of fleet performance, identifying patterns, inefficiencies, and areas of improvement. The conversion of this raw data into clear, digestible reports enables fleet managers to monitor operations more effectively, reduce costs, and make informed decisions about maintenance, routes, fuel consumption, and driver performance.

The power of data analytics extends beyond mere reporting, as advanced analytical techniques can predict future trends and outcomes, allowing for proactive rather than reactive management. Predictive analytics, for instance, can forecast maintenance needs, estimate fuel consumption patterns, and identify potential vehicle malfunctions before they result in costly repairs or breakdowns. By leveraging machine learning algorithms and historical data, these systems can provide actionable insights that improve fleet reliability and safety. Reporting tools, integrated with dashboards, allow fleet operators to visualize key metrics in real-time, providing them with up-to-date information on vehicle health, fuel efficiency, and driver behavior. This real-time feedback loop ensures that fleet managers are not only aware of current operations but are also equipped to make strategic adjustments that optimize overall fleet performance.

Figure 3. Data Analytics and Reporting: Turning Raw Data into Insights

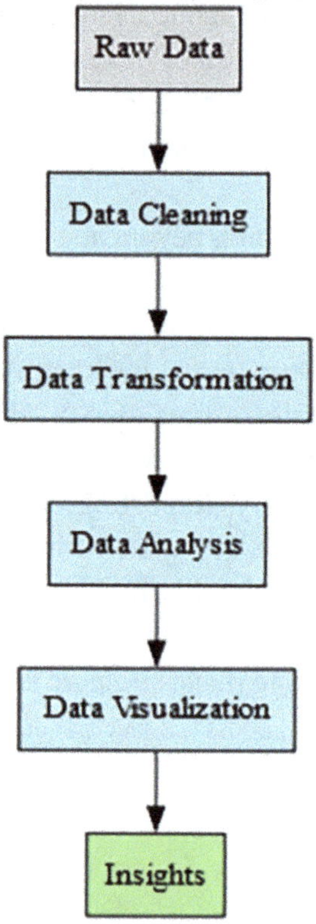

IoT and GPS Integration in Fleet Management

Understanding IoT Architecture for Fleet Management

The architecture of an IoT-based fleet management system is designed to integrate a variety of components that work together to provide real-time data on vehicle performance and location. At its core, the IoT system is comprised of sensors embedded in the vehicles, capable of monitoring multiple parameters such as fuel levels, engine health, tire pressure, speed, and driver behavior. These sensors collect data continuously, which is then transmitted to a central system via communication networks such as GSM, Wi-Fi, or cellular networks. The collected data is sent to a cloud-based platform where it is processed and analyzed. This cloud-based architecture provides scalability, allowing fleet managers to manage large numbers of vehicles efficiently and access real-time information from anywhere, at any time. The IoT sensors provide not only crucial operational data but also enable fleet managers to maintain a high level of control over each vehicle's health, ensuring timely maintenance and minimizing the risk of unexpected breakdowns.

Incorporating GPS technology into the IoT framework enhances the system's capability by providing accurate, real-time location data for each vehicle. The GPS module works in conjunction with the IoT sensors, enabling fleet managers to track the movement, speed, and routes of their vehicles at all times. This integration allows for better route optimization, reducing fuel consumption and improving overall fleet efficiency. The data generated by both IoT sensors and GPS modules is stored in the cloud, where it is processed and presented to fleet operators through dashboards or mobile applications. The seamless communication between IoT devices and GPS ensures that fleet managers receive comprehensive insights into vehicle operations, allowing them to make informed decisions regarding routing, maintenance schedules, fuel usage, and driver performance.

Limitations and Failure Scenarios in IoT-Based Fleet Management Systems

While IoT-based fleet management systems offer significant advantages in terms of real-time monitoring and operational efficiency, they are also subject to several limitations and potential failure scenarios that must be carefully considered. One of the primary challenges is network dependency, as these systems rely heavily on continuous connectivity through GSM, cellular, or internet networks. Any disruption in network availability can lead to delays in data transmission, resulting in incomplete or outdated information for fleet managers.

Another critical limitation is sensor reliability. IoT sensors deployed in vehicles may experience calibration issues, hardware degradation, or environmental interference, leading to inaccurate data collection. Faulty sensor readings can mislead decision-making processes, particularly in areas such as fuel monitoring and predictive maintenance.

GPS-related inaccuracies also pose a concern, especially in urban environments with high-rise buildings (urban canyon effect) or in remote areas with weak satellite signals. These inaccuracies can affect route optimization and real-time tracking precision.

Data security and privacy risks are additional challenges in IoT-enabled systems. As large volumes of sensitive operational data are transmitted and stored in cloud platforms, these systems become potential targets for cyberattacks, unauthorized access, and data breaches. Without robust encryption and authentication mechanisms, the integrity of fleet data may be compromised.

Furthermore, system scalability and integration complexity can impact performance as fleet size increases. Managing large-scale fleets requires efficient data handling, processing capabilities, and seamless integration with existing enterprise systems, which may not always be achievable with basic IoT architectures.

Addressing these limitations through robust system design, fault-tolerant mechanisms, and enhanced security protocols is essential for ensuring the reliability and effectiveness of IoT-based fleet management solutions in real-world applications.

The Role of GPS in Fleet Positioning and Monitoring

Figure 4. The Role of GPS in Fleet Positioning and Monitoring

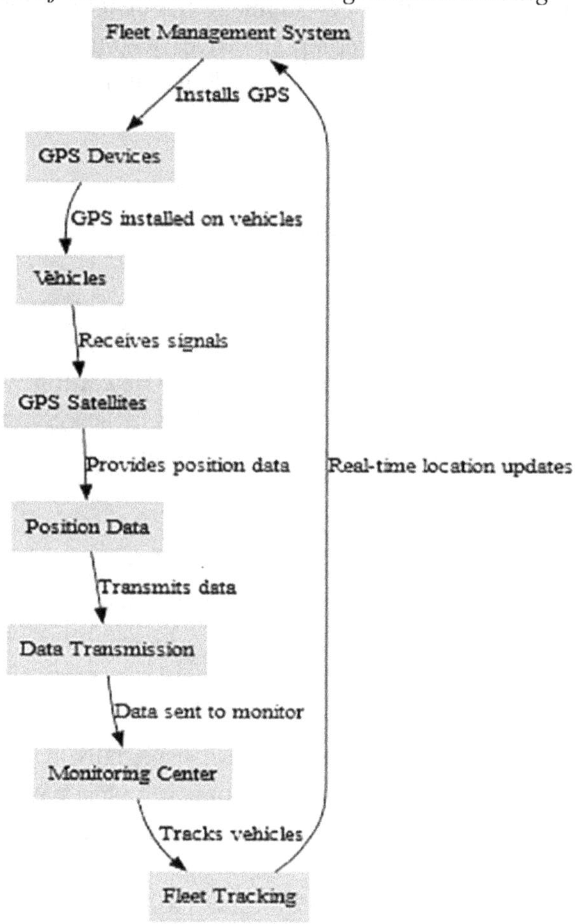

GPS plays a crucial role in fleet management by providing precise, real-time positioning and tracking of vehicles, which is fundamental for optimizing fleet operations. Through GPS technology, fleet managers are able to monitor the exact location of each vehicle, track its route, and ensure that it adheres to the planned schedule. This continuous flow of location data allows for better decision-making regarding route optimization, reducing delays and minimizing fuel consumption by selecting the most efficient paths. GPS also enhances the ability to track vehicles in real-time, making it easier to respond to unforeseen circumstances such as traffic disruptions or route deviations, and rerouting vehicles to maintain operational efficiency.

In addition to route optimization, GPS enables the monitoring of important metrics related to vehicle performance and behavior. By analyzing the location data in conjunction with other vehicle parameters such as speed, fuel consumption, and engine diagnostics, fleet managers can gain deeper insights into driver performance, vehicle health, and operational efficiency. GPS tracking also helps in ensuring compliance with safety regulations by providing data on speed limits and unauthorized stops.

Combining IoT and GPS: Achieving Real-Time Fleet Visibility

The integration of IoT and GPS technologies in fleet management systems plays a critical role in achieving real-time visibility and enhancing the overall efficiency of fleet operations. GPS technology allows for precise, continuous tracking of vehicle locations, providing fleet managers with up-to-the-minute data on vehicle movements, routes, and speeds. This real-time tracking capability enables managers to optimize routes, avoid traffic delays, and ensure that vehicles are operating within expected parameters. When combined with IoT sensors, which monitor various vehicle conditions such as fuel levels, engine health, and tire pressure, the system offers a comprehensive view of fleet performance. By capturing data on both the location and the health of the vehicles, fleet managers can make informed decisions that optimize performance, reduce downtime, and prevent unnecessary fuel consumption.

The synergy between IoT and GPS enables fleet managers to monitor and respond to operational issues in real time, providing insights that were previously unavailable. IoT devices collect vast amounts of data from each vehicle and transmit it to a central cloud-based system for processing and analysis. This seamless integration of GPS and IoT data allows for a more granular understanding of fleet performance, from pinpointing inefficient routes to identifying problematic driving behaviors that may lead to fuel wastage or vehicle wear and tear. Real-time fleet visibility not only supports more effective decision-making but also enhances the safety and security of fleet operations.

Case Studies on Successful IoT and GPS Integration in Fleets

Case studies of successful IoT and GPS integration in fleet management offer valuable insights into the practical applications and benefits of these technologies.

One such example is the implementation of an IoT-based fleet management system by a major logistics company, which integrated GPS tracking with real-time vehicle health monitoring. By using GPS to track the location of each vehicle, fleet managers could optimize routes, reduce idle times, and improve delivery efficiency. IoT sensors embedded in the vehicles monitored critical factors such as tire pressure, fuel consumption, and engine performance, sending alerts for any irregularities. This integration allowed the company to proactively address maintenance issues before they led to breakdowns, ultimately reducing downtime and improving the overall reliability of the fleet.

Figure 5. Case Studies on Successful IoT and GPS Integration in Fleets

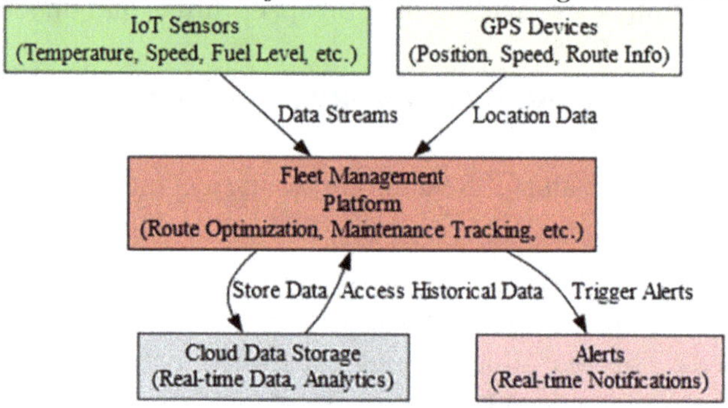

Another case study demonstrates how a transportation service provider improved fuel efficiency and driver safety through the combined use of GPS and IoT. By installing fuel sensors along with GPS tracking devices, the company was able to monitor real-time fuel consumption and identify patterns of fuel waste, such as excessive idling or inefficient driving habits. The system provided actionable insights into driving behavior, allowing the company to implement targeted driver training programs. Over time, this led to a significant reduction in fuel costs and an improvement in driver safety scores. The integration of IoT and GPS not only enhanced operational efficiency but also contributed to the company's sustainability goals by reducing its carbon footprint and lowering fuel-related emissions.

While the presented case studies demonstrate the effectiveness of IoT and GPS integration in improving fleet efficiency, several real-world challenges must also be considered for practical deployment. One of the major challenges is handling heterogeneous fleet environments, where vehicles of different types, ages, and technological compatibility levels must be integrated into a unified system. This often leads to interoperability issues and inconsistent data quality. Another key challenge is infrastructure dependency, particularly in regions with limited network coverage. In such scenarios, real-time data transmission becomes unreliable, affecting tracking accuracy and decision-making. Additionally, environmental conditions such as extreme temperatures, dust, and vibrations can impact sensor performance and longevity, leading to increased maintenance requirements.

Driver behavior adaptation also presents a practical challenge, as the effectiveness of IoT-based recommendations depends on driver compliance with suggested practices such as optimized routing and fuel-efficient driving. Resistance to change or lack of proper training can limit system effectiveness. The cost of implementation and system maintenance remains a barrier, especially for small and medium-scale fleet operators. Initial investment in IoT devices, cloud infrastructure, and analytics platforms may be significant, requiring careful cost-benefit analysis. Addressing these real-world challenges is essential to ensure the successful adoption and long-term sustainability of IoT and GPS-based fleet management systems across diverse operational environments.

Figure 6. Proposed IoT and GPS-Based Fleet Management System Architecture

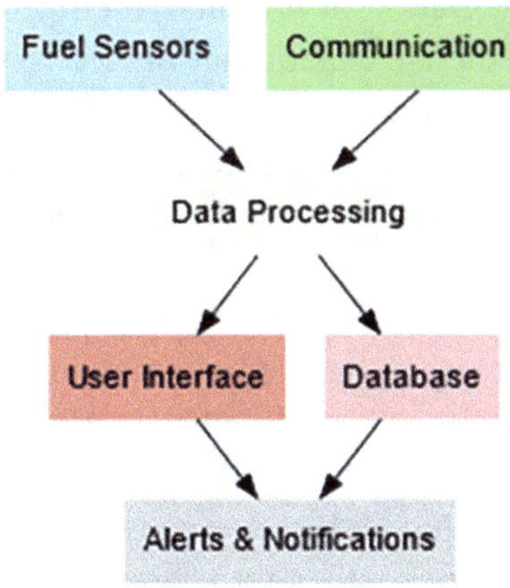

The proposed fleet management system architecture consists of multiple interconnected layers that enable real-time data acquisition, transmission, processing, and decision-making. At the vehicle layer, IoT sensors continuously monitor parameters such as fuel levels, engine performance, tire pressure, and speed. These sensors work in conjunction with GPS modules to provide precise location and movement data. The collected data is transmitted through the communication layer using technologies such as GSM, 4G/5G networks, or lightweight protocols like MQTT, ensuring reliable and efficient data transfer. This data is then received by the cloud layer, where it is stored, processed, and managed using scalable cloud infrastructure.

At the analytics layer, advanced data processing techniques, including machine learning algorithms, are applied to extract meaningful insights such as fuel consumption patterns, predictive maintenance alerts, and route optimization strategies. Finally, the processed information is presented to fleet managers through user interfaces such as dashboards or mobile applications, enabling real-time monitoring and informed decision-making.

FUTURE TRENDS IN IoT AND GPS TECHNOLOGY FOR FLEET MANAGEMENT

The future of IoT and GPS technology in fleet management is poised to revolutionize how fleets are monitored and operated. With advancements in communication technologies such as 5G, IoT devices are expected to become more efficient, enabling faster data transmission and real-time processing. This will allow fleet managers to receive immediate updates on vehicle location, performance, and condition, leading to quicker decision-making and a more responsive approach to fleet management. The growing integration of artificial intelligence (AI) with IoT and GPS will also enhance the predictive capabilities of fleet systems, enabling more accurate forecasting of vehicle maintenance needs, fuel consumption, and overall fleet performance. These advancements will lead to smarter, more autonomous systems capable of handling increasingly complex data and providing insights that were previously unavailable.

In addition to the technological advancements in connectivity and data processing, future trends will also focus on improving the sustainability and efficiency of fleet management systems. With the rise of electric vehicles (EVs) and hybrid fleets, IoT and GPS technologies will evolve to better manage the unique challenges of these vehicles, such as battery monitoring and energy consumption tracking. Real-time integration with renewable energy sources for charging infrastructure will allow for more efficient energy use within fleets, reducing operational costs and carbon footprints. Furthermore, the use of blockchain technology for secure and transparent data sharing within fleet networks is expected to enhance the overall security and reliability of fleet management systems.

Fuel Monitoring and Efficiency Optimization

Technologies Behind Fuel Monitoring Systems

Fuel monitoring systems rely on a variety of advanced technologies to provide real-time insights into fuel consumption and optimize fleet efficiency. These systems are primarily powered by sensors that are installed in the vehicle's fuel tank or fuel lines. The most commonly used sensors include ultrasonic sensors, capacitive sensors, and fuel flow meters, each designed to measure different aspects of fuel usage. Ultrasonic sensors, for instance, work by emitting sound waves to measure the fuel level in the tank, providing accurate data without any physical contact with the fuel. Capacitive sensors, on the other hand, use changes in capacitance to detect fuel levels, while fuel flow meters track the volume of fuel being consumed by monitoring the flow rate. These sensors transmit real-time data to the central system, where it is processed and analyzed to monitor fuel efficiency, detect anomalies, and identify patterns in fuel consumption.

The integration of these sensors with IoT technology allows for continuous data transmission to cloud-based platforms where advanced analytics tools are applied. Machine learning algorithms can process the large volumes of data collected, identifying trends in fuel consumption and optimizing routes or driving behaviors to improve fuel efficiency. In some systems, fuel monitoring is coupled with GPS tracking, providing fleet managers with a comprehensive view of both fuel levels and vehicle locations, enabling them to identify inefficient routes or excessive idling. Fuel monitoring systems also feature alert mechanisms to notify fleet managers of any unusual fuel consumption or potential issues such as fuel theft, leakage, or misfueling.

Figure 7. Technologies Behind Fuel Monitoring Systems

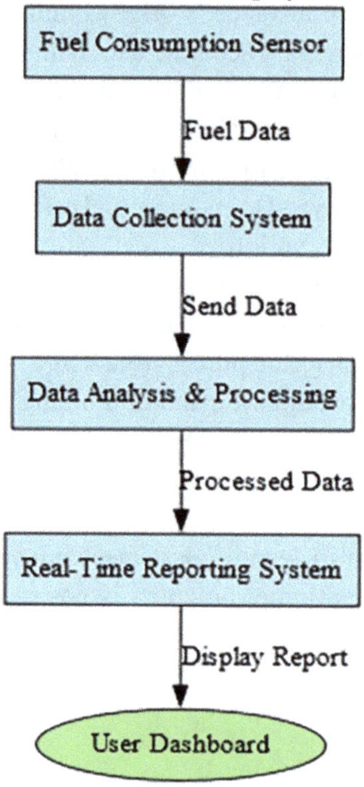

Identifying Fuel Theft and Leakage Through IoT Sensors

Fuel theft and leakage are significant issues in fleet management, as they directly impact operational costs and overall fleet efficiency. IoT-based fuel monitoring systems are increasingly being deployed to address these challenges by providing real-time monitoring of fuel levels and detecting abnormal fluctuations that may indicate theft or leakage. IoT sensors installed in fuel tanks continuously measure the fuel level, and when integrated with other vehicle data, they can provide detailed insights into fuel consumption patterns. Any sudden drop in fuel levels, which deviates from expected consumption patterns based on driving behavior and route history, triggers alerts to fleet managers. This real-time monitoring allows for immediate action to be taken, preventing significant losses and minimizing the risk of fraudulent activities.

The precision of IoT sensors in detecting fuel theft and leakage has greatly improved, with advanced systems capable of identifying small discrepancies that may otherwise go unnoticed. These sensors work by analyzing the rate of fuel consumption against various factors such as vehicle load, speed, and driving conditions. By combining data from multiple sensors, the system can differentiate between normal fuel usage and unauthorized fuel extraction or leakage. The integration of GPS data further enhances this capability by providing accurate location tracking of the vehicle, allowing fleet managers to verify whether the vehicle has been idling for longer than expected or has deviated from its designated route.

Real-Time Fuel Consumption Monitoring and Analysis

This section builds upon the previously discussed fuel monitoring mechanisms by focusing on the analytical interpretation of fuel consumption data and its role in predictive decision-making. Real-time fuel consumption monitoring has become a critical component of modern fleet management systems. By utilizing IoT-enabled sensors installed in vehicles, fleet operators can continuously track the fuel levels and consumption patterns across their entire fleet. These sensors provide valuable data, including fuel usage per vehicle, fuel efficiency trends, and sudden fluctuations in consumption, such as fuel theft or leakage. This constant stream of data is transmitted to a central cloud platform where it is processed and analyzed to generate actionable insights. Real-time monitoring ensures that fleet managers can detect issues immediately, allowing them to make adjustments before inefficiencies escalate, thus minimizing fuel waste and reducing operational costs.

The ability to analyze fuel consumption data in real-time further enhances the optimization of fleet operations. Advanced analytics tools can process this data to identify trends and patterns that may not be immediately apparent, such as the impact of specific routes, driving behaviors, or even weather conditions on fuel efficiency. By integrating these insights into a comprehensive reporting system, fleet managers can adjust driving practices, optimize routes, and schedule vehicle maintenance to improve overall fuel economy. In addition, predictive algorithms can be employed to forecast fuel needs, enabling more accurate refueling schedules and reducing downtime.

Figure 8. Real-Time Fuel Consumption Monitoring and Analysis

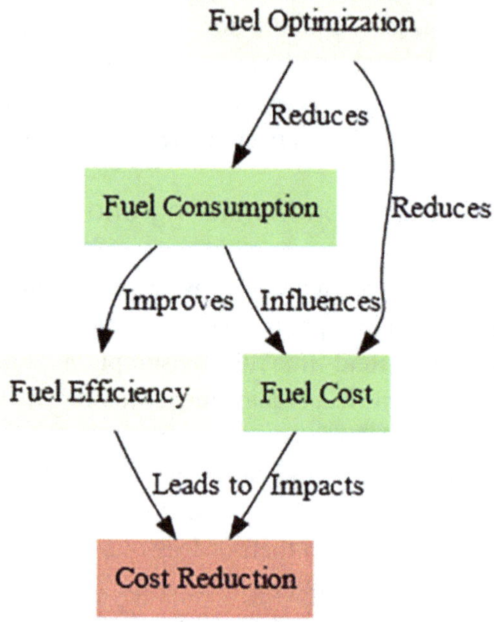

Fuel Efficiency Optimization Algorithms in Fleet Management

Fuel efficiency optimization algorithms play a crucial role in reducing fuel consumption and operational costs in fleet management systems. By analyzing real-time data from various sensors, such as fuel level sensors, GPS, and vehicle performance monitors, these algorithms can detect inefficiencies in fuel usage, such as excessive idling, harsh acceleration, or suboptimal routes. Through advanced data analysis, algorithms can provide fleet managers with recommendations to optimize driving behavior, adjust routes, or implement changes in vehicle operation that reduce fuel consumption. These recommendations can be integrated into a fleet management dashboard, providing actionable insights that directly impact fuel efficiency, leading to significant cost savings over time.

As the fleet management sector continues to evolve, the integration of machine learning and artificial intelligence into fuel efficiency optimization algorithms is expected to bring even greater benefits. By using historical data and real-time inputs, machine learning models can predict fuel consumption trends, identify patterns of waste, and offer tailored solutions for each vehicle in the fleet. These predictive models enable more precise forecasting of fuel needs, allowing for proactive management of fuel resources and better decision-making.

Cost Reduction through Effective Fuel Management

Building on fuel monitoring and analytical insights, this section emphasizes the economic impact of optimized fuel management, particularly in reducing operational costs and improving overall fleet profitability. Effective fuel management is essential for reducing operational costs in fleet management, as fuel represents one of the largest expenses for fleet operators. By implementing fuel monitoring systems, fleet managers can gain real-time insights into fuel consumption patterns, enabling them to identify inefficiencies and optimize usage. These systems track fuel levels, detect fuel theft, and highlight discrepancies in fuel usage that could indicate maintenance issues, such as leaks or engine malfunctions. With accurate, real-time data on fuel consumption, fleet operators can implement strategies to reduce waste, such as optimizing routes to minimize fuel consumption or adjusting driver behavior to encourage more efficient driving practices. This level of precision ensures that fuel is used in the most efficient way possible, contributing directly to cost reduction.

Figure 9. Cost Reduction through Effective Fuel Management

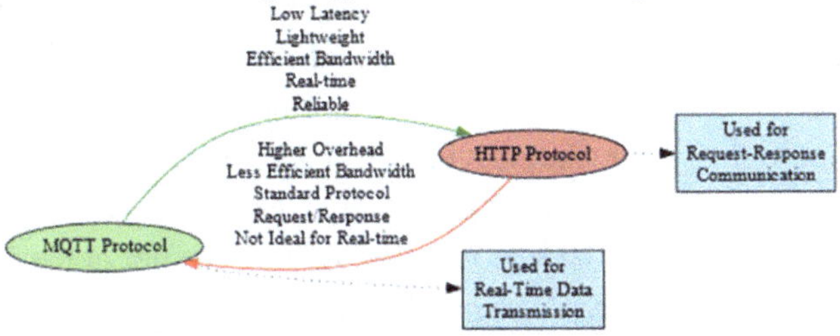

The integration of fuel monitoring systems within a larger IoT-enabled fleet management framework allows for enhanced visibility and control over fuel usage. Fleet managers can analyze data trends to identify high-consumption vehicles or drivers who are prone to inefficient practices, such as excessive idling or rapid acceleration. By addressing these behaviors, fleet managers can implement targeted training programs or enforce stricter driving policies, further enhancing fuel efficiency. Furthermore, the automation of fuel data collection and reporting reduces the administrative burden and potential for human error, allowing fleet managers to focus on strategic decision-making rather than manual monitoring. Overall, effective fuel management not only leads to direct cost savings but also promotes long-term sustainability by reducing a fleet's carbon footprint and ensuring more responsible fuel usage.

Data Collection, Transmission, and Management

Real-Time Data Acquisition from Vehicles and Sensors

Real-time data acquisition from vehicles and sensors is the cornerstone of modern fleet management systems, enabling fleet operators to continuously monitor the performance and health of their vehicles. IoT sensors embedded in the vehicles collect critical data such as location, speed, fuel levels, engine health, and driver behavior. These sensors capture a wealth of information that is crucial for decision-making, allowing fleet managers to track vehicle performance in real time. The data acquisition process is designed to be seamless, with sensors transmitting data at regular intervals, ensuring that fleet managers are always up to date on the status of their fleet. The real-time nature of the data allows for immediate responses to any irregularities, such as fuel theft or a vehicle breakdown, reducing response times and minimizing downtime.

Once the data is captured by the sensors, it is transmitted to a central system for processing and analysis. The data acquisition process is typically powered by reliable communication technologies, such as GSM, GPRS, or Wi-Fi, which ensure that data is transmitted with minimal delay and high accuracy. This transmission process is crucial for enabling fleet managers to make informed decisions based on current, real-time data rather than relying on outdated or incomplete information. The real-time nature of the data also facilitates more effective monitoring of fleet operations, ensuring that any issues, whether related to fuel consumption, vehicle performance, or safety concerns, are addressed promptly.

Data Transmission Protocols: MQTT vs HTTP for Fleet Management

Data transmission protocols are critical for the efficiency and reliability of fleet management systems, as they govern how data collected from IoT sensors and GPS devices is communicated to cloud servers for analysis and decision-making. Two of the most commonly used protocols in fleet management systems are MQTT (Message Queuing Telemetry Transport) and HTTP (Hypertext Transfer Protocol). MQTT is a lightweight messaging protocol specifically designed for low-bandwidth, high-latency, or unreliable networks. It enables efficient communication between devices and cloud platforms by using a publish/subscribe model, making it well-suited for real-time applications in fleet management. The protocol ensures minimal data loss and reduced transmission time, providing fleet managers with near-instantaneous updates on vehicle status, fuel consumption, and other performance metrics. MQTT's ability to handle numerous devices simultaneously and its small message sizes make it ideal for managing large fleets with numerous IoT sensors.

Figure 10. Data Transmission Protocols: MQTT vs HTTP for Fleet Management

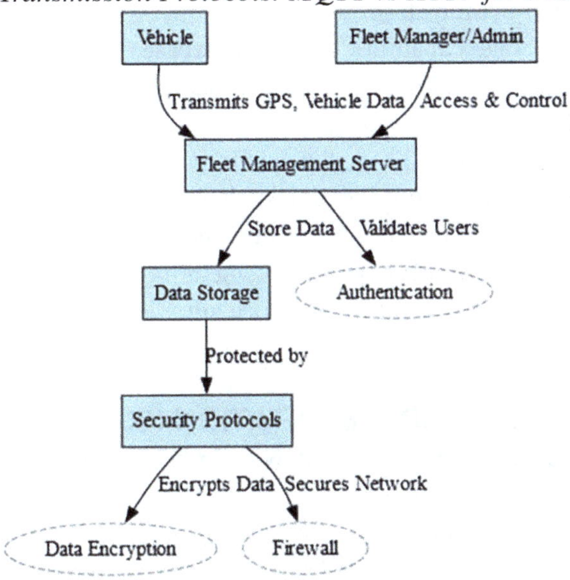

On the other hand, HTTP, a more traditional protocol, is widely used for web-based applications due to its simplicity and reliability. While HTTP is highly effective for request-response transactions, it is less efficient in real-time, low-latency environments compared to MQTT. HTTP requires a continuous connection for sending data, which can lead to higher overheads and latency, particularly when dealing with a large volume of data or frequent updates from numerous vehicles. Despite this, HTTP remains a robust choice for certain fleet management systems where complex data retrieval and large payloads are involved, as well as when devices are directly interfacing with web-based platforms for reporting or visualization purposes.

Cloud Databases and Their Role in Fleet Data Storage

Cloud databases play a crucial role in the efficient storage and management of the vast amounts of data generated by IoT sensors and GPS devices in fleet management systems. As vehicles are equipped with multiple sensors that continuously collect data on parameters such as location, fuel consumption, engine health, and driving behavior, cloud databases provide a scalable solution to store this real-time information. The cloud's flexibility allows for the seamless storage of large volumes of data without the need for on-premise infrastructure, reducing the costs and complexities associated with physical storage. This centralization of data ensures that fleet managers have easy access to historical and real-time information, which can be critical for making informed decisions related to fleet performance, maintenance scheduling, and fuel optimization.

In addition to storage capabilities, cloud databases also facilitate efficient data retrieval and processing, enabling advanced analytics and reporting. Data from various fleet vehicles are aggregated and stored in cloud databases, where they can be organized, indexed, and categorized for easy access. This setup allows fleet operators to retrieve insights on-demand, such as fuel consumption trends, vehicle performance, and maintenance needs. With cloud storage, fleet managers can also benefit from enhanced data security features, including encryption, redundancy, and access controls, ensuring that sensitive information is protected.

CHALLENGES AND IMPLEMENTATION ISSUES IN IoT-BASED FLEET MANAGEMENT SYSTEMS

The advantages of IoT and GPS-based fleet management systems, several practical challenges and implementation issues must be addressed to ensure effective deployment. One of the primary challenges is system integration, where different hardware components, communication protocols, and software platforms must work seamlessly together. In heterogeneous fleet environments, integrating legacy systems with modern IoT infrastructure can be complex and resource-intensive. Another critical issue is scalability. As fleet size increases, the volume of data generated grows significantly, requiring robust cloud infrastructure and efficient data processing mechanisms. Without proper scalability planning, system performance may degrade, leading to delays in analytics and decision-making.

Network dependency also presents a major limitation, as continuous connectivity is required for real-time data transmission. In regions with poor network coverage, data loss or delays may occur, affecting system reliability. Additionally, IoT devices deployed in vehicles are exposed to harsh environmental conditions such as temperature variations, dust, and vibrations, which can impact sensor accuracy and durability. Data security and privacy remain ongoing concerns, particularly when sensitive operational data is transmitted over networks and stored in cloud platforms. Ensuring secure communication through encryption, authentication, and access control mechanisms is essential to prevent unauthorized access and cyber threats.

The cost of implementation, including hardware installation, cloud infrastructure, and maintenance, may be a barrier for small and medium-sized fleet operators.

Data Privacy and Security in Fleet Management Systems

In fleet management systems, data privacy and security are paramount concerns due to the sensitive nature of the information being collected, transmitted, and stored. As fleet operations increasingly rely on IoT devices and GPS technologies to gather real-time data, the amount of information being exchanged grows exponentially. This data, which can include vehicle location, fuel consumption, driver behavior, and maintenance records, is highly valuable but also vulnerable to cyber threats. Without adequate protection, fleet operators risk exposing themselves to data breaches, unauthorized access, and potential misuse of sensitive information. Securing this data requires robust encryption protocols, secure communication channels, and continuous monitoring to ensure that it remains protected at all stages, from data collection to transmission and storage.

Figure 11. Data Privacy and Security in Fleet Management Systems

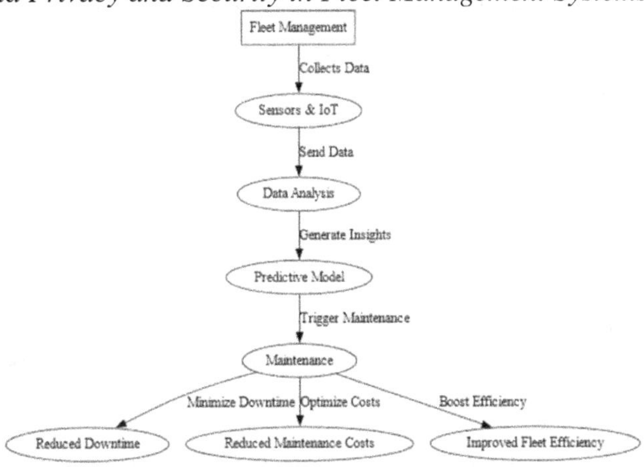

Data security in fleet management systems also extends to the cloud platforms where much of the information is stored and analyzed. These platforms must implement strict access controls, ensuring that only authorized personnel can view or modify sensitive data. Security features such as multi-factor authentication (MFA) and role-based access controls (RBAC) help mitigate risks by limiting access to only those with a legitimate need to know. Furthermore, fleet management systems should be designed with data integrity in mind, using methods such as data hashing and secure transmission protocols to prevent tampering or manipulation. As fleets become more connected and dependent on real-time data for decision-making, investing in comprehensive data security measures is essential to maintain trust, protect assets, and comply with relevant data protection regulations.

A conventional security measures, the growing scale of IoT-enabled fleet systems introduces new challenges related to scalability and system resilience. As fleet size increases, the volume of data generated by sensors and GPS devices grows exponentially, requiring efficient data processing, storage, and real-time analytics capabilities. Without scalable cloud architectures and optimized data handling mechanisms, system performance may degrade, leading to delays in decision-making and reduced operational efficiency. Scalability also impacts network bandwidth and latency, particularly in large fleets operating across geographically distributed regions. Ensuring seamless communication between thousands of connected devices requires the adoption of lightweight communication protocols and edge computing solutions to reduce dependence on centralized cloud systems.

From a security perspective, large-scale deployments expand the attack surface, making fleet management systems more vulnerable to cyber threats such as Distributed Denial of Service (DDoS) attacks, data interception, and unauthorized device access. Therefore, advanced security mechanisms such as end-to-end encryption, intrusion detection systems, blockchain-based data integrity solutions, and zero-trust architectures should be considered to enhance system robustness. A scalable and secure fleet management system must therefore balance performance, reliability, and protection against evolving cyber threats, ensuring continuous and trustworthy operation in complex real-world environments.

Data Management for Scalable Fleet Operations

Data management is a critical component of scalable fleet operations, as it ensures that the vast amounts of data generated by vehicles and IoT sensors are organized, stored, and processed efficiently. In a fleet management system, data from various sensors, including GPS trackers, fuel monitors, and engine diagnostics, needs to be collected in real-time, transmitted to a centralized platform, and stored securely for analysis. As fleets expand, the volume of data grows exponentially, making it crucial to implement robust data management systems that can handle this increase without compromising performance or reliability. Efficient data management strategies, such as data indexing, categorization, and archiving, enable fleet managers to access relevant information quickly and effectively, even with a large fleet of vehicles. This allows for informed decision-making across all levels of fleet operations, from daily scheduling to long-term strategic planning.

As fleet operations scale, it is essential to ensure that the data management infrastructure remains flexible and adaptable to handle new vehicles, additional sensors, and growing data streams. Cloud-based storage solutions are particularly valuable in this context, as they offer the scalability needed to accommodate increasing data volumes. These platforms provide the flexibility to scale up or down based on the fleet's needs, while ensuring that data is stored securely and remains easily accessible. Data management for scalable fleet operations also involves integrating data from diverse sources, such as telematics systems, fuel sensors, and maintenance logs, into a unified platform. This integration allows fleet managers to gain a comprehensive view of the entire fleet's performance, identify inefficiencies, and optimize operations without the need for multiple disconnected systems.

Predictive Maintenance and Performance Monitoring

The Need for Predictive Maintenance in Fleet Management

Predictive maintenance is an essential component of modern fleet management, as it allows for the proactive identification and resolution of potential issues before they lead to costly breakdowns or operational disruptions. Traditional maintenance approaches, which often rely on fixed schedules or reactive repairs, can result in unnecessary downtime, unexpected repair costs, and a reduction in vehicle lifespan. In contrast, predictive maintenance leverages real-time data collected from IoT sensors and GPS tracking systems to monitor the condition of vehicles continuously. By analyzing this data, fleet managers can detect early signs of wear and tear, such as irregular engine performance or abnormal fuel consumption, and schedule maintenance activities only when needed, reducing both repair costs and unnecessary maintenance intervals.

The adoption of predictive maintenance in fleet management offers several significant benefits. First, it helps prevent costly emergency repairs by identifying problems in their early stages, often before they cause significant damage. This early detection not only extends the lifespan of the fleet but also improves the overall reliability of the vehicles. Second, it reduces downtime, as maintenance is performed during scheduled intervals rather than in response to unexpected failures. As a result, fleet operators experience fewer disruptions to their operations, ensuring better fleet availability and performance. Lastly, predictive maintenance enhances safety by addressing potential issues before they can result in hazardous breakdowns or accidents.

Figure 12. The Need for Predictive Maintenance in Fleet Management

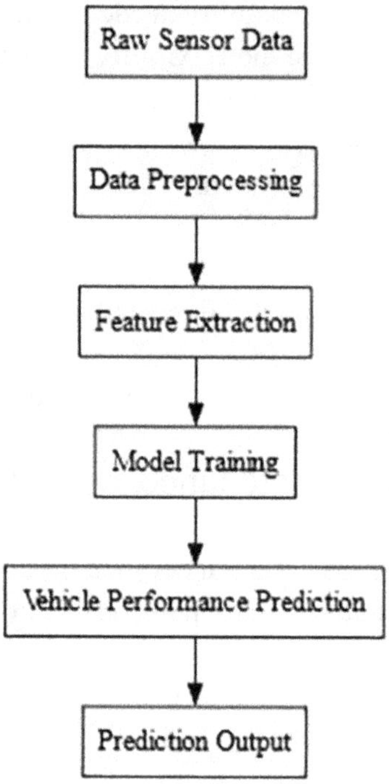

How IoT and GPS Enable Proactive Maintenance Scheduling

IoT and GPS technologies play a critical role in enabling proactive maintenance scheduling within fleet management systems. By continuously monitoring key vehicle parameters such as engine temperature, tire pressure, fuel consumption, and overall vehicle performance, IoT sensors provide real-time data that can be analyzed to identify potential issues before they escalate into costly breakdowns. These sensors capture a wealth of information that, when integrated with GPS tracking, allows fleet managers to correlate vehicle performance data with specific routes, driving behaviors, and environmental conditions. This comprehensive data set enables a more accurate assessment of vehicle health, facilitating predictive maintenance decisions based on actual usage rather than traditional time-based schedules.

The ability to predict maintenance needs based on real-time data transforms how fleet operators manage vehicle upkeep. IoT-enabled systems can generate alerts for potential maintenance issues, such as excessive engine wear, low tire pressure, or abnormal fuel consumption, which can be addressed before causing significant downtime. This shift from reactive to proactive maintenance helps minimize disruptions to fleet operations and reduces the risk of unexpected breakdowns that can lead to costly repairs. With GPS tracking data, maintenance scheduling can also be optimized by aligning vehicle servicing with the most efficient routes or downtimes, ensuring minimal impact on fleet availability.

Analyzing Sensor Data for Vehicle Performance Prediction

Figure 13. Analyzing Sensor Data for Vehicle Performance Prediction

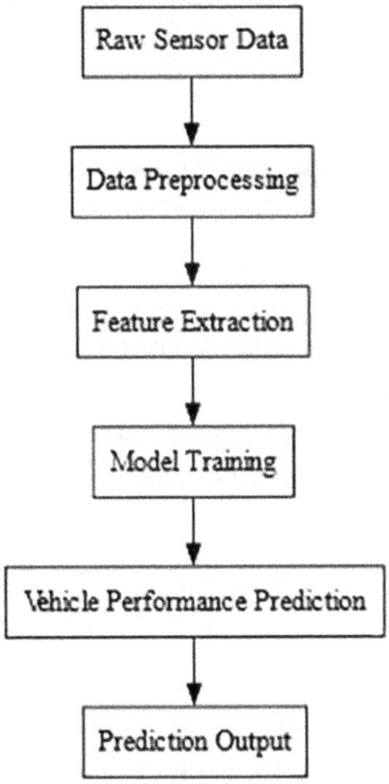

Analyzing sensor data plays a crucial role in predicting vehicle performance and enabling predictive maintenance in fleet management systems. Sensors embedded in vehicles continuously monitor various parameters such as engine temperature, tire pressure, fuel consumption, and brake performance. By collecting real-time data from these sensors, fleet managers can identify early signs of wear and tear or potential malfunctions before they lead to significant breakdowns. Machine learning algorithms process this data, recognizing patterns that indicate when a vehicle is likely to experience issues, allowing maintenance actions to be scheduled proactively. This predictive approach minimizes unexpected downtime and extends the life of the vehicle fleet, ultimately leading to lower maintenance costs and improved operational efficiency.

The ability to analyze sensor data effectively enhances decision-making by providing fleet managers with actionable insights into vehicle health. For instance, if sensor data shows that a vehicle's engine is consistently running at higher temperatures, it may signal an impending failure of the cooling system. Predicting such issues early enables fleet managers to perform targeted repairs before the vehicle experiences a catastrophic failure. The continuous flow of sensor data also allows for the optimization of vehicle performance over time by fine-tuning engine performance, reducing fuel consumption, and improving overall driving efficiency. By integrating sensor data with predictive maintenance strategies, fleet management systems can shift from reactive maintenance to a proactive, data-driven approach that optimizes both the performance and longevity of the fleet.

Reducing Fleet Downtime Through Predictive Analytics

Predictive maintenance plays a crucial role in reducing fleet downtime by allowing fleet managers to address potential issues before they lead to vehicle breakdowns. By leveraging data from IoT sensors embedded in the vehicles, predictive analytics can monitor various components such as engine health, tire pressure, fluid levels, and battery performance in real-time. Through the analysis of historical data and ongoing sensor inputs, algorithms can detect early warning signs of potential failures or performance degradation. This proactive approach enables fleet managers to schedule maintenance or repairs during non-peak times, minimizing the impact on operations and reducing the likelihood of unexpected vehicle failures that could result in costly delays or disruptions.

The ability to predict maintenance needs also helps to extend the lifespan of fleet vehicles, as timely interventions prevent minor issues from evolving into major, expensive repairs. Predictive maintenance not only improves the reliability of the fleet but also contributes to better resource management. By optimizing the timing of maintenance activities, fleet operators can avoid unnecessary maintenance tasks and reduce vehicle downtime. This leads to increased vehicle availability, allowing the fleet to operate more efficiently and reducing the costs associated with unscheduled repairs. Through continuous performance monitoring and the use of predictive analytics, fleet operators can ensure that their vehicles remain in optimal condition, ultimately enhancing fleet performance and driving long-term cost savings.

CASE STUDIES ON THE IMPACT OF PREDICTIVE MAINTENANCE IN FLEET MANAGEMENT

Case studies on the implementation of predictive maintenance in fleet management highlight the significant impact of this approach on reducing operational downtime and maintenance costs. One such case study involved a large logistics company that integrated predictive maintenance technology into their fleet management system. By leveraging IoT sensors and data analytics, the company was able to monitor the health of their vehicles in real-time, identifying potential issues before they resulted in vehicle failure. This proactive approach led to a notable reduction in unplanned maintenance and vehicle breakdowns, resulting in improved fleet uptime and efficiency. The predictive system used historical data, real-time monitoring, and machine learning algorithms to forecast when a vehicle would require maintenance, enabling the company to schedule repairs during off-peak times and avoid costly emergency fixes.

Another case study focused on a public transportation company that employed predictive maintenance to optimize vehicle performance and extend the lifespan of its fleet. By collecting data on various vehicle components, such as engines, brakes, and tires, the company was able to predict when certain parts were nearing the end of their useful life. This allowed them to replace or repair parts proactively, preventing costly and time-consuming breakdowns that could disrupt service. The predictive maintenance system also helped the company to optimize the frequency of maintenance activities, ensuring that vehicles were serviced at the most opportune times, thereby minimizing downtime and maximizing service availability. As a result, the transportation company saw a significant reduction in maintenance costs and an increase in the reliability of its fleet, demonstrating the tangible benefits of integrating predictive maintenance into fleet management practices.

CONCLUSION

The integration of IoT and GPS technologies into fleet management systems has ushered in a new era of operational efficiency, cost reduction, and enhanced fleet performance. Traditional fleet management systems often struggled with issues such as fuel inefficiency, maintenance delays, and poor vehicle tracking. However, the advent of IoT-based solutions and GPS tracking has addressed these challenges by providing real-time data on critical fleet parameters, including vehicle health, fuel consumption, and driver behavior. This technological convergence has not only optimized fleet operations but also reduced operational costs and minimized the environmental impact of fleet activities.

Through continuous monitoring and analysis, fleet managers are now equipped with unprecedented visibility into their fleet's performance, allowing for data-driven decisions that enhance operational efficiency. IoT sensors provide valuable insights into fuel usage, engine conditions, tire pressure, and more, which can be analyzed to detect inefficiencies, prevent fuel theft, and schedule maintenance based on actual vehicle needs rather than arbitrary intervals. By combining IoT with GPS, fleet managers can track the exact location of their vehicles in real time, optimizing routes to save fuel, reduce wear and tear, and improve overall fleet productivity.

Predictive maintenance, fueled by IoT data, has proven to be one of the most impactful benefits of this integrated system. By identifying potential issues before they lead to failures or breakdowns, fleet operators can reduce downtime, extend the life of their assets, and avoid costly emergency repairs. This proactive maintenance approach not only contributes to better fleet reliability but also results in substantial cost savings over time.

The cloud-based platforms that support these systems have revolutionized how data is collected, processed, and visualized. Real-time dashboards allow fleet managers to instantly access performance metrics, track fuel consumption trends, and monitor vehicle health. Advanced data analytics, combined with machine learning, further enhance decision-making capabilities, enabling predictive modeling and optimizing fleet operations in ways previously unattainable.

The future of fleet management lies in the continued evolution of these technologies. As the adoption of electric vehicles (EVs) and autonomous driving continues to grow, the integration of IoT and GPS will be critical in managing these new types of fleets. Furthermore, innovations in AI, machine learning, and 5G connectivity will drive even greater advancements in fleet management systems, offering further opportunities for optimization, sustainability, and cost-efficiency.

While the proposed IoT and GPS-based fleet management framework demonstrates significant advantages in terms of real-time monitoring, fuel optimization, and predictive maintenance, certain limitations must be acknowledged. The effectiveness of the system depends heavily on network availability, sensor reliability, and seamless integration across heterogeneous fleet environments. Additionally, challenges related to data security, system scalability, and implementation cost may impact real-world deployment. Addressing these limitations through robust system design, advanced security mechanisms, and scalable architectures remains an important direction for future research.

REFERENCES

Abdirahman, A. A., Hashi, A. O., Dahir, U. M., Elmi, M. A., & Hashim, S. Z. M. (2025). Enhanced vehicle tracking: A GPS-GSM-IoT approach. *International Journal of Computing and Digital Systems*, *17*(1), 1–11. DOI: 10.12785/ijcds/1571012134

Alam, T., & Shakil, K. A. (2023). A review of IoT applications in fleet and transportation management systems. *Sustainable Computing : Informatics and Systems*, *38*, 100857.

Ali, M., Kamal, M. D., Tahir, A., & Atif, S. (2021). Fuel consumption monitoring through COPERT model: A case study for urban sustainability. *Sustainability (Basel)*, *13*(21), 11614. DOI: 10.3390/su132111614

Babiyola, A., Saillaja, V., Shally, S. P., & Omkumar, S. (2023). Development of an IoT-based integrated system for fleet management in real time. In Proceedings of the 2nd International Conference on Edge Computing and Applications (ICECAA) (pp. 1265–1269). IEEE.

Chandana, S., Dhanyashree, C. J., Ashwini, K. L., Harini, R., Premkumar, M., & Abualigah, L. (2023). Fuel automata: Smart fuel dispenser using RFID technology and IoT-based monitoring for automotive applications. *Journal of Autonomous Intelligence*, *6*(1), 78–92. DOI: 10.32629/jai.v6i1.682

Chaganti, P. C. V., Manitha, P. V., & Nithya, M. (2024). Fleet management using IoT and machine learning techniques. In Proceedings of the 5th International Conference on Intelligent Communication Technologies and Virtual Mobile Networks (ICICV) (pp. 498–503). IEEE.

Degadwala, S., Upadhyay, R., Upadhyay, S., Dave, S. S., Mahida, D., & Vyas, D. (2023). Enhancing fleet management with ESP8266-based IoT sensors for weight and location tracking. In Proceedings of the 3rd International Conference on Innovative Mechanisms for Industry Applications (ICIMIA) (pp. 13–17). IEEE.

Devi, J. Y., Murki, K., Hanmanthu, B., Ram, A., & Kamran, M. (2023). Detection of fuel monitoring system through IoT sensors for sustainability. E3S Web of Conferences, 430, 01095.

Desai, S., Suthar, R., Yadav, V., Ankar, V., & Gupta, V. (2022). Smart bus fleet management system using IoT. In Proceedings of the International Conference on Emerging Research in Electronics, Computer Science and Technology (ICERECT) (pp. 1–6). IEEE.

Farahpoor, M., Esparza, O., & Soriano, M. C. (2023). Comprehensive IoT-driven fleet management system for industrial vehicles. *IEEE Access : Practical Innovations, Open Solutions*, *12*, 193429–193444. DOI: 10.1109/ACCESS.2023.3343920

Gubbi, J., Buyya, R., Marusic, S., & Palaniswami, M. (2022). Internet of Things (IoT): A vision, architectural elements, and future directions. *Future Generation Computer Systems*, *29*(7), 1645–1660. DOI: 10.1016/j.future.2013.01.010

Hapsari, A. W., Prastowo, H., & Pitana, T. (2021). Real-time fuel consumption monitoring system integrated with Internet of Things (IoT). Kapal: Jurnal Ilmu Pengetahuan dan Teknologi Kelautan, 18(2), 88–100.

Infrastructure, S. T. (2024). Internet of Things (IoT) solutions for smart transportation infrastructure and fleet management. *Journal of Propulsion Technology*, *45*(4), 1492–1509.

Jimoh, A. A. (2025). Development and implementation of IoT-based vehicle tracking system. Journal of Engineering Research and Development.

Kamolsin, C., Rungkul, W., Limpalasuk, C., Pensiri, F., Ratanavilisagul, C., & Visutsak, P. (2023). Vehicle management system: Fuel, mileage, cost, and maintenance tracker. In *Proceedings of the International Conference on Cyber Management and Engineering (CyMaEn)* (pp. 283–288). IEEE. DOI: 10.1109/CyMaEn57228.2023.10051089

Khan, R., Khan, S. U., Zaheer, R., & Khan, S. (2021). Future Internet: The Internet of Things architecture, applications, and challenges. *IEEE Communications Surveys and Tutorials*, *16*(1), 234–250.

Lin, J., Yu, W., Zhang, N., Yang, X., Zhang, H., & Zhao, W. (2022). A survey on Internet of Things: Architecture, enabling technologies, security, and privacy. *IEEE Communications Surveys and Tutorials*, *19*(4), 2287–2326.

Mathi, S. C., Jambula, S. R., Surat, D. A., & Maridinapalli, M. C., PV, M., & Sailaja, V. (2024). Development of an intelligent fuel monitoring and theft detection system with GPS and GSM integration. In Proceedings of the International Conference on Inventive Computation Technologies (ICICT) (pp. 2095–2102). IEEE.

Meenambika, A., Karuppasamy, P., & Vithyaa, T. (2023). Fleet Guard: Advanced fleet management and vehicle monitoring system. *Journal of Electronic Design Technology*, *14*(2), 18–24.

Moumen, I., Rafalia, N., Abouchabaka, J., & Aoufi, M. (2023). Real-time GPS tracking system for IoT-enabled connected vehicles. E3S Web of Conferences, 412, 01095.

Nagesh, A. S., Aparna, C., & Rao, M. C. S. (2025). A GPS-enabled fuel sensor-based vehicle tracking system for fleet management using the Internet of Things. Scalable Computing. *Practice and Experience*, *26*(1), 34–47.

Patil, D., Mane, S., Biradar, S., Rankhamb, S., & Bhonsle, M. (2023). Fleet management mobile application using GPS shortest path. In *International Conference on Data Intelligence and Cognitive Informatics* (pp. 205–213). Springer.

Prabhu, S. B., Deepa, K., & Nithya, M. (2022). IoT-enabled fuel level monitoring and automatic fuel theft detection system. In *Proceedings of the 13th International Conference on Computing Communication and Networking Technologies (ICCCNT)* (pp. 1–7). IEEE. DOI: 10.1109/ICCCNT54827.2022.9984515

Punith, M. S., Nithya, M., & Deepa, K. (2022). IoT-enabled smart fleet management. In Proceedings of the IEEE International Conference on Cybernetics, Cognition and Machine Learning Applications (ICCCMLA) (pp. 256–260). IEEE.

Rathnayaka, R. P. D. T., Ekanayake, K. V. J. P., Rathnayake, H. U. W., & Jayetileke, H. R. (2021). Fleet management with real-time data analytics. In Proceedings of the 2021 6th International Conference on Information Technology Research (ICITR) (pp. 1–6). IEEE.

Rudrusamy, B., Teoh, H. C., Pang, J. Y., Lee, T. H., & Chai, S. C. (2023). IoT-based vehicle monitoring and driver assistance system framework for safety and smart fleet management. *International Journal of Integrated Engineering*, *15*(1), 391–403. DOI: 10.30880/ijie.2023.15.01.035

Teja, P. S., Reddy, N. S. S., Pandugu, P. K., & Vangari, P. (2023). Smart vehicle monitoring and tracking system. E3S Web of Conferences, 391, 01099.

Tulloh, R., Ramadan, D. N., & Hadiyoso, S., Rohmattullah, & Rahmana, Z. (2021). Fuel truck tracking for real-time monitoring system using GPS and Raspberry Pi. In Proceedings of ICEBEHI 2020 (pp. 29–41). Springer. DOI: 10.1007/978-981-33-6926-9_3

Yamini, N., Kumar, R. B., Mohamed, J. S., & Iyengar, N. C. S. (2023). Smart IoT-based system for monitoring and detecting fuel theft and fuel indication for refilling process. *International Journal of Human Computations & Intelligence*, *2*(5), 247–255.

Zhang, Y., Chen, X., & Li, L. (2023). Intelligent fleet management using IoT and machine learning: A comprehensive review. *IEEE Internet of Things Journal*, *10*(5), 4123–4145.

Chapter 5
Geomatics Applications in Rooftop Solar PV Potential Mapping:
A Case Study at The Papua New Guinea University of Technology

Jeremiah Renagi
http://orcid.org/0009-0003-9825-7532
Papua New Guinea University of Technology, Papua New Guinea

Tingneyuc Sekac
http://orcid.org/0000-0001-5972-7149
Papua New Guinea University of Technology, Papua New Guinea

Sujoy Kumar Jana
http://orcid.org/0000-0003-0901-5316
Papua New Guinea University of Technology, Papua New Guinea

ABSTRACT

PNG is rapidly embracing solar PV technology, especially in rural and peri urban areas, aligning with the government's ambitious target of achieving 100% renewable electrification by 2050. This study investigates rooftop solar potential at the PNGUoT using UAV photogrammetry integrated with GIS. The analysis generated rooftop solar energy maps, visually supporting quantitative estimations. Results revealed that three sampled rooftops could produce 90.32 MWh annually, far exceeding their combined demand of 21.49 MWh, leaving 68.83 MWh surplus for storage or export. This demonstrates that rooftop PV can fully meet residential electricity needs while generating excess energy. The study highlights UAV photogrammetry and GIS as cost

DOI: 10.4018/979-8-3373-7941-8.ch005

efficient, reliable, and scalable tools for suburban scale solar assessment, offering a replicable model for campus wide and national renewable energy deployment. It underscores the importance of renewable energy awareness and knowledge dissemination to drive PNG's transition from fossil fuels, supporting policy, planning, and the 2050 electrification vision.

INTRODUCTION

Papua New Guinea's ongoing electricity crisis remain a significant constraint to the economic and socio-economic development of this resource-abundant nation (PNG Power Limited, & National Energy Authority, 2024). Whilst only 13-15% of the nation (PNG) have access to grid-supplied electricity, the remaining population turns to other alternatives such as solar lanterns, solar home systems, local mini-grids, and thermal generation etc. Furthermore, more than 25% of PNG's total population (31%) still do not have access to any form of electricity (Rahman & Koo, 2024. The PNG government's interest in addressing the adverse impacts of climate change positions solar energy and technology at the forefront of providing reliable, affordable and accessible electricity whilst combating the negative effects of thermal generation. However, the implementation of several Solar PV initiatives in PNG has had mixed successes, proving to be more challenging (Aiau et al.,2016). Thus, the transition from fossil fuel generation to climate-friendly energy sources requires an effective planning mechanism that deals with input from multiple stakeholders and is guided by continuous collection and analysis of site-specific data by local agencies, which results in a more coordinated approach (Rawali et al.,2019; Kezazy, et al., 2025).

Globally, GIS-based modelling approaches have become increasingly popular with the availability of Lidar data (Song et al.,2025; Eitel. et al., 2016; Mandla et. al., 2021) giving analysts the flexibility of systematically assessing the behaviour of solar radiation over multiple building rooftop surfaces. GIS is essential for spatial analysis and solar radiation modelling, UAV photogrammetry provides high-resolution 3D surface data through image-based reconstruction, and solar PV concepts relate to the conversion of solar radiation into electricity and estimation of energy potential. This approach starts by developing 3D models or using existing models from topographic data and calculating the solar radiation incident on the existing surface with considerations such as rooftop geometry (slope, aspect, elevation) and surrounding obstructions.

This approach enables the automatic computation of key rooftop PV characteristics, such as the slope, orientation, and total area of the rooftop, leading to accurate estimations of PV potential and adhering to the 5 key principles of assessing the solar PV potential of rooftops (Gassar et al., 2021; Long et al.,2025):

1. Total Rooftop Area
2. Available Rooftop Area
3. Available Solar Radiation
4. PV Potential Output
5. Economic Potential

However, in data-scarce regions and developing countries, the cost of running a LiDAR flight mission is not cost-effective for solar radiation mapping at smaller built-up scales. The emergence of Unmanned Aerial Vehicle (UAVs) photogrammetry, integrated with spatial analysis tools, provides a fast and effective method for data collection, processing, and analysis, essential for the data-driven implementation of solar initiatives. The UAV-DSM method uses a photogrammetric software that processes a set of overlapping images to generate 3D topographic data, in particular a Digital Surface Model (DSM), essential for solar radiation estimation algorithms. Due to the limited capabilities of these UAV's, the approach is suitable for small built-up areas such as university campuses (Wongwan et al.,2023.

LiDAR-based modelling is widely used at city and regional scales, where high-density laser point clouds generate precise 3D Digital Surface Models (DSM) that capture rooftop geometry, slope, aspect, and shading from trees or nearby buildings, enabling highly accurate solar radiation estimation and suitability mapping. satellite-derived solar mapping uses multi-source remote sensing data (e.g., high-resolution imagery and geostationary satellites) combined with GIS to assess solar irradiance and rooftop availability over large geographic extents, making it scalable for national or global studies. AI-driven PV estimation has emerged as a cutting-edge approach, where machine learning and deep learning models automatically detect rooftops, segment suitable areas, estimate installed PV systems, and even predict energy generation using integrated spatial and temporal datasets.

When estimations are conducted using computer models, the results must be validated to establish a degree of certainty before using the data for decision-making. This includes validating solar radiation models against ground-based weather station data using appropriate statistical methods (Singh, A., 2023 or comparing the solar PV potential estimates with existing PV system data (Fuentes et al., 2020; Radosevic et al.,2022). Researchers and analysts around the globe apply various validation techniques. However, the availability of data and the scale of the study determine the preferred and applicable method.

Currently, within Papua New Guinea, there are several renewable energy initiatives regarding solar PV (Bosip, 2023), in particular, the PPLs rooftop solar pilot project at the Mineral Resource Authority (MRA) headquarters, NCDC, which aims to cut annual energy expenditures by around 60% (MRA, 2022). Despite the growing use of GIS and RS tools and techniques for rooftop solar potential assessments globally, there are no available studies in PNG that use UAV and photogrammetric methods to estimate the PV potential of building rooftops.

This study aims to investigate the application of UAV photogrammetric methods and the integration of GIS to estimate the PV potential of residential rooftops in the Papua New Guinea University of Technology (PNGUoT) campus. In an urban context where rooftop space is abundant, the application of rooftop PV systems exploits unused roof space, unobstructed vision of the sky and is less vulnerable to vandalism, reinforcing the importance of the built environment for reducing emissions associated with energy efficiency (Wolfs, 2017; Yao et al.,2023; Yin et al.,2025).

MATERIALS USED AND METHODOLOGY

Background of the Study Area

The Papua New Guinea University of Technology campus is in the located in Lae District and covers approximately 220 hectares. The campus relies solely on the national grid (PPL). However, several academic and administrative buildings, as well as the student dormitories, are connected to diesel backup generators, which are costly both financially and environmentally. The staff residential area, which occupies a significant portion of the campus, comprises approximately 367 buildings of varying designs and sizes, subdivided into five distinct residential zones. Additionally, the residential buildings lack access to backup energy sources, making them vulnerable to the adverse effects of blackouts.

The study site in this case covers a small portion of the residential zone (1) and is defined in the WGS 84 / UTM Zone 55S coordinate system (EPSG:32755), with Easting values ranging from 499,085.818 m to 499,408.595 m and Northing values ranging from 9,262,537.185 m to 9,262,823.674 m, covering an area of 48502 sqm. A total of 23 residential buildings of diverse rooftop designs within the study site were assessed. The complex arrangement of buildings and the varied roof geometries necessitate a GIS-based modelling approach to produce realistic, validated estimates of solar PV potential.

Figure 1. The Study Site Map

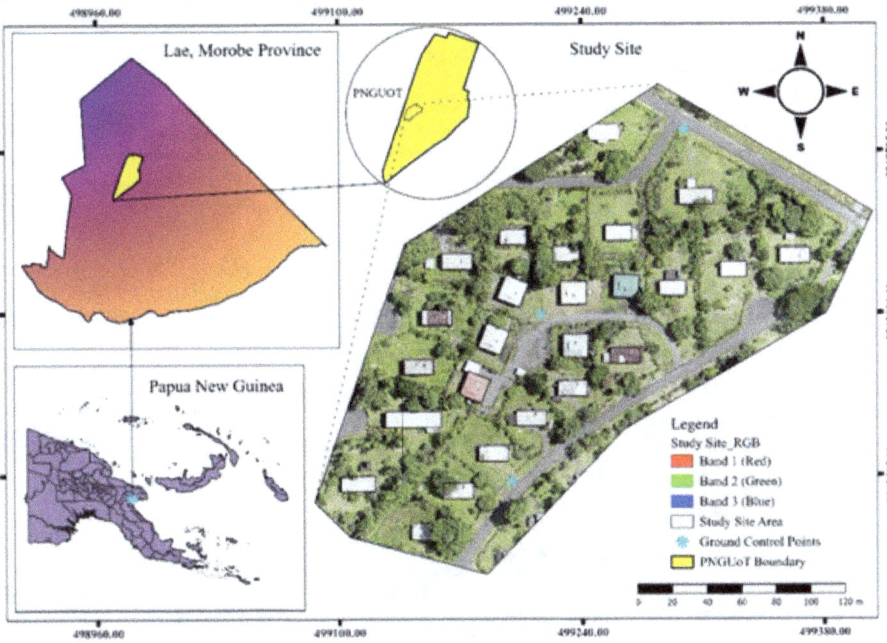

Data Source and Pre-processing

The datasets used in this research were primarily obtained through primary data collection methods, including Unmanned Aerial Vehicle (UAV) photogrammetric surveys, structured interviews and questionnaires and meteorological observations from a local ground weather station. A DJI Mavic 2 Pro drone equipped with a Hasselblad L1D-20c RGB camera was used to capture raw, overlapping images of the study site for photogrammetric processing. Before the flight mission was conducted, three (3) ground control points were established in an approximately uniform order, and their precise coordinates were measured using a GNSS RTK base setup at BM 32 and a rover. The precise coordinates of these points are shown in Table 1.

Table 1. Established Ground Control Points in the Study site

GCP Tagged Name	E	N	Z
GCP 1 - Imila	9262688.864	499216.2522	130.315
GCP 2 - Ramu	9262591.53	499199.0869	129.1714
GCP 3 - Warangoi	9262771.22	499142.4734	130.5557
Benchmark 32 (BM32)	9262864.032	499861.978	133.276

Figure 2. Setting of RTK Base Station A) DJI Mavic 2 Pro drone equipped with a Hasselblad L1D-20c RGB camera (5472 × 3648 resolution). B) Measuring 3D Coordinates using GNSS RTK Rover. C) Setting up GNSS RTM Base at BM32. D) HI-Target V200 RTK GNSS Base and Rover

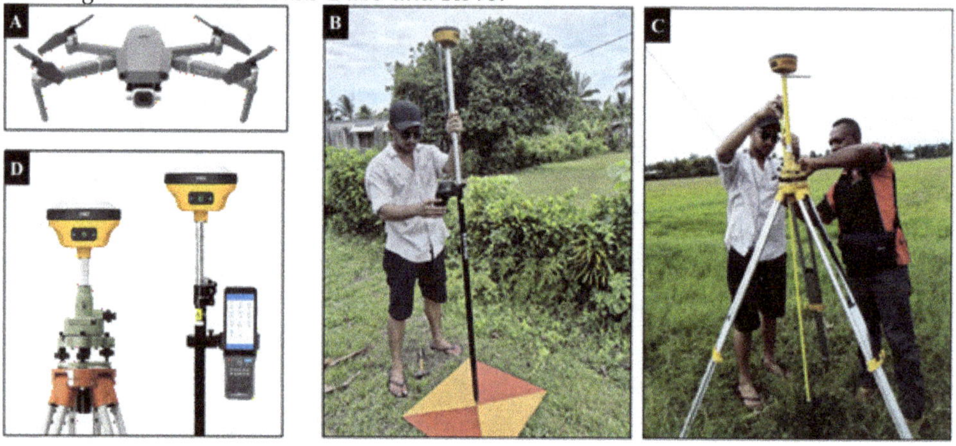

After the flight mission was completed, a total of 161 overlapping images were processed using Pix4Dmatic v1.71.0, a professional photogrammetric software package. This phase aimed to derive three (3) topographic outputs essential in modelling rooftop solar radiation, i.e., i) Digital Surface Model, ii) Orthomosaic iii) Dense Point Cloud layer. The resulting DSM and Orthomosaic achieved an average Ground Sampling Distance (GSD) of 2.2 cm, providing sufficient spatial resolution to digitise the rooftop boundary and facilitate accurate rooftop area extraction. Apparent distortions on rooftop surfaces were identified through careful observation of the DSM. These distortions may be attributed to highly reflective rooftops; vegetation cover over rooftops or insufficient image overlap. Despite these anomalies, rooftop surfaces for the selected houses were manually corrected and remodelled by triangulating extracted 3D coordinates of the rooftop edges.

Figure 3. Rooftop Condition A) Irregular rooftop surface B) Highly reflective rooftop surface in raw imagery C) Extended view of rooftop in a 3D point cloud environment (Cloud Compare)

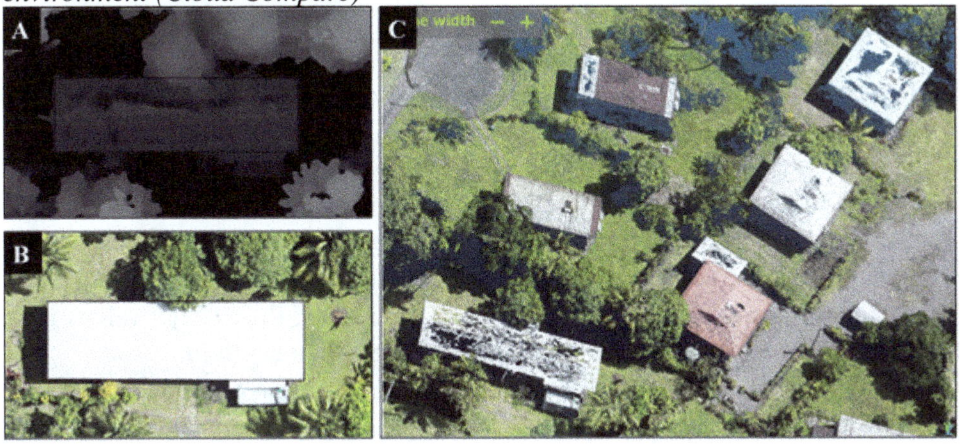

In order to statistically validate the solar radiation model, ground-based solar data were needed and collected from the UNITECH Agricultural Farm meteorological station. The raw data is logged every 2 hours, and solar insolation is measured in kWp, which was later converted to kWh to standardise with GIS solar radiation model estimations. The available data ranged from 'January 2025 to August 2025', hence, limiting a full-year analysis.

The final required dataset to complete the planned analysis was collected through questionnaires from the PNG Power Head Office in Lae MP. The data were obtained in billing format over one year for three select households within the study site. The billing data were aggregated, and the average monthly power bill expenditure was converted to kWh to proxy the average monthly electricity demand.

METHODOLOGY

Digitizing Building Footprints & Extracting Rooftop Vertex Points

The first step in reconstructing the 3D surfaces was digitizing the building rooftops. Firstly, the flat rectangular rooftops were digitized using the rectangular polygon guide tool. The same tool was employed on the sloped rooftops due to the rectangular footprint geometry; thus, multiple rectangular polygons were used to capture the distinct points and break lines of the rooftop. Each polygon was attributed to its corresponding building address (street name and house number). After all the rooftops were digitized, the *extract vertices tool* was used to extract rooftop vertices and create a rooftop vertex point layer in preparation for sampling and assigning elevation values.

Dense Point Cloud Sampling

The dense point cloud layer was derived from photogrammetric processing and was used as the basis for extracting elevation (z) values for the rooftop vertex points. The georeferenced dense point clouds generated in Pix4Dmatic (v.1.7.1) were imported into Cloud Compare (v.2.13.2), a 3D point cloud environment, where individual rooftop vertices were manually inspected, and their height values were sampled and assigned to their corresponding vertex point in QGIS.

For flat rooftops, the sampled elevation values were taken from one rooftop corner, and the same value was applied to the other vertex points within its respective footprint. For sloped rooftops, the sampling required careful inspection at each rooftop vertex point, due to the distorted points and dense tree cover. Hence, in instances where rooftop surfaces were severely distorted or where no dense point cloud data existed, the elevation value was adopted from a nearby or identical building (rooftop design).

Figure 4. A) Building footprints digitised in rectangular segments and extracted rooftop vertex points. B/C) Manually sampling 3D coordinates of rooftop edges for the corresponding vertex point

After the completion of this phase, the initial 2D rooftop vertex point layer now comprised a z-value column that stored elevation values of the respective rooftop edges. Subsequently, they were transformed into 3D points, ready to interpolate a 3D surface.

TIN Interpolation Tool

The process of triangulation connects a set of discrete elevation points to form a set of overlapping triangles representing a continuous terrain surface. The Delaunay triangulation is one of the commonly used algorithms. A TIN's fidelity comes from its ability to ensure that no other vertices lie within the circumcircle of any triangle, resulting in high-precision terrain modelling. This method effectively improves the geometric uniformity and structural stability of the TIN. Furthermore, this method provides the flexibility of manually adding supplementary points into a TIN model, thereby allowing the analyst to reconstruct the TIN surface to their preference (He et al., 2026).

The 3D point layer, which represents all rooftop edges, was the main input into the TIN interpolation tool. After the first operation of the tool, all flat rooftops were modelled correctly; however, several sloped rooftops at the base showed incorrect geometries due to the proximity of other rooftop points.

To correct this error, supplementary 3D points were manually added along the vertices of the affected rooftop, thereby avoiding the interference from other rooftop points. The final TIN rooftop model represents the exact geometry of the existing rooftops; however, it omits the presence of other surface features such as terrain and vegetation etc.

Finally, manual validation was conducted by comparing elevations between smooth rooftop surfaces in the original DSM and the TIN rooftop model, ensuring all differences fell within the acceptable tolerance defined for this study (±0.05 – 0.10 m).

Figure 5. Addition of supplementary points to correct the initial TIN interpolated surface

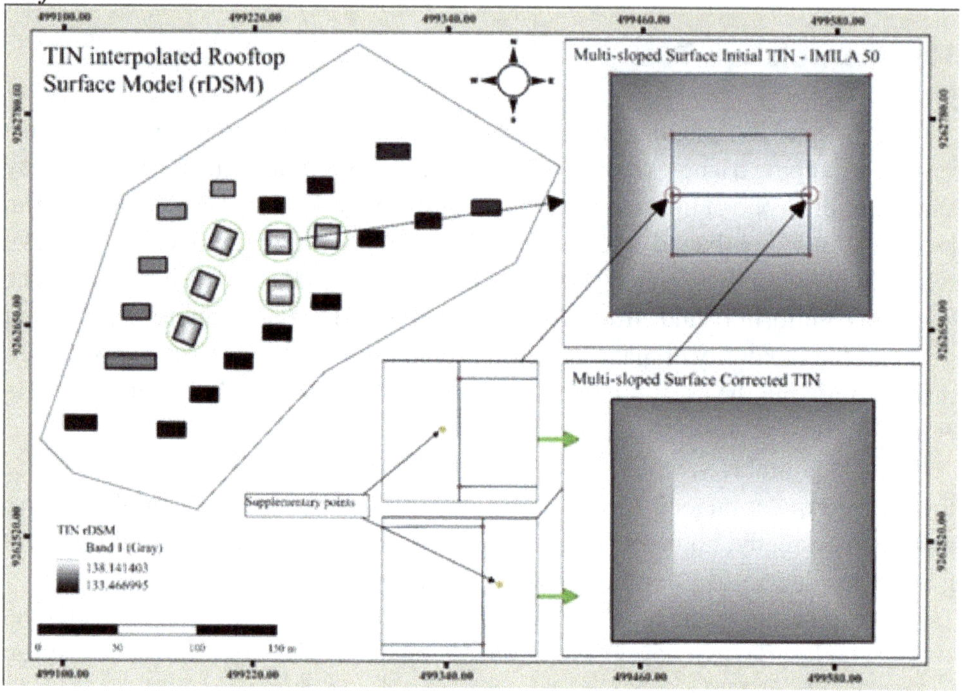

Rooftop Solar Radiation Modelling

In order to determine the potential power generation of a solar PV system, the amount of solar resource on the surface must be estimated. The ArcGIS Area Solar Radiation Tool is integrated into a GIS that should use view shed analysis result and diffuse/direct radiation to calculate the radiation at each raster cell of a DSM. These calculations are based on an upward-looking hemispherical view shed algorithm that evaluates surrounding raster cells within the DSM to determine which portions of the sky are visible or blocked by terrain, buildings, or vegetation by computing sky visibility across multiple azimuth and zenith angles. The model quantifies direct, diffuse and reflected radiation for every cell (Fu, 1999).

The parameters were specified in the ASR tool dialogue box, enabling considerations regarding climatic, spatial and temporal factors. All modelling parameters were defined within the ASR tool's parameter dialogue box. The modelling period was set to the Whole Year, with monthly intervals enabled to generate individual solar radiation maps for each month. Atmospheric parameters, including transmissivity and diffuse proportion, were configured as a uniform sky. Sky resolution and z-factor settings were also applied to maintain the default values.

The building footprint polygons were used to guide the solar radiation calculations done only on the raster cells within the rooftop polygons. For each of the twelve rooftop solar radiation maps, the average solar radiation (kWh/month) for each rooftop was extracted using the Zonal Statistics as Table tool and organised within an Excel spreadsheet, clearly depicting the estimated average solar radiation each rooftop receives each month.

Mean Absolute Percentage Error

Validating forecasted data against ground-truth data plays an important role in establishing a degree of certainty regarding the GIS-based solar radiation model. The Mean Absolute Percentage Error (MAPE) is a statistical measure used to evaluate the accuracy of forecasting or predictive models and expresses prediction errors as a percentage, showing how far off forecasts are from actual values on average.

The computation of absolute errors for monthly forecasted data and mean percentage error was performed in a Microsoft Excel spreadsheet. The first task was importing both the forecasted GIS data and the calibrated weather station for each month in kWh/m2. Secondly, the absolute error for the monthly forecasts was computed, followed by the absolute percentage error for each month.

Table 2. Weather station data against GIS-based forecasted data

(kWh/m2/month)	Jan	Feb	Mar	Apr	May	Jun	Jul	Aug
Farm_WS Data	171.89	135.38	159.60	141.27	131.48	112.82	122.03	120.89
Forecasted Data	157.57	146.92	161.52	144.66	132.42	110.79	128.25	144.17

Finally, the mean of the percentage errors of each month represents the overall error associated with the modelled solar radiation. Furthermore, the data available for appraisal of this study covered only January 2026 to August 2026. Hence, only the corresponding solar radiation monthly estimates were validated.

$$MAPE = (100\%/n)\sum_{t=1}^{n}\left|\frac{A_t - F_t}{A_t}\right|$$

Whereas

A_t = observed monthly average daily solar radiation from the meteorological dataset

F_t = model-predicted monthly average daily radiation from the ASR output
n = number of months evaluated

Table 3. Interpretation of MAPE Results

MAPE (%)	Interpretation
< 10	Highly accurate forecasting
10–20	Good forecasting
20–50	Reasonable forecasting
> 50	Inaccurate forecasting

Rooftop PV Potential Computation

Once the average solar radiation for each rooftop per month is computed, these values are then integrated into a system-specific calculation of basic parameters to determine the PV Potential per rooftop. This study adopted the constant value method applied to the Asian Development Bank (ADB) headquarters in Manila (Paton, C. P., 2014). This approach followed a two-step calculation using preselected system parameters and the individual rooftop area to determine the potential installation capacity of a rooftop, followed by computing the potential PV generation for each rooftop.

Potential Installed Capacity

The potential installed capacity describes the theoretical PV system size that can be installed on a specified rooftop, measured in kilowatts peak (kWp) and is computed using equation 2 and the parameter values in Table 8. This approach follows the constant value method that employs assumptions for parameters regarding the rooftop suitable area. This case study employed system parameter values from an existing PV setup within the UNITECH campus. The panel in this case has a module area (Am) of 2.59m2 and a rated capacity (Cm) of 550Wp. The roof-cover ratio (RCR) refers to the fraction of the total roof area that the modules will cover. Due to the variety of rooftop designs, a constant value was applied to flat rooftops (80%) and sloped rooftops (50%).

The roof area (Ar) is the total rooftop area minus the area of any obstacles on the roof. Post-inspection of the high-resolution orthophotograph indicated that all rooftops comprised a single solar hot water system, measured at approximately 5 m2, which was subtracted from the total area of each rooftop to determine the AR.

$$C_r = \left(\frac{C_m}{1000}\right) \times \left(\frac{A_r \times RCR}{A_m}\right)$$

Table 4. Input Parameters and Values of Equation 2

Parameters	Input Value	Unit	Description
CM	550Wp	Wp	Rated capacity of specified PV module converted to (kWp)
RCR	Slope-Flat	Dimensionless	Roof Cover Ratio
AR	Varies per rooftop	m2	Available roof area or installation of solar modules
Am	2.59	m2	Area of the specified PV module

Potential Rooftop PV Generation

The completion of this phase achieves the third (3) objective of this study and informs us how much energy the PV system will potentially generate if installed. Hence, the technical potential is determined by the product of the potential installed capacity of a rooftop (Cr) by the respective average solar radiation estimates, and the adopted derating factor, which is set based on a standard value. This calculation is done in an Excel spreadsheet for each month for each building rooftop by substituting the respective values into its corresponding parameters.

$$E = C_r \times GHI_a \times D$$

E: Energy yield
C_r: Potential installed capacity
GHI_a: Global Horizontal Irradiation (x)
D: Derate factor (0.75)

Table 5. Software Used

Software Name	Version	Task Performed
Pix4Dmatic	1.71.0	● Generate DSM ● Generate Orthomosaic ● Generate dense point clouds
QGIS	3.40.1	● Digitising building footprints ● Extracting Rooftop vertices ● Creating 3D point layer ● Generating TIN rooftop surfaces ● Extracting mean rooftop solar radiation values
Cloud Compare	2.13.2	● Sampling z-values for rooftop vertex points
ArcGIS Pro	3.8	● Solar radiation modelling – Area Solar Radiation Tool
Microsoft Word	Microsoft 365	● Dissertation write-up
Microsoft Excel		● Processing household demand data ● Processing weather station data ● Processing extracted monthly rooftop solar radiation data ● Performing MAPE analysis ● Performing PV Estimation ● Generating Tables and Graphs for Analysis

RESULTS AND DISCUSSION

As outlined in section (Rooftop Solar Radiation Modelling), twelve thematic maps were generated by the Area Solar Radiation tool in ArcGIS Pro 3.8. Each map represents the average solar radiation the rooftops receive that month, respectively. These monthly rooftop solar radiation maps provide a clear visual understanding of how sun angles and roof orientation interact throughout different time periods to influence variations in the availability of solar energy.

Solar Radiation Modelling Results

The rooftop solar radiation map for January, along with the extracted monthly solar radiation values in Figure 6, indicates an average aggregated potential of 157.57kWh/m2. Furthermore, the distribution of solar radiation on flat rooftops is uniform, whilst sloped rooftops indicate a significant portion of incident radiation on south-facing rooftops over other oriented segments. There is an apparent balance of solar radiation distribution on sloped rooftops in February based on visual observation. Additionally, the average aggregate experiences a drop in potential to 146.92kWh/m2. Following this drop, the highest monthly average (161.52kWh/m2) for the site is achieved in March. This month also sees a shift in solar radiation intensity from south-oriented rooftop segments towards North-facing segments.

As indicated by Figure 6, the average potential for each month follows a downhill trajectory from March to April (144.66kWh/m2) down to May (132.42kWh/m2), where the average solar radiation of the site reaches its annual lowest of 110.79kWh/m2 in June. During this period, north-oriented rooftops receive more solar energy than other oriented surfaces, including flat rooftops.

Leading into the second half of the year, the intensity of solar radiation starts to shift back to south-oriented segments. The average monthly solar radiation of the site gradually increases from July (128.25kWh/m2) to September and is depicted clearly in the distribution of solar radiation, which is fair around sloped rooftops, whilst intensity is greater on flat rooftops. There is a slight decrease in intensity on all rooftops in September from 159.12kWh/m2 to 158.10kWh/m2 in October, and then down to November. After which, the rooftop aggregates solar radiation increases in November (153.84 kWh/m2) through to December (155.12 kWh/m2). Furthermore, there is a noticeable increase towards south-facing surfaces. These spatial and temporal variations form an important foundation for evaluating rooftop solar PV potential, as they allow for more accurate identification of high-performing roof segments and more realistic estimates of annual solar energy production.

Figure 6. Forecasted average monthly solar radiation

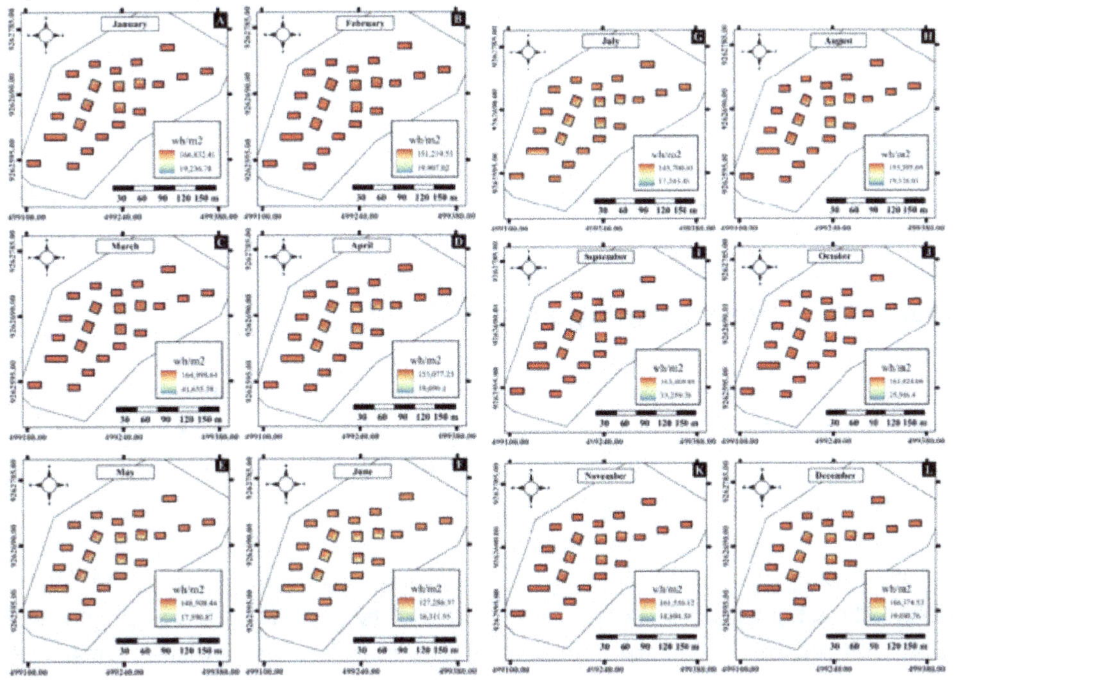

Figure 7. Monthly Rooftop Solar Radiation Maps for A. January B. February C. March D. April E. May F. June G. July H. August I. September J. October K. November L. December

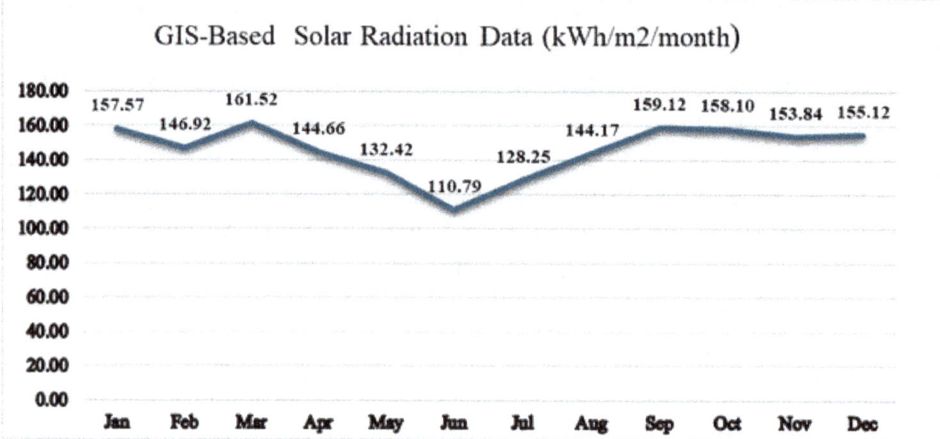

These findings are consistent with previous studies that emphasize the importance of roof orientation, tilt, and seasonal sun-path dynamics in determining rooftop solar potential. For instance, GIS-based solar assessments by Gassar & Cha, 2021 and Fuentes et al.,2020 demonstrate that equatorial and tropical regions exhibit bi-directional seasonal shifts in optimal roof orientation, unlike temperate regions where one dominant aspect prevails. Similarly, studies using LiDAR and UAV-derived DSMs confirm that flat rooftops generally show uniform radiation distribution, while sloped roofs exhibit strong directional dependency due to shading and incidence angle effects.

Mean Absolute Percentage Error Results

The MAPE analysis results indicate a strong correlation between the monthly weather station data and the GIS-based modelled data. The months of January and February achieve an absolute percentage error of 8.33% and 8.53%. The absolute percentage errors for March (1.2%), April (1.2%), May (2.4%), and June (0.7%) all fall below 3%, indicating a strong correlation between the data sources for those months. Analysis for July obtained a 5.1% absolute percentage error, and August obtained the highest absolute percentage error of 19.25%. Thus, the average across all months yields a mean absolute percentage error of 5.92%, which is considered excellent forecasting. Hence, most of the absolute percentage errors for each month fall under the recommended excellent forecasting range, with only August, which falls within the good forecasting range. These findings are consistent with previous studies in solar radiation modelling, where MAPE values below 10% are widely considered indicative of high model accuracy. For example, studies by Fuentes et al., 2020 and Radosevic et al., 2022 using UAV and GIS-based approaches reported similar accuracy ranges, emphasizing that integration of high-resolution spatial data significantly improves model reliability. Likewise, broader solar modelling literature suggests that errors typically increase during months with higher atmospheric instability, supporting the observed spike in August.

Table 6. MAPE Computation Results

Month	Actual	Forecasted	Absolute Error	Absolute Percentage Error	Forecasting Accuracy
Jan	171.89	157.57	14.32	8.33	Highly Accurate
Feb	135.38	146.92	11.54	8.53	Highly Accurate
Mar	159.60	161.52	1.92	1.20	Highly Accurate
Apr	141.27	144.66	3.39	2.40	Highly Accurate
May	131.48	132.42	0.94	0.71	Highly Accurate
Jun	112.82	110.79	108.76	1.80	Highly Accurate
Jul	122.03	128.25	6.22	5.10	Highly Accurate
Aug	120.89	144.17	23.27	19.25	Good

Solar PV Potential Calculation and Supply/Demand Results

The rooftop solar PV potential for each building presented in this phase informs the building occupant/owner how much electricity can be generated if the specified PV panels were placed based on the parameters substituted into equation 2. Hence, based on the temporal nature of this study, this approach gives a monthly interval of PV potential per rooftop.

Firstly, all rooftop PV potentials are aggregated into monthly intervals, indicating that the PV potential follows a similar trend to solar radiation. Hence, in March and September, the building rooftops achieved the highest PV potential among the other months, 68334.84kWh and 67322.93kWh. The rooftop PV potential aggregated achieves an annual total of 395.88 MWh. The results obtained indicate a very high potential for rooftop solar PV installations based on the sampled rooftops.

The integration of PV potential estimates with visual maps can inform solar engineers and planners on the optimal placement of the PV panels and design of the PV system. Notably, the rooftop area has direct implications and distinguishes the PV potential between the rooftops. Additionally, the reductions imposed on sloped rooftops also reduce the potential, even though they (Imila 50 – 56) have a greater area than the flat rooftops (Ramu Buildings). These findings are strongly supported by previous studies in rooftop solar assessment. For example, Gassar & Cha, 2021 and Wongwan et al., 2023 emphasize that roof area and usable surface fraction are the most critical determinants of PV capacity, often outweighing minor differences in solar irradiance. Similarly, research using UAV and LiDAR-derived models (e.g., Fuentes et al., 2020) shows that PV potential is highly sensitive to geometric constraints and rooftop segmentation, particularly in heterogeneous urban environments. The seasonal peaks in PV output aligning with solar radiation maxima also corroborate global findings that PV generation in tropical regions follows a bimodal distribution, unlike temperate regions with a single summer peak.

Figure 8. A) Annual Rooftop PV Potential per Household. B) Monthly Rooftop PV Potential vs monthly average electricity demand of Imila 53 .C) Monthly Rooftop PV Potential vs monthly average electricity demand of Ramu 47. D) Monthly Rooftop PV Potential vs monthly average electricity demand of Warangoi 23

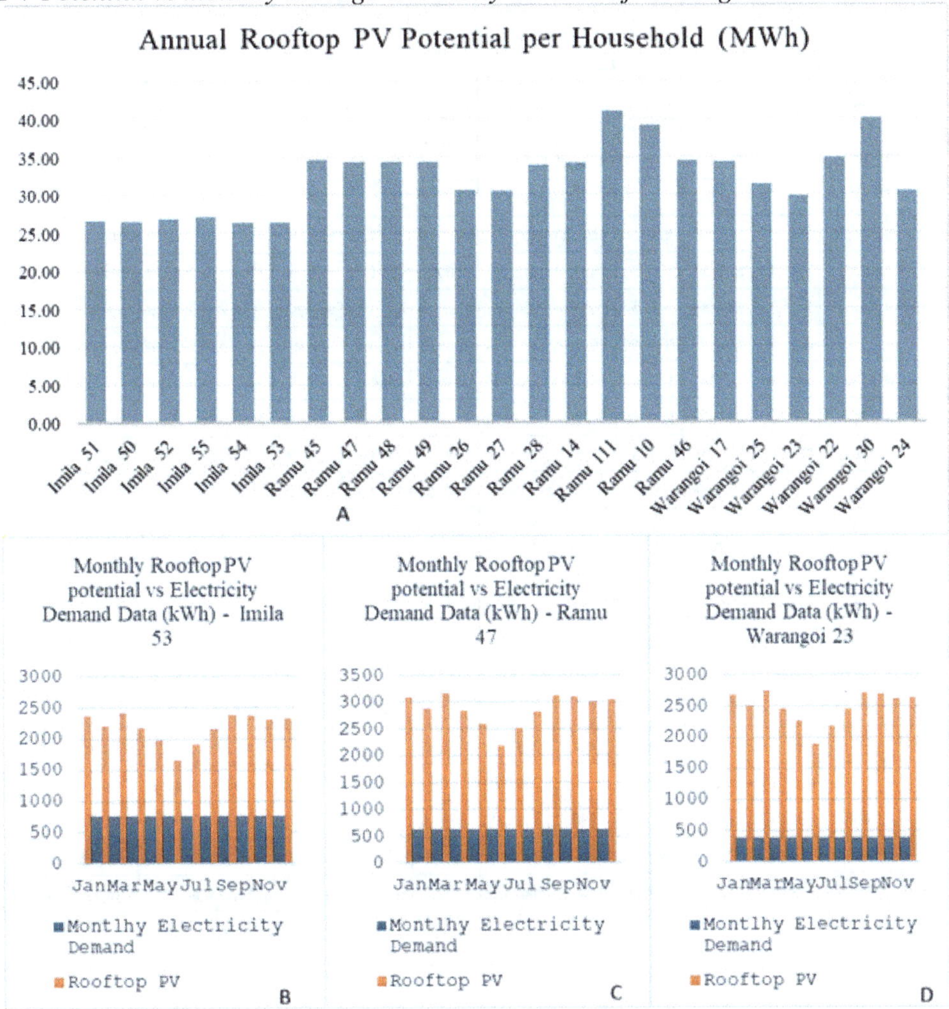

Table 7. Monthly Rooftop PV Potential Estimates for Each Household

Address	Ar(m2)	Jan	Feb	Mar	Apr	May	June	July	Aug	Sept	Oct	Nov	Dec
Imila	**kWh**												
50	204.74	2374.75	2214.49	2435.08	2181.97	1998.31	1673.1	1935.83	2174.84	2399.28	2382.99	2318.69	2338.05
51	205.84	2394.18	2233.1	2455.13	2197.94	2010.27	1681.58	1946.61	2189.8	2418.53	2403.09	2337.82	2356.2
52	207.87	2415.45	2251.69	2474.79	2216.44	2029	1698.54	1965.36	2208.88	2438.06	2422.71	2358.27	2378.36
53	203.29	2361.51	2203.02	2422.63	2169.34	1984.48	1660.2	1921.69	2161.44	2386.67	2370.88	2306.02	2323.92
54	203.29	2361.65	2202.92	2422.14	2168.52	1983.39	1659.17	1920.58	2160.5	2386.06	2370.66	2306.11	2324.1
55	209.70	2438.17	2273.33	2499.3	2239.08	2050.25	1716.42	1986.05	2231.65	2462.42	2446.19	2380.55	2400.57
Ramu													
10	176.56	3511.31	3273.8	3598.56	3222.84	2950.07	2468.09	2857.32	3211.88	3545.22	3522.7	3428.28	3456.96
14	155.08	3072.56	2864.88	3149.36	2820.84	2582.36	2160.59	2501.24	2811.35	3102.77	3082.76	2999.94	3024.95
26	139.27	2745.34	2559.65	2813.54	2519.94	2306.87	1929.84	2234.37	2511.46	2771.86	2754.26	2680.44	2702.82
27	138.82	2738.26	2553.33	2807.09	2514.46	2301.99	1926.01	2229.7	2506.04	2765.63	2747.58	2673.58	2695.78
28	154.09	3051.89	2845.63	3128.23	2801.93	2565.06	2146.11	2484.47	2792.51	3081.95	3062.06	2979.76	3004.59
45	156.65	3103.25	2893.54	3180.89	2849.06	2608.15	2182.09	2526.2	2839.46	3133.83	3113.61	3029.92	3055.15
46	156.65	3101.78	2892.19	3179.41	2847.7	2606.84	2180.88	2524.92	2838.09	3132.36	3112.16	3028.48	3053.69
47	155.84	3084.28	2875.88	3161.47	2831.6	2592	2168.34	2510.51	2822.01	3114.69	3094.61	3011.41	3036.46
48	155.71	3080.75	2872.19	3156.74	2826.67	2586.98	2163.89	2505.54	2816.9	3109.81	3090.47	3007.88	3033.11
49	156.09	3087.97	2879.11	3164.53	2833.75	2593.83	2169.71	2512.27	2824.08	3117.5	3097.99	3014.96	3040.15
111	297.98	3690.09	3440.13	3780.99	3386.01	3099.41	2593.41	3001.97	3374.48	3724.86	3701.56	3602.76	3633.1
17	155.89	3089.23	2880.39	3166.34	2836.01	2596.2	2172.17	2514.64	2826.45	3119.49	3099.43	3016.21	3041.38
22	157.91	3130.98	2919.35	3209.22	2874.47	2631.47	2201.71	2548.81	2864.81	3161.75	3141.36	3056.98	3082.48
23	135.61	2673.81	2493.23	2741.04	2455.35	2247.96	1880.88	2177.38	2447.16	2700.56	2682.9	2610.64	2632.33
24	139.45	2743.14	2558	2812.19	2518.72	2305.52	1928.27	2233.01	2510.16	2770.58	2752.62	2678.38	2700.53
25	143.09	2820.17	2630.11	2892.02	2590.9	2372.18	1984.48	2297.71	2582.32	2849.43	2830.34	2753.64	2776.28
30	180.59	3594.41	3351.43	3684.15	3299.77	3020.71	2527.31	2925.8	3288.63	3629.62	3606.29	3509.45	3538.72
Total	**3990.00**	**66664.94**	**62161.37**	**68334.84**	**61203.34**	**56023.28**	**46872.81**	**54261.99**	**60994.91**	**67322.93**	**66889.2**	**65090.15**	**65629.7**

Table 8. Forecasted Solar PV Supply vs Electricity Demand Data

Building Address	Average Monthly Electricity Demand (kWh)	Annual Electricity Demand (MWh)	Annual PV Potential (MWh)	Surplus Energy (MWh)
Imila 53	769.81	9.24	26.27	17.03
Ramu 47	622.04	7.46	34.30	26.84
Warangoi 23	398.63	4.78	29.74	24.96
Total		21.49	90.32	68.83

The final analysis involved comparing individual solar PV estimates against household electricity demand data. Due to data availability, only three buildings, i.e., Imila 53, Ramu 47 and Warangoi 23, were selected as shown with their respective supply and demand values in Table 8.

The first house sampled for comparison (Imila 53) has an average monthly electricity demand of 769.81kWh. Hence, the forecasted rooftop PV estimation indicates that there is enough potential to supply the building's monthly energy demand. The annual PV potential (26.27MWh) exceeds the annual electricity demand of 9.24MWh by 17.03MWh.

Similarly, the annual rooftop PV potentials for Ramu 47 and Warangoi 23 exceed their annual electricity demand by 26.84MWh and 24.96MWh. Additionally, the average monthly electricity demand for both buildings is well covered by their monthly forecasted PV potential. These results are consistent with previous studies on rooftop solar PV potential, which have shown that in high solar irradiance regions, rooftop systems often exceed residential demand. For instance, studies by Radosevic et al.,2022 and Paton, 2014 demonstrate that urban rooftop PV systems in tropical and subtropical regions can generate two to four times the household electricity demand, particularly when optimized for available roof space. Similarly, GIS-based assessments in Asia and Europe have highlighted that rooftop PV systems frequently achieve energy self-sufficiency with surplus generation, especially in low- to medium-density residential areas. The findings also align with global research emphasizing that distributed solar systems can significantly reduce dependence on centralized fossil-fuel-based generation, supporting broader energy transition goals.

Economic feasibility analysis of rooftop solar PV in Papua New Guinea (PNG) involves assessing whether the financial benefits outweigh the initial and operational costs. A cost–benefit analysis considers capital investment (solar panels, inverters, installation), maintenance costs, and savings from reduced electricity bills or diesel fuel substitution, which is particularly relevant in PNG where backup diesel generation is common. From a policy and regulatory perspective, the PNG government has set ambitious targets for achieving near-universal electrification through renewable energy by 2050, promoting solar adoption through national energy policies and pilot projects, although regulatory frameworks for grid integration, net metering, and incentives are still evolving and can affect investment attractiveness. When compared with alternative renewable energy sources, such as hydropower, biomass, or wind, rooftop solar PV offers advantages in decentralization, scalability, and lower installation time, especially for rural and peri-urban areas; however, it may face limitations related to intermittency and storage requirements. Hydropower, while more stable, requires large infrastructure and high upfront investment, whereas solar PV provides a flexible and rapidly deployable solution aligned with PNG's dispersed settlement patterns and energy access challenge.

Global solar adoption trends show that solar energy has become the fastest-growing component of the global energy mix, driven by falling technology costs, strong policy support, and the urgency of climate mitigation. Recent data indicate that global solar capacity additions reached record levels, with around 655 GW expected in 2025 and rapid year-on-year growth, including a 64% surge in installations in early 2025. Solar now plays a central role in the global energy transition, contributing significantly to meeting rising electricity demand and even surpassing fossil fuels in some periods, while renewables collectively account for nearly half of global power capacity. A notable trend is the rise of distributed and rooftop solar systems, supported by policies such as net metering and decentralized energy planning, alongside the integration of storage technologies. Linking these trends to energy transition debates in Papua New Guinea (PNG), the findings suggest that PNG is well-positioned to adopt a decentralized solar pathway due to its high solar potential, dispersed population, and limited grid infrastructure. However, unlike leading solar markets (e.g., China, India, and Germany) that benefit from strong policy frameworks and large-scale investments, PNG still faces challenges related to financing, regulatory support, and technical capacity.

CONCLUSION AND RECOMMENDATIONS

The PNG-National Energy Policy 2017-2027 stated that one of the main challenges of solar energy development was the lack of awareness of the potential opportunities and economic benefits offered by solar technologies. The integration of Geographic Information System (GIS) and Remote Sensing (RS) techniques provides an efficient alternative to conventional in-situ measurements, such as pyranometer readings, which are often labour-intensive and impractical for multiple building structures over extended periods.

This study demonstrated the application of Unmanned Aerial Vehicle (UAV) photogrammetric techniques and GIS-based modelling approaches in determining the technical feasibility of installing rooftop PV systems, using 23 residential buildings of various roof designs in the Papua New Guinea University of Technology (PNGUoT) campus and a specific PV module in a case study.

Due to surface modelling limitations attributed to photogrammetric processing, a manual method of sampling and assigning height values to building footprint vertices was performed to create 3D rooftop vertices, which were subsequently interpolated using the TIN tool to model rooftop surfaces. This approach was taken to address the distorted rooftops.

The solar radiation model generated by ArcGIS 3.8 (ASR Tool) was successfully validated against the AgriFarm weather station data, achieving a MAPE of 5.92%, classified as very good forecasting. Thus, it is important to note that specific climatic variables were not applied in this study. The monthly rooftop solar maps generated support the PV estimations, illustrating the variation in solar radiation over different surfaces and can act as a visual aid, informing solar planners to make decisions and design the PV systems specific to the site.

The annual rooftop PV potential for the 3 buildings included in the final comparison between potential supply and demand indicated that all the estimated rooftop PV potential has the capacity to cater for 100% of their demand, while leaving with an excess of 68.83 MWh to be either shared, stored or exported back to the grid pending the design of the system.

The first recommendation of this study is to improve UAV photogrammetric outputs through enhanced data collection tools and GCP coverage. The accuracy of rooftop surface reconstruction can be improved by using a multispectral drone. Secondly, this study recommends that future work should integrate the use of measured energy output from existing or nearby rooftop PV systems. Comparing GIS-based PV estimates with logged inverter or smart-meter data would provide a stronger empirical basis for validating the Area Solar Radiation (ASR) modelling outputs. This would enable the identification of any model biases and improve confidence in predicting long-term PV generation.

Finally, this study recommends expanding the analysis to include all academic, administrative, and service buildings, which would provide a complete solar PV energy potential profile for the entire PNGUoT campus. A full-scale assessment would support strategic renewable energy planning, allowing the university to identify priority buildings for solar installation, estimate overall campus energy independence, and potentially create a roadmap for future PV deployment.

REFERENCES

Aiau, S. S., Kavi, M., Pirapaharan, K., Hoole, P. R. P., Anyi, M., & Hoole, S. R. H. (2016). PNG's energy sector and estimation of renewable energy resources in morobe Province, Papua New Guinea: Solar and wind power for New Umi township. [JTEC]. *Journal of Telecommunication, Electronic and Computer Engineering*, *8*(12), 39–46.

Bosip, P., & Fletcher, L. (2023). PNG's energy policy and practice for climate and people: Building on what works. Jubilee Australia Research Centre. https://celcorblog.wordpress.com

Eitel, J. U., Höfle, B., Vierling, L. A., Abellán, A., Asner, G. P., Deems, J. S., Glennie, C. L., Joerg, P. C., LeWinter, A. L., Magney, T. S., Mandlburger, G., Morton, D. C., Müller, J., & Vierling, K. T. (2016). Beyond 3-D: The new spectrum of lidar applications for earth and ecological sciences. *Remote Sensing of Environment*, *186*, 372–392. DOI: 10.1016/j.rse.2016.08.018

Fu, P., & Rich, P. (1999). *The Solar Analyst 1.0 user manual*. Helios Environmental Modelling Institute.

Fuentes, J. E., Moya, F. D., & Montoya, O. D. (2020). Method for estimating solar energy potential based on photogrammetry from unmanned aerial vehicles. *Electronics (Basel)*, *9*(12), 1–21. DOI: 10.3390/electronics9122144

Gassar, A. A. A., & Cha, S. H. (2021). Review of geographic information systems-based rooftop solar photovoltaic potential estimation approaches at urban scales. *Applied Energy*, *291*, 116817. Advance online publication. DOI: 10.1016/j.apenergy.2021.116817

He, Z., Gan, S., & Yuan, X. (2026). A Terrain-Constrained TIN Approach for High-Precision DEM Reconstruction Using UAV Point Clouds. *Journal of Imaging*, *12*(1), 8. DOI: 10.3390/jimaging12010008 PMID: 41590892

Kezazy, H. E., & Hilmi, Y. (2025). Promoting the Energy Transition Throughout Dealing with the Climate Change Issue. In *Digital Technology for an Innovative Energy Transition: Perspectives and Opportunities* (pp. 77–93). Springer Nature Switzerland. DOI: 10.1007/978-3-031-89657-6_4

Long, Y., Xu, X., & Huo, Z. (2025). Urban rooftop photovoltaic potential model: A study on assessment methods and model framework. *Energy and Building*, *345*, 116138. DOI: 10.1016/j.enbuild.2025.116138

Mandla, V. R., Chokkavarapu, N., & Peddinti, V. S. S. (2021, April). Role of drone technology in sustainable rural development: opportunities and challenges. In *International Conference on Unmanned Aerial System in Geomatics* (pp. 301-318). Cham: Springer International Publishing Mineral Resource Authority. (2022). Minerals for life: MRA launches solar power system.

Paton, C. P. (2014). *Handbook for rooftop solar development in Asia*. Asian Development Bank.

PNG Power Limited, & National Energy Authority. (2024). Papua New Guinea National Energy Access Transformation Project: Environmental and social management framework (P173194).

Radosevic, N., Liu, G. J., Tapper, N., Zhu, X., & Sun, Q. (2022). Solar energy modeling and mapping for the sustainable campus at Monash University. *Frontiers in Sustainable Cities*, *3*, 745197. Advance online publication. DOI: 10.3389/frsc.2021.745197

Rahman, A. S., & Koo, B. B. (2024). *Papua New Guinea energy access diagnostic report based on the multi-tier framework*. World Bank. DOI: 10.1596/41626

Rawali, M., Bruce, A., Raturi, A., Spak, B., & Macgill, I. (2019). Electricity access challenges and opportunities in Papua New Guinea (PNG). UNSW. http://hdl.handle.net/1959.4/unsworks_65345

Singh, A. (2023). Assessment of potential rooftop solar PV electricity at a suburban scale. University of Wollongong. https://ro.uow.edu.au/theses1

Song, Y., Helmholz, P., Su, F., Zhou, C., Lotfata, A., Zohar, M., Leonardo, M. G., & Sila-Nowicka, K. (2025). Advancing geospatial methods for addressing global resource and sustainability challenges. *Resources, Conservation and Recycling*, *223*, 108517. DOI: 10.1016/j.resconrec.2025.108517

Wolfs, T. (2017). *A GIS-based approach to estimate the geographic potential for rooftop solar PV: Developing a database at building-level for Denmark*. Aalborg University.

Wongwan, W., Pleerux, N., Thanomsat, N., & Moukomla, S. (2023). Assessment of solar rooftop potential using a digital surface model based on a photogrammetric survey by an unmanned aerial vehicle. *Environment Asia*, *16*(3), 68–81. DOI: 10.14456/ea.2023.36

Yao, H., & Zhou, Q. (2023). Research status and application of rooftop photovoltaic Generation Systems. *Cleaner Energy Systems*, *5*, 100065. DOI: 10.1016/j.cles.2023.100065

Yin, H., Krayenhoff, E. S., Voogt, J., Heusinger, J., & Aliabadi, A. A. (2025). Simulation of Tilted Rooftop Photovoltaic Panels at City Scale: Novel Measurements, Model Development, and Application in WRF. *Journal of Geophysical Research: Atmospheres, 130*(9), e2024JD043133.

Chapter 6
Geomatics in Urban Expansion Mapping for Sustainable Urban, Warehouse, and Logistics Planning

Nelly Rose Gham
Papua New Guinea University of Technology, Papua New Guinea

Sujoy Kumar Jana
http://orcid.org/0000-0003-0901-5316
Papua New Guinea University of Technology, Papua New Guinea

Tingneyuc Sekac
http://orcid.org/0000-0001-5972-7149
Papua New Guinea University of Technology, Papua New Guinea

Bikash Ranjan Moharana
http://orcid.org/0000-0001-8208-6191
Papua New Guinea University of Technology, Papua New Guinea

ABSTRACT

Alotau, the capital town of Milne Bay Province, is experiencing rapid urbanization, leading to significant pressure on its limited land resources and demanding improved planning strategies. The urban expansion from 2005 to 2025 was analysed using Geographic Information Systems (GIS), remote sensing techniques and land use land cover (LULC) maps, and found a significant improvement in urbanization i.e.,

DOI: 10.4018/979-8-3373-7941-8.ch006

~148%. The expansion was predominantly confined to low-elevation, low-gradient terrain, resulting in the conversion of vegetated and barren land covers, and was strongly regulated by topographic constraints. The LULC maps with high classification accuracy (> 85%), providing reliable, data-driven evidence of Alotau's growth patterns. These findings offer critical insights for town planners and policy makers, underscoring the urgent need for strategic and sustainable urban planning to guide Alotau's future development via. manufacturing firms, logistics planning etc., in a manner that is both resilient and environmentally conscious.

1 INTRODUCTION

Urban planning is a complex and multifaceted field that requires careful consideration of various factors, including demographic trends, infrastructure development, transportation systems, and environmental sustainability (Cheshmehzangi, 2025; Adesina et al., 2024; Sekac et al., 2017). Geospatial analysis, using Geographic Information Systems (GIS), has become an essential tool in urban planning, enabling policymakers and planners to make informed decisions based on spatially referenced data. In this article, we will explore the world of geospatial analysis for urban planning, delving into GIS techniques, tools, and applications for informed decision-making (Prabha et al., 2025; Mansourihanis et al., 2025; Anucharn et al., 2025; Andrew et al., 2023; Poi et al., 2018).

Alotau's town itself grew by 3.570 km^2 in 2011 (PNG NSO), yet more than half of the expansion lacks zoning records. Alotau, the capital town of Milne Bay Province is situated on the coastal area of the forested foothills that extends from the Owen Stanley Range in an eastward direction. The town has a population of 11,857 (2011 census). The town serves as a commercial and administrative hub, attracting the people from the rural areas seeking economic opportunities. The town has been experiencing rapid changes in terms of urbanization in the recent years. As more people move into town and the development of infrastructures increases, changes in the land use and settlement patterns are becoming more obvious. Alotau's urbanization mirrors Papua New Guinea's broader challenges such as unplanned growth, coastal squeeze, and terrain limited expansion (UN-Habitat, 2020). Most of the town's suitable land for urbanization is being occupied, making it to expand into the nearby communities. This study fills a critical gap because while prior PNG research focused on Port Moresby and Lae, Alotau's unique topography and tourism driven economy demand localized analysis. This town is a recipient of urban expansion due to the increase in infrastructures in the town itself and expanding into the nearby communities. The expansion is accompanied by various problems including unstructured planning, the increase in settlements and overcrowding.

This study utilizes Remote Sensing and Geographic Information Systems (GIS) technologies to examine the urban expansion of Alotau town, identify the patterns and drivers of change, and generate information that can be used to support better land use planning and decision-making in the future. This study also holds critical importance as it provides the first comprehensive, data-driven analysis of urban expansion in Alotau, Milne Bay Province, Papua New Guinea, using GIS and Remote Sensing techniques. By mapping land use changes for the years 2005, 2015 and 2025, the research will deliver actionable insights for sustainable urban planning, addressing Alotau's rapid unregulated urban growth. It also advances methodologies for tropical urban mapping, offering a replicable model for similar Pacific towns. Ultimately, this study will bridge the gap between spatial data and policy-making, ensuring Alotau's growth is both strategic and sustainable. The aim of the study is to utilize and GIS and Remote Sensing techniques to quantify and model urban sprawl patterns (2005 – 2025) through spatially explicit LULC change analysis.

The specific objectives are

(i) To do change detection study over Alotau town and the surrounding communities.
(ii) Analyse and map out these changes to detect the extent of urban expansion of the town and need for individual zones to meet the public demand.
(iii) To do terrain analysis to identify its influence on the pattern of urban spread.
(iv) Identify the factors contributing to the changes and suggest recommendations for a sustainable urban development and optimum route planning for reducing logistics cost and time.

2 LITERATURE SURVEY FOR URBAN EXPANSION STUDIES

Urbanization is world-wide phenomena mostly caused by population growth, economic developments and the migration of people from the rural areas to towns and cities that has been increasing in the past decades. Kisamba and Li's (2023) study (as cited in Slimani and Raham, 2023) defines urbanization as the development and expanding of urban centres at the cost of rural areas, reflecting the trends in lifestyles and social impacts. Urban expansion is a commonly increasing issue in many parts of the world especially in developing countries where urban areas usually grow at a quick pace without proper planning.

This literature review discusses how urban expansion has been studied, focusing on the methods researchers use especially Remote Sensing and Geographic Information Systems (GIS). The aim is to understand the concepts, tools and challenges involved in order to come up with a strong approach for studying urban expansion in Alotau Town. By exploring relevant studies and methods, this review will help set the stage for the methods used in this research.

The physical expansion of towns, cities and metropolitan regions into the neighboring rural areas is known as urban expansion (Abadura & Sokido, 2024). The study of urban expansion is crucial for policymakers, planners and environmentalists to ensure sustainable urban development. Studies such as Seto et al., (2011) and Angel (2023) have highlighted the need for an effective tracking, monitoring and managing urban growth.

In Papua New Guinea, small towns like Alotau are currently going through noticeable changes especially in the changes in the Land Use Land Cover (LULC) that is mainly caused by the expansion of urban boundaries into the nearby rural communities. Understanding the pattern and spatial dimensions of urban expansion helps decision-makers and planners to come up with sustainable development strategies.

2.1 Role of GIS and Remote Sensing in Urban Expansion Studies

The integration of Remote Sensing (RS) and Geographic Information Systems (GIS) make available the space and tools for mapping, monitoring and quantifying urban growth. Remote sensing data particularly satellite images (e.g., Landsat Imagery and Sentinel), is usually used in urban expansion studies. Ancillary data such as administrative boundaries, topographical sheets and census data further supports these analyses (Gour et al., 2014; Stemn and Agyapong, 2014; Katyambo and Ngigi, 2017; Hegazy and Kaloop, 2015).

In PNG, few studies have explored this approach, however most studies have been done in the country's main cities and less to none in the small towns. GIS enhances remote sensing by enabling multi-criteria analysis, spatial modelling and decision support.

2.2 Methodologies and Analysis Techniques for Urban Expansion

Selecting the appropriate data sources and conducting preprocessing are the most critical steps in remote sensing studies. Remote sensing data particularly satellite imagery or aerial photographs is being widely used in several studies including urban expansion mapping. Previous studies have utilized several data collection methods to assess urban expansion mapping including satellite imagery such as Landsat satellite imagery and Sentinel satellite imagery and other ancillary data such as administrative boundaries, topographical sheets and census data. The common primary source data across the different studies was the Landsat satellite images from different time frames mostly acquired from the United States Geological Survey (USGS) archives. The type of sensors used for the images varied between different studies. To support the analysis of the study, ancillary data such as topographical sheets, aerial photographs and high-resolution imagery from Google Earth and Google Maps are collected and used as reference data. Other datasets include the census data and the administrative map boundaries of the study areas.

After the data collection the satellite images have to go through some pre-processing session before they are used for further analysis. This stage is crucial because it converts raw satellite data into a valid form for analysis. It includes processes such as band stacking, image resampling, mosaicking, image sub setting, atmospheric correction, geometric correction, radiometric correction etc, which can be done using multiple Remote Sensing and GIS software such as ERDAS IMAGINE, ArcGIS, QGIS etc. To improve the reliability and accuracy of the study, there has to be a deep level of rigorousness. Gour et al. (2014) emphasized the importance of some of the image improvements techniques in their study such as data scaling and histogram equalization, stating that they improve the satellite image quality making it more clear for interpretation. For a place like Alotau, where cloud cover and terrain can affect the image quality, preprocessing is necessarily important.

Different methods were used by researchers to analyse urban expansion. One common approach is the image classification, where the land cover types are grouped into several classes such as built-up, bare land, water, vegetation and other necessary classes. Image classification can be done in two ways, supervised image classification and the unsupervised image classification. In supervised classification, the commonly used algorithm is the Maximum Likelihood method which uses training data to guide the process and is considered to be the usually more accurate. Unsupervised classification does not require training data and is less reliable especially in areas with mixed land uses. The commonly used algorithm for an unsupervised classification is the ISODATA method. Gour et al. (2014) used both supervised and unsupervised classification methods in their study utilizing the Maximum Likelihood and ISODATA algorithms to classify the pre-processed images. The final output of both classifications gave them the land use land cover image of their study area to investigate the changing scenario.

To measure changes overtime, researchers use techniques like comparing classified maps from different years. Two commonly used indices are the Normalized Difference Vegetation Index (NDVI) and the Built-Up Index (BUI). Gour et al. (2014) did NDVI calculations on their study area using the formula NDVI = (NIR – RED) / (NIR + RED) and investigate the land use or land cover change. They also calculated the BUI "BUI = (MIR – NIR) / (MIR + NIR)" and stated that it is a good indicator for monitoring the changes of land use land cover especially in terms of built-up areas. GIS tools then help to analyse the spatial patterns and link them with other features such as roads or elevation.

Accurate land-use or land-cover classification is essential for sustainable urban planning, particularly in supporting decisions related to industrial zoning, warehouse development, and logistics infrastructure planning. Spatial information derived from reliable classification helps planners identify suitable locations for logistics hubs, distribution centers, and warehouse facilities, while minimizing conflicts with residential areas and environmentally sensitive zones. Where training data are limited, traditional classifiers can still provide dependable spatial information that contributes to sustainable urban development and strategic warehouse and logistics planning (Rodrigue, 2020; Muduli et al., 2020; Das et al., 2023).

After classification and change detection analysis, it is crucial to check how accurate the results are. In other words, it is important to validate the results. This is usually done by comparing the result with known data, using a tool called the confusion or error matrix. This process gives us overall accuracy and shows how well each land type classes was classified. The Kappa Coefficient is also used to measure the agreement or disagreement between the map and the reference data. Stemn and Agyapong (2014) performed error matrix calculations of their two different land cover maps (1991 and 2008) and got a reasonably high accuracy assessment, concluding that the data are sufficient for urban growth detection.

In places such as Alotau where there is lack of ground truth data, Google Earth or even local knowledge can be used to validate the results. Acquiring a good accuracy (>85%) is essential because small errors may lead up to wrong conclusions and misinformation on how the town is expanding.

2.3 Literature Gaps and Methodological Challenges

Even though technology has evolved in the recent years, there are still many challenges faced in the mapping of urban expansion especially in smaller towns in developing countries. Tropical urban studies often underrepresent cloud cover challenges (Wang et al., 2020). Also, because of the limitations and lack of technical skills, there may not be many studies in local areas or less reliable local data. These challenges require researchers to come up with creative solutions such as using participatory mapping or combining different data sources.

Based on what has been reviewed, this study uses Landsat Satellite Imagery for the years 2005, 2015 and 2025, supervised classification and GIS based spatial analysis to study how Alotau town has expanded. Landsat images from the USGS archives were chosen because of its historical data and decent resolution. Studies have shown that Maximum Likelihood Classification (MLC) can outperform some machine learning algorithms when training datasets are limited, as the method relies on statistical probability distributions rather than large training samples (Hegazy & Kaloop, 2015). The MLC approach also advantageous particularly in mixed urban environments, where land-cover classes such as built-up areas, vegetation, bare soil, and transportation infrastructure often occur in highly fragmented and heterogeneous patterns. While NDVI suffices for vegetation tracking, BUI is prioritized for Alotau due to its mixed urban-rural fringe. Post classification comparison was used to detect changes because it is effective and straightforward. Classification accuracy was checked using the confusion matrix with the help from Google Earth for validation. GIS analysis techniques were also used to overlay urban growth on physical features such as the slope to study the influence of the terrain in urban spread.

This approach fits well with what the literature suggests and has been adjusted to match the local context of Alotau.

3 METHODOLOGY

The study area is Alotau Town which is a small town and the capital of Milne Bay Province and the surrounding communities that is situated in the eastern part of Papua New Guinea (shown in figure 1). It has Latitude and Longitude of -10°18'54" S and 150°27'26" E (WGS 84 coordinate reference system) and an elevation of 11 meters (36 ft) above sea level.

Figure 1. Geographic location of Alotau Town

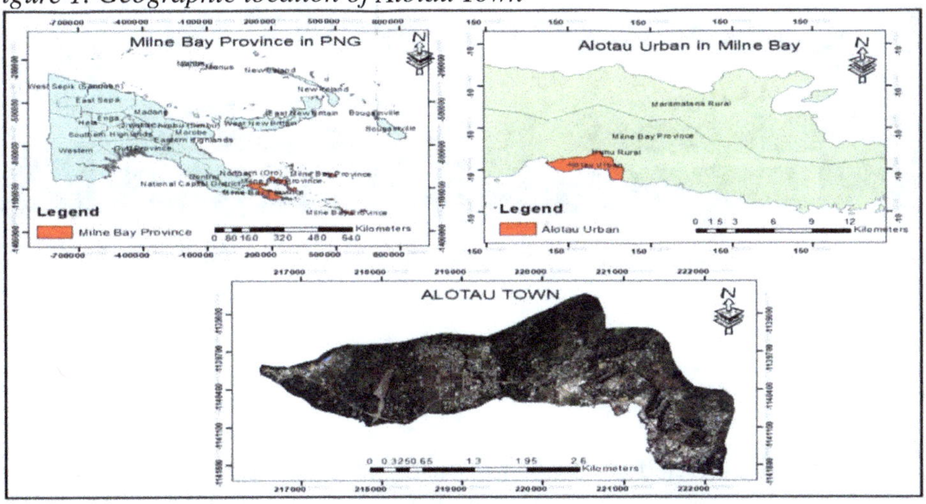

The methodology was designed to meet the research objectives by utilizing Remote Sensing (RS) and Geographic Information Systems (GIS) to analyse the spatial and temporal changes in Land Use Land Cover (LULC). The three major components used to derive the outcome of the research were data acquisition, data pre-processing and the urban expansion analysis. This methodological design (flow chart shown in figure 2) ensured a systematic and objective approach to identifying, quantifying and interpreting the urban expansion in Alotau town using widely accepted remote sensing and GIS methods.

Figure 2. Methodological flowchart

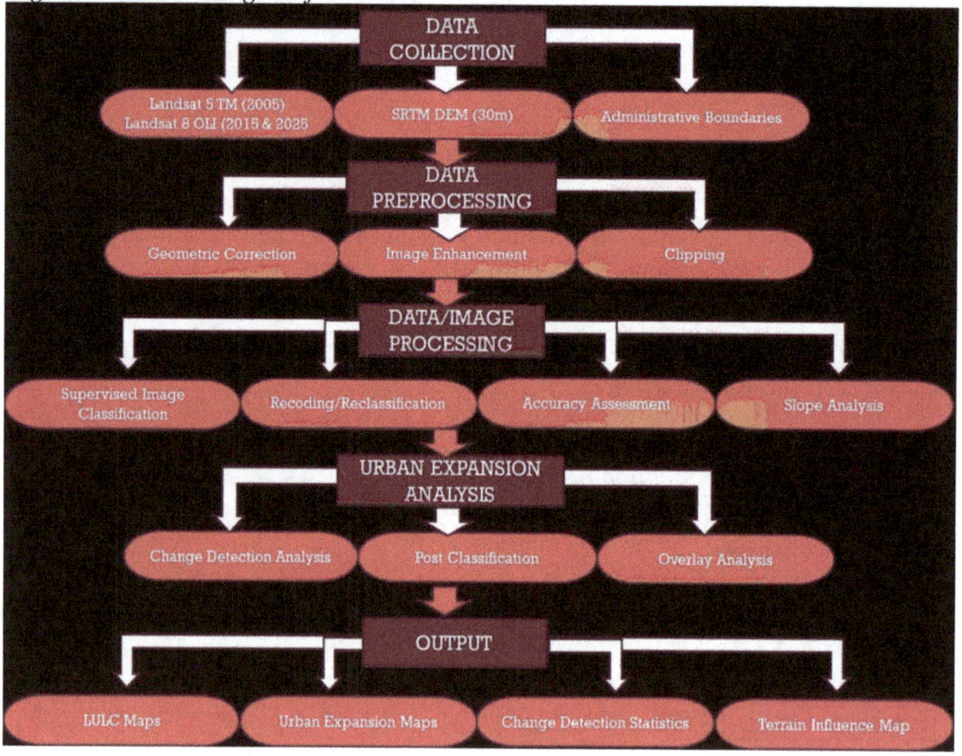

3.1 Data Collection

The data are collected from remote sensing satellite imagery, that was freely available online. The main datasets included Landsat Satellite Images, a Digital Elevation Model (DEM) and supporting ancillary data such as the administrative boundary shape file, the road networks data and the population census data. The data type and sources are given in the table 1.

Table 1. Data type and sources

YEAR	DATA TYPE	RESOLUTION	DATA SOURCE
2005	Landsat 5 TM	30m	USGS
2015	Landsat 8 OLI	30m	USGS
2025	Landsat 8-9 OLI	30m	USGS
	Administrative Boundary Shapefile		NSO
	Road Networks		Open Street Map
	SRTM DEM	30m	USGS

The software used for the analysis are as follows;

1. ArcGIS 10.8 (for preprocessing, clipping, projection, change detection analysis)
2. QGIS 3.42.2 (for downloading and clipping road networks)
3. ERDAS IMAGINE 8.5 (for classification and accuracy assessment)
4. MS Excel (for generation of graphs)

3.2 Image Preprocessing

Before image classification and further analysis, the satellite imagery was pre-processed to improve its quality and ensure accurate change detection results. The Landsat images for the years 2005, 2015 and 2025 underwent several preprocessing steps to correct atmospheric disturbance and geometric errors, and improve image quality to prepare them for image classification. Preprocessing steps include pan sharpening with band 8 from the satellite data to enhance the satellite images for better interpretation. Geo-referencing of the satellite images was done for correct spatial alignment. The clipping tool in ArcGIS was used to extract the study area.

3.3 Image Classification

Supervised image classification was used to classify the individual images into three land cover classes namely urban, non-urban and inland water. Training samples for the various land cover types were selected with the help of Google Earth. The Maximum Likelihood algorithm was used to classify the images according to the covariance and variance of the spectral response patterns of each pixel in the image. To ensure the accuracy of the classification, an accuracy assessment was conducted for each image. This gave validation and confidence to the following attached results. The accuracy assessment includes the computation of an overall accuracy as well as the Kappa statistics.

3.4 Urban Expansion Analysis

Post classification change detection was performed by comparing the classified images from the different years to identify the transition of land use land cover. The post classification process is recommended to be effective however it relies entirely on the accuracy assessment of the classified images for a true and well-informed results.

The urban expansion analysis focused on the identification of how built-up areas have changed spatially and temporally over a defined period. The 2005, 2015 and 2025 built-up images were overlaid to obtain an urban expansion image. For further analysis of urban expansion, the urban growth image was overlaid with the slope of the town to highlight the influence the terrain had on the urban expansion spread. The goal of the urban expansion analysis is to understand both the rate and pattern of the expansion, which can inform sustainable urban planning.

This analysis is crucial for understanding the dynamics of the town's development and for guiding future planning efforts that reduces environmental and social impacts.

4 RESULT AND DISCUSSIONS

This session presents the outcomes of the study based on the implemented methodology. The application of GIS and remote sensing techniques produced the following results: classified LULC maps, urban expansion maps, change detection statistics and terrain influence maps. The results demonstrate the significant spatial changes in land use patterns in Alotau town between 2005 and 2025.

4.1 Classified "Land Use Land Cover" (LULC) Maps

The three classified LULC maps for 2005, 2015 and 2025 as shown in figure 3, provide a clear visual narrative of the transformation of the town. The most obvious trend observed is the consistent and rapid conversion of non-urban land (primarily vegetation and bare-land) into built-up areas. In 2005, the urban footprint was largely concentrated around the town's focal point and the port area covering 111.15 hectares. This pattern is a characteristic of a small, centralized urban hub. By 2015, the urban area had more than doubled to 260.46 hectares, revealing a clear pattern of linear expansion along the major road networks. This shift indicates that transportation routes became the primary areas for development, facilitating urban sprawl into the surrounding communities. By 2025, the built-up area had reached 275.47 hectares, with significant expansion towards the western and southern parts of the town.

Figure 3. Land use land cover change for Alotau Urban (2005, 2015 & 2025)

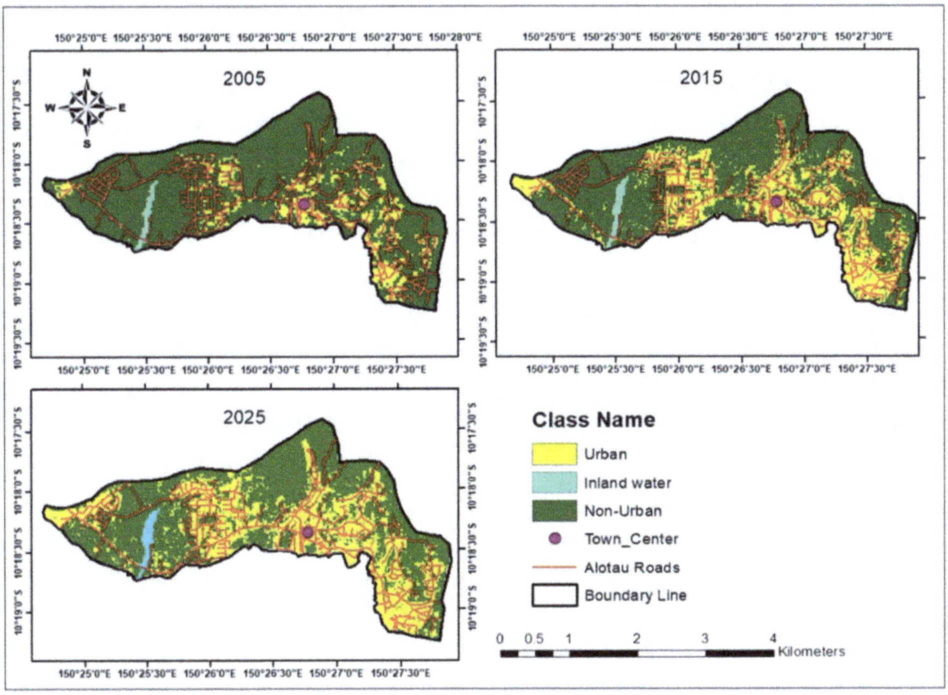

This growth pattern directly results from the town's rising population and its central role as the commercial and administrative hub of Milne Bay Province. As the town spread into vegetated or bare-land areas, a major concern regarding the availability of easy access land for buildings arises, squeezing between the coast and the steep terrain. The LULC maps clearly show that urban development has been the single most powerful force reshaping the town's landscape over the last 20 years, most times replacing natural areas.

The Classified LULC results showed that:

- In 2005, Alotau was smoothly covered by vegetation and bare-land with limited built-up areas concentrated near the town center.
- By 2015, built-up areas increased along the road networks causing a decrease in vegetation and bare-land.
- By 2025, built-up areas expanded significantly, especially towards the western and southern parts of the town.

4.2 Classification Accuracy Assessment

Further analysis of the urban expansion solely depends on the accuracy of the classified LULC maps, as given in table 2. The accuracy assessment displayed overall accuracy values of 93.48% (2005), 94.00% (2015) and 88.00% (2025), and Kappa coefficients all above 0.81, indicating strong results. These values meet and exceed the >85% benchmark set in the methodology, confirming that the classification process was done correctly and the resulting maps and analysis that follows after are highly reliable representation of reality.

The high Kappa values, particularly for 2015 (0.8955), indicate an excellent agreement between the classified maps and the reference data from Google Earth, significantly beyond what would be expected by chance. With such a high level of accuracy, we can be highly confident in the urban growth areas I identified and the change statistics I calculated. The slight dip in accuracy for the 2025 image could be attributed to increased spectral complexity from mixed land uses at the urban-rural fringe, a common challenge in rapidly expanding towns. Nevertheless, the accuracy remained well within the acceptable range for robust urban expansion analysis.

Table 2. Accuracy assessment results for the classified LULC images of Alotau Urban

YEAR	OVERALL ACCURACY (%)	KAPPA STATISTICS
2005	93.48	0.8196
2015	94.00	0.8955
2025	88.00	0.8130

4.3 Urban Expansion Map

The urban expansion maps (refer to figure 4) were developed to visualize the growth and spread of built-up areas overtime. The maps illustrate the spatial extent and direction of urban growth, clearly showing the expansion of urban development into the surrounding rural communities. The urban area expanded by 149.31 ha and 64.4175 ha in between 2005-2015 and 2015-2025 respectively.

Figure 4. Urban expansion map of Alotau (a) between 2005-2015, (b) between 2015-2025

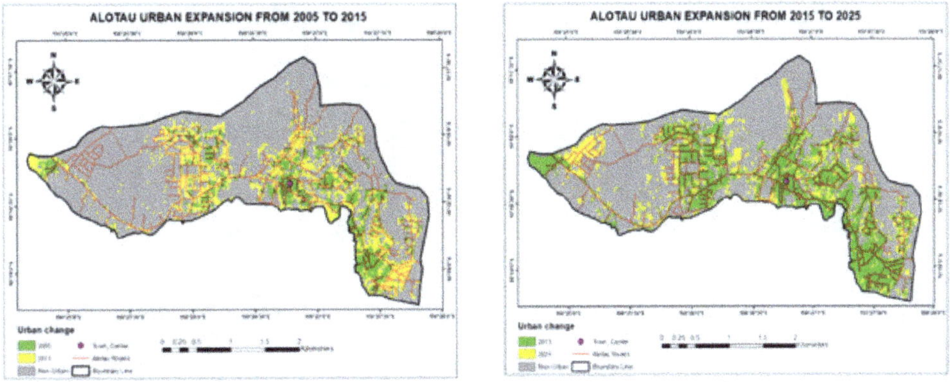

(a) (b)

The maps revealed that new built-up areas occurred mainly along the major road networks suggesting that accessibility was a primary driver. The results also showed a notable directional expansion towards the western and southern parts of the town between 2015 and 2025. These spatial patterns reflect Alotau's role as the commercial and administrative hub of Milne Bay Province. The growth along the road networks is evidence of people seeking connection to the town's economic centre. The directional spread of built-up areas towards the western and southern parts indicates the need for land suitability, highlighting the coastal squeeze as the town bumps up against its geographical limits. The overall urban expansion of Alatau was recorded approx. 164.3175 ha from 2005 to 2025, shown in figure 5.

Figure 5. Overall urban expansion map for Alotau from 2005 to 2025.

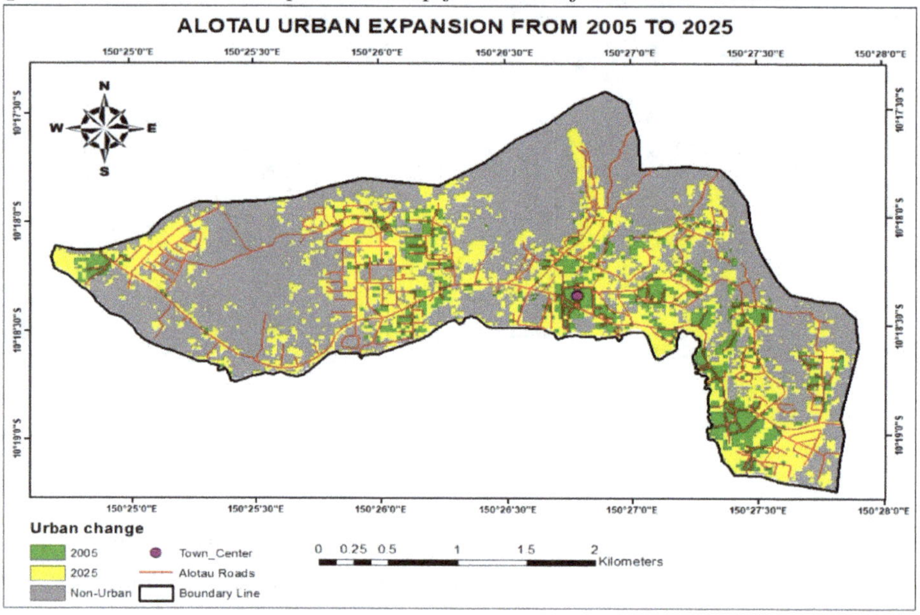

4.4 Change Detection Statistics

Change detection statistics were produced to quantify the land cover changes from 2005 to 2025 given in Table 3, 4 and 5. The statistics show the area in hectares gained or lost by each land cover class over the study period. A change matrix was also created (Table 6) to detail the specific conversions between classes, such as non-urban to urban. The change detection statistics shows a distinct phase of growth. From 2005 to 2015, urban expanded rapidly, with the urban area growing by 149.31 hectares which is an increase of over ~ 134%. The change matrix (refer table 6) shows that 168.31 hectares of non-urban land were converted to urban during this decade, indicating a massive and direct transformation of the natural landscape to accommodate growth.

Table 3. Alotau urban area cover and changes in hectares for the years 2005-2015

LULC Class	Area in hectares		
	2005	2015	Change
Urban	111.15	260.46	149.31
Non-Urban	721.53	572.58	-148.95
Inland Water	9.63	9.27	-0.36

Table 4. Alotau urban area cover and changes in hectares for the years 2015-2025

LULC Class	Area in hectares		
	2015	2025	Change
Urban	275.46	339.89	64.43
Non-Urban	546.91	479.92	-66.99
Inland Water	9.63	12.18	2.56

Table 5. Overall Alotau urban area cover and changes in hectares for the years 2005-2025

LULC Class	Area in hectares		
	2005	2025	Change
Urban	111.15	339.89	228.74
Non-Urban	721.53	479.92	-241.61
Inland Water	9.63	12.18	2.55

Table 6. Change matrix for Alotau urban for the years 2005-2015 and 2015-2025

For years 2005 – 2015					For years 2015 - 2025				
From/ To	Urban	Non-Urban	Inland Water	Total (2005)	From/ To	Urban	Non-Urban	Inland Water	Total (2015)
Urban	91.62	19.53	0	111.15	Urban	228.17	46.86	0.42	275.45
Non-Urban	168.39	551.7	1.44	721.53	Non-Urban	111.61	432.02	3.28	546.91
Inland Water	0.45	1.35	7.83	9.63	Inland Water	0.11	1.03	8.48	9.62
Total (2015)	260.46	572.58	9.27	842.31	Total (2025)	339.89	479.91	12.18	831.98

The results for 2015-2025 also shows same trend of urbanization i.e. increase of 64.43 hectares in urban area. The change percentage in this decade is low compare to 2005-2015. The reason being for this decrease may due to policy-driven land use causing the eviction of two settlements in the town. Despite this decrease, the overall trend from 2005 to 2025, the town's urban area grew by 228.74 hectares which is ~ 205%. The change matrix for 2005-2015 effectively identifies the dynamics of change, showing that while 91.62 hectares of the original 2005 urban remained stable, the vast majority of new urban land was created at the expense of non-urban areas. This provides crucial., evidence-based data on the rate and source of urban expansion, which is essential for managing future growth and avoiding unregulated development.

4.5 Terrain Influence Map

The terrain influence analysis confirms that topography plays an important part in the distribution of built-up areas in Alotau urban. The findings show that there is a relationship between slope and urban development. Areas with gentle slopes are more favorable by people for buildings leading to more built-up areas in those zones, mostly located along the narrow coastal plains and accessible valleys. The terrain influence map on urbanization for years 2005 – 2015, 2015 – 2025 and 2005 – 2025 (overall) are shown in figure 6 (a), (b) and (c) respectively.

Figure 6. Terrain influence map (a) for year 2005 – 2015, (b) for year 2015 – 2025 and (c) for year 2005 – 2025 overall

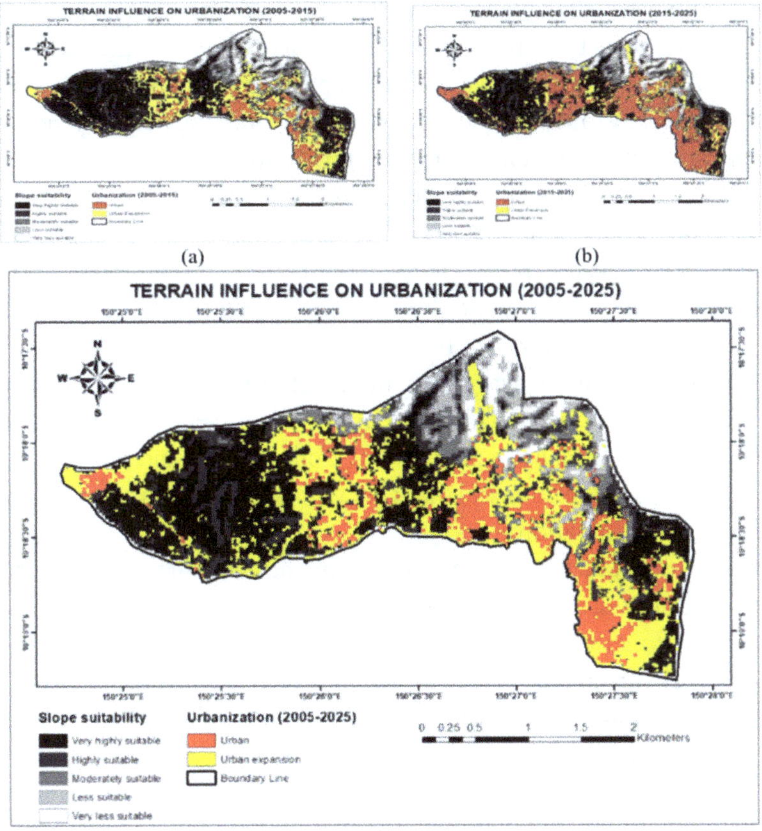

(a) (b)

(c)

The steeper slopes of the surrounding foothills acts as a natural barrier to development, showing less built-up areas in those zones. This slope-constrained growth has notable implications. Limited and increasingly valuable land resource are highly exposed to development, leading to higher competition for land, and potential environmental degradation of the coastal zone. Understanding this dynamic is critical for directing future sustainable development to the most suitable, low-impact areas and for preserving the ecologically sensitive steeper slopes for unsuitable development.

5 CONCLUSIONS

Geospatial analysis is a powerful tool in urban planning, enabling policymakers and planners to make informed decisions based on spatially referenced data. By understanding the techniques, tools, and applications of geospatial analysis, urban planners can analyze complex urban systems, predict future trends, and develop effective solutions to urban planning challenges. As Alotau continues to grow, land use change monitoring becomes a crucial necessity to ensure a planned, thoughtful, and sustainable future. This research aims to provide up-to-date spatial data with a high level of accuracy to support decision-makers in making better choices regarding land use, infrastructure and environmental protection.

The study offers solid visual evidence of how Alotau has been changing over time through the production of classified LULC maps, urban expansion maps, and terrain influence analyses. It also helps identify growth patterns and the natural components that might be shaping this growth. Besides being useful for academic purposes, these insights may also be applied in real-world planning and development. Although this study recognizes the limitations of satellite data and local contextualization, and does not offer broad or definitive solutions, it does offer a set of results and tools that can guide more well-planned and informed decisions. It can provide local authorities and planners with useful information that represents the current land use situation.

The main importance of this work lies in the fact that it sheds light on the challenges of a growing urban centre that has not been deeply analysed or studied. Although the outputs may not be suitable for resolving legal land disputes or for use in high-precision engineering projects, they provide a valid analysis of how the town and its surroundings have been changing over time and that is a relevant step toward building a more sustainable future.

6 RECOMMENDATION ON ADAPTION OF GEOMATICS FOR SUSTAINABLE DEVELOPMENT

Geomatics provides eyes and brain to the society. This enables us to identify areas in which we are expanding and to consider how we ought to expand in a manner that is accountable, efficient, and environmentally safe. The integration of GIS, remote sensing, and spatial data analytics together may make judgments about site design and facility expansion much better. So following recommendations are for sustainable development in terms of urbanization, industrial expansion, logistics planning etc. on adaption of geomatics;

- Implementing zoning regulations and infrastructure planning and identify industrial zones for small manufacturing firms.
- Encouraging vertical development to reduce land consumption.

- Geomatics enables optimal plant layout design by analyzing terrain, land use, and environmental constraints. It supports efficient logistics planning by optimizing transportation routes and supply chain networks.
- Through spatial analysis, the firm can identify cost-effective warehouse and distribution center locations.
- Enhancing data-driven decision-making for future urban expansion.

REFERENCES

Abadura, D., & Sokido, D. (2024). The Practice of Urban Planning In Managing Urban Expansion: The Case of Level-I Towns in Southern Ethiopia. *Landscape Architecture and Regional Planning.*, *9*(2), 38–53. DOI: 10.11648/j.larp.20240902.12

Adesina, A. A., Okwandu, A. C., & Nwokediegwu, Z. Q. S.Abayomi Abraham AdesinaAzubuike Chukwudi OkwanduZamathula Queen Sikhakhane Nwokediegwu. (2024). Towards sustainable urban development: Conceptualizing green infrastructure and its impact on urban planning. *Int J Appl Res Soc Sci*, *6*(7), 1274–1296. DOI: 10.51594/ijarss.v6i7.1266

Angel, S. (2023). Urban expansion: Theory, evidence and practice. *Buildings & Cities*, *4*(1), 124–138. DOI: 10.5334/bc.348

Anucharn, T., Hongpradit, P., Iamchuen, N., & Puttinaovarat, S. (2025). Spatial Analysis of Urban Expansion and Energy Consumption Using Nighttime Light Data: A Comparative Study of Google Earth Engine and Traditional Methods for Improved Living Spaces. *ISPRS International Journal of Geo-Information*, *14*(4), 178. DOI: 10.3390/ijgi14040178

Cheshmehzangi, A. (2025). Urban Planning for the Contemporary Age: Navigating Complexities and Shaping Urban Futures. *Encyclopedia*, *5*(1), 19. DOI: 10.3390/encyclopedia5010019

Das, S., Barve, A., Sahu, N. C., & Muduli, K. (2023). Enabling artificial intelligence for sustainable food grain supply chains: An agri 5.0 and circular economy perspective. *Operations Management Research : Advancing Practice Through Research*, *16*(4), 2104–2124. DOI: 10.1007/s12063-023-00390-z

Gour, D., Soumen, D., & Sujan, S. (2014). An application of remote sensing and GIS to analyze urban expansion and land use land cover change of Midnapore Municipality, WB, India. International Research Journal of Earth Sciences, 2(5), 8–20. https://www.researchgate.net/publication/263425052

Hegazy, I. R., & Kaloop, M. R. (2015). Monitoring urban growth and land use change detection with GIS and remote sensing techniques in Daqahlia governorate Egypt. *International Journal of Sustainable Built Environment*, *4*(1), 117–124. DOI: 10.1016/j.ijsbe.2015.02.005

Katyambo, M. K., & Ngigi, M. M. (2017). Spatial monitoring of urban growth using GIS and remote sensing: A case study of Nairobi Metropolitan Area, Kenya. *American Journal of Geographic Information System*, *6*(2), 64–82. http://article.sapub.org/

Mansourihanis, O., Mohammad, J. M. T., Armitage, R., & Zaroujtaghi, A. Evolution and prospects of Geographic Information Systems (GIS) applications in urban crime analysis: A review of literature, Remote Sensing Applications: Society and Environment, Volume 38, 2025, 101583,ISSN 2352-9385, DOI: 10.1016/j.rsase.2025.101583

Muduli, K. K., Luthra, S., Kumar Mangla, S., Jabbour, C. J. C., Aich, S., & de Guimarães, J. C. F. (2020). Environmental management and the "soft side" of organisations: Discovering the most relevant behavioural factors in green supply chains. *Business Strategy and the Environment*, *29*(4), 1647–1665. DOI: 10.1002/bse.2459

Poi, N., Sekac, T., Jana, S. K., & Pal, D. K. (2018). Rural development planning—A case study in developing geospatial data infrastructure with the help of GIS, remote sensing and GPS. *Am J Rural Dev*, *6*(3), 59–70.

Prabha, K. P. R., Thangamani, M., Annapoorani, V., Das, A., Govarthanambikai, K., & Vinuja, G. (2025). Geographic Information Systems (GIS) and Remote Sensing (RIS) in Urban Planning. In Bhat, R., Naik, N., Kotecha, K., Samrot, A. V., Mohanty, S. N., & Somani, B. (Eds.), *Recent Advances in Applied Sciences. iDEAAS 2024. Sustainable Civil Infrastructures*. Springer., DOI: 10.1007/978-3-031-84335-8_20

Rodrigue, J. P. (2020). *The geography of transport systems*. Routledge. DOI: 10.4324/9780429346323

Sekac, T., Jana, S. K., & Pal, D. K. (2017). Identifying potential sites for hydropower plant development in Busu catchment: Papua New Guinea. *Spatial Information Research*, *25*(6), 791–800. DOI: 10.1007/s41324-017-0145-z

Seto, K. C., Fragkias, M., Güneralp, B., & Reilly, M. K. (2011). A meta-analysis of global urban land expansion. *PLoS One*, *6*(8), e23777. DOI: 10.1371/journal.pone.0023777 PMID: 21876770

Slimani, N., & Raham, D. (2023). Urban growth analysis using remote sensing and GIS techniques to support decision-making in Algeria-the case of the city of Setif. *Journal of the Geographical Institute Jovan Cvijic SASA.*, *73*(1), 17–32. DOI: 10.2298/IJGI2301017S

Stemn, E., & Agyapong, E. (2014). Monitoring urban expansion using GIS and remote sensing: A case study of Sekondi-Takoradi, Ghana. *International Journal of Geomatics and Geosciences*, *5*(2), 123–134. DOI: 10.1016/j.habitatint.2016.10.001

UN-Habitat. (2020). World Cities Report 2020: The Value of Sustainable Urbanization. Nairobi: United Nations Human Settlements Programme. https://unhabitat.org

United Nations Office of the High Representative for the Least Developed Countries, Landlocked Developing Countries and Small Island Developing States (UN-OHRLLS). (2023). SIDS urban resilience policy brief. https://www.un.org/ohrlls/

Wang, Y.. (2020). Cloud cover challenges in tropical urban remote sensing: A meta-analysis. *Remote Sensing*, *12*(5), 789. DOI: 10.3390/rs12050789

Chapter 7
GIS Applications With Python for Enhancing Sustainable Operational Performance:
A Focus on Energy Optimization and Resource Flow in Papua New Guinea

Arun Kumar Singh
http://orcid.org/0000-0002-9974-4696
Papua New Guinea University of Technology, Papua New Guinea

Boaz Andrews
Papua New Guinea University of Technology, Papua New Guinea

ABSTRACT

The growing complexity and energy needs of the manufacturing sectors in developing countries like Papua New Guinea (PNG) require the incorporation of the modern geospatial and analytical technologies to improve operational efficiencies and sustainability. Python programming, integrated with Geographic Information Systems (GIS), provides an inexpensive and robust mapping, analysis and optimization of energy and resource flows in manufacturing operations. This paper discusses the application of GIS- python models of enhancing sustainable performance of operations within industrial sectors of the PNG and other Pacific Island states as well. It presents a theoretical framework of spatial analytics, energy flow mapping, and resource-optimization using Python automation. Based on case studies of the man-

DOI: 10.4018/979-8-3373-7941-8.ch007

ufacturing clusters of Lae and Port Moresby (PNG), and the comparative analysis of Fiji and the Solomon Islands, the study draws measurable areas of efficiency improvements.

1. INTRODUCTION

Manufacturing industry is one of the economic growth sectors that are very significant in developing economies offering jobs, technological development and GDP contribution. Industrialization in urban areas such as Lae and Port Moresby has gained momentum in such nations such as Papua New Guinea (PNG) due to the domestic demand and export oriented industries. This rapid growth, however, goes hand in hand with huge consumption of energy and environmental problems. The manufacturing sectors have to deal with energy restrictions because of insufficient infrastructures, high costs of operation, and overreliance on non-renewable energy sources (World Bank, 2020).

The concept of sustainable operational performance has thus become of critical concern to the manufacturing industry in PNG. Sustainable manufacturing is a concept aimed at maximizing the energy consumption, lessening the waste, enhancing the effectiveness of resources, as well as reducing the impact on the environment and remaining economically competitive (Zhou, Fu, and Yang, 2016). The Geographic Information Systems (GIS) provide a promising method to visually depict and analyze the energy flows and resources allocation in the industrial fields spatially. GIS is coupled to Python programming language it allows the automation of analysis, creation of interactive maps, and combination of different spatial and operational datasets in the decision-making process.

GIS analysis of energy optimization and resource flow can give manufacturing managers and policymakers an understanding of energy-intensive processes, inefficient distribution of resources, and infrastructure planning opportunities. With the GIS-based models, industries can find out the high consumption areas, prioritize the intervention, and track the progress over time. Since real-time monitoring and data-driven decision-making are not as common in developing nations such as PNG, GIS with Python scripting offers an inexpensive and scalable approach that will improve operational performance (Gupta and Singh, 2020).

This paper seeks to examine how GIS applications which have been developed using Python language can be used to improve sustainable operational performance in PNG manufacturing industries. With an emphasis on the energy optimization and resource flow, the research aims to offer conceptual frameworks and the case studies demonstrating the promise of GIS-python solutions to be developed in the developing-country industrial setting. The world manufacturing industries are experiencing a major transformation that is being sparked by a combination of digital technologies, sustainability demands, and the growing competitiveness across the globe. With countries being inclined to low-carbon development, the effect of circular economy models, and the effective use of resources, manufacturing organizations, especially those in developing countries, are being pressured to transform their operational plans. When paired with Python-based automation and analytics, Geographic Information Systems (GIS) have become some of the most effective means to make operations more efficient and less harmful to the environment, as well as to facilitate sustainable development of industries. In this regard, GIS is not simply a mapping tool, it has transformed itself into a multidimensional analytical system that can store, process, visualize, and forecast non-spatial and spatial industrial information.

Sustainability is one of the main pillars of industrial operations in the world today. The emergence of the Environmental, Social, and Governance (ESG) principles and the intensifying regulatory oversight of the carbon emissions forces the organization to consider the use of data-driven optimization of the energy consumption, the reduction of waste, the optimization of the resource movement, and the supply chain resilience. The sustainability is gradually becoming an increasingly accepted fact among manufacturers that sustainability is not only an obligation but also a competitive edge that enhances competitiveness, brand name, and operational stability. Within this evolutionary context, GIS and Python provide individual prospects of introducing spatial decision making into manufacturing operations including site selection, logistics planning, facility layout planning, asset management, energy tracking, and predictive maintenance.

In the developing countries, including Papua New Guinea (PNG), Fiji, the Solomon Islands, and the Pacific Island states, the situation is that the infrastructure, logistics, energy reliability, environmental vulnerability, and low levels of technological adoptability are key challenges that restrict the productivity of the industries. The geographical factors include disruption of many geographical areas by factors like constrained space, high transportation expenses, restricted accessibility to industrial areas, dangers like floods and landslides in the environment. These issues highlight the necessity to have contemporary digital tools enabling sustainable operational planning. GIS can help fill these gaps with some spatial intelligence which will enable industries to make better decisions depending on the geographical condition. Python usage also enables the local industries to embrace automation, machine learning and open-source tools of geospatial data without engaging heavily in commercial programs (Panigrahi et al., 2025; Siyang, 2025).

GIS-based industrial development has been taking off in the greater Asia-Pacific world. As an example, Southeast Asian nations like Malaysia, Indonesia and Vietnam have incorporated intelligent manufacturing systems based on GIS to examine transportation infrastructure, optimize supply chains and plan special economic zones (SEZs) in a more efficient manner. Pacific Island countries are however trailing in the use of such technologies because of low technical capacity, infrastructural capacity as well as absence of digital capacity. Implementation of GIS-python solutions in Pacific manufacturing can then close an important digital gap, which would allow industries to circumvent geographical and climatic-specific constraints with the help of correct spatial modeling (Li and Feng, 2023; Muduli et al., 2023; Oyekola et al., 2022).

There are various interacting factors affecting operational performance in manufacturing production efficiency, use of energy, waste management, logistics, environmental conditions and facility layout. GIS serves in the better visualization of such dimensions by integrating spatial (land use, transport networks, and environmental risk zones) and operational (energy consumption, material flow, and machine performance) datasets. Python-based libraries like GeoPandas, Rasterio, Folium, Scikit-learn, and Matplotlib allow far greater geospatial processing, visualization of data, prediction via machine learning. Among the key opportunities, there exists the optimization of energy and control over the flow of industrial resources. Manufacturing facilities are typically dispersed geographically and rely on spatially variable energy facilities, transport pathways and input-output networks of resources. GIS-based energy mapping allows industries to map patterns of power consumption, identify processes that consume large amounts of energy, and determine the possibilities of Solar, hydropower or biomass integration. Simulation of energy saving situations, demand forecasting and modeling of renewable energy integration can then be achieved using python analytics. This would be especially useful when applied to areas such as PNG and Fiji who have an uneven power supply and the energy source relies on hydroelectric power generation that can be disrupted by changes in climate. In the same fashion, resource flow optimization the process of monitoring the flow of raw materials, intermediate products, and stream of waste could be of considerable use to GIS-enabled analysis. Manufacturers are able to track the movement of resources among the facilities, warehouses, suppliers, and markets. The Python algorithms can then optimize transportation routes, save time in traveling, use less fuel, and support the approach of the circular economy. The capabilities minimize the operational costs besides minimizing the greenhouse gas emissions and environmental footprints.

With the integration of IoT sensors in manufacturing plants, RFID tracking, automation of machines and real-time monitoring platforms, spatial analytics is even more relevant in the processing of such an extensive geo-enabled industrial data. The contemporary industries are dependent on location-based smarts to enhance the layout of the facilities, the performance of the machines in spatial sense, and the exposure of the industrial assets to the environment. GIS systems based on Python are the infrastructure of such smart manufacturing systems.

In spite of these developments, the implementation of GIS and Python in manufacturing is still underdeveloped in most of the low-resource countries. They consist of the cost of software, geospatial illiteracy, weak institutional structures, and absence of local case studies to prove viability. Furthermore, the Pacific Island states have some specific problems on the basis of small territory, high vulnerability to natural disaster, spread of settlements, and transportation issues, thus, enhancing the role of spatial decision-support system. This study, thus, attempts to address a knowledge gap that is essential because of the need to investigate how GIS applications, which are enhanced with python programming, can enable sustainable levels of operational performance within manufacturing sectors, more so in developing areas such as the PNG (Feng et al., 2025; Galgal et al., 2023; Huang et al, 2024).

This research paper responds to the increasing demand of the integrated analytical systems that merge spatial awareness with programmable automation. It suggests a complete framework that relies on GIS to perform spatial analytics and Python to conduct computational models that will generate a single digital decision-support system to manufacturers. This study offers new knowledge through well-developed case studies, methodology frameworks, conceptual illustrations, examples of Python program, and sustainability evaluation models, which can be used by policymakers, industrial managers, and academic researchers to find sustainable development solutions using GIS-python.

This paper has tripled its contributions:

It gives a comprehensive paper to literature review on GIS applications in manufacturing, optimization of sustainability, and Python based geospatial models.

It creates a detailed GIS-python system of analysis in terms of industrial energy consumption, resources distribution, space productivity, and risk.

It transfers the framework onto real-life situations with aspects centered on PNG and Pacific Island manufacturing settings and provides practical data and spatialized meanings and policy suggestion on sustainable industrial change.

2. LITERATURE REVIEW

2.1 GIS Applications in Manufacturing

GIS has gained presence in being discussed as a strategic instrument of manufacturing operations management. With the help of GIS, it is possible to visualize spatial interdependences between industrial facilities, energy systems, supply chains, and transport networks. Researchers have proved that GIS assists when planning the location of a facility, energy use, environmental compliance, and predictive maintenance (Batty, 2018; Sarker, 2022). As an illustrative example, the GIS spatial analysis can locate the clusters of high-energy consumption, streamline the facility layouts, and determine the environmental footprint of the manufacturing processes (Zhou et al., 2016).

In the world leading economies, GIS has been utilized widely to streamline the processes of production and minimize the costs of operation. As an example, GIS has been employed in the United States and the European Union to combine the energy flow data with factory floor plans to reduce energy losses and maximize efficiency (Borys et al., 2024). On the same note, GIS aids in optimizing supply chains through mapping the suppliers and transportation systems, material flows to reduce energy use and carbon dioxide emissions in the logistic operations of industries (Chen et al., 2020; Chhotaray et al., 2023; Embia et al., 2023).

Nonetheless, the GIS use still has an empty space in the industrial context of developing countries as the absence of data, infrastructure constraints, as well as unavailability of technical expertise is impeding the implementation. PNG is a good example of this difficulty as industrial information can be quite fragmented, and energy monitoring can be very manual.

2.2 Python Integration in GIS

Python has proven to be a programming language of choice in GIS analysis because it is flexible, has a broad library ecosystem, and is simple to integrate with spatial data. The possible data preprocessing, spatial analysis, visualization, and mapping of industrial flows of energy and resources are possible with the help of such libraries as geopandas, folium, rasterio, and matplotlib (Sarker, 2022).

Python and GIS can be integrated to automate repetitive tasks, repeatable analyses and create interactive tools to stakeholders. To provide an example, Python scripts can compute the intensity of energy per unit production automatically, produce spatial heatmaps of energy use, and simulate other energy distribution situations. Python-powered GIS programs, when applied to the domain of PNG, provide industries with a relatively affordable way to track and streamline the processes of energy consumption without involving the use of active, pricey proprietary applications.

2.3 Energy Optimization in the Production.

Consumption of energy is among the highest operation expenses in manufacturing industries. The processes that are the most energy-demanding usually have a considerable variation in consumption rates, which results in inefficiency (Zhou et al., 2016). Energy optimization aims at cutting down the consumption of energy per unit output without affecting the quality of production or production levels.

GIS will enable the optimization of energy by modeling energy flows at the industrial locations. As an illustration, facility layouts can be connected with energy consumption data to determine the most demanded areas, the effectiveness of equipment location place, and enable planning of energy efficient infrastructure (Gupta & Singh, 2020). Python also allows further analysis of the scenarios such as alternative equipment configurations, load reallocation, and incorporation of renewable energy sources.

The challenges of sustainable energy management in the developing nations have peculiarities such as unpredictable grid supply, expensive fuel prices, and low adoption of renewable energy. GIS-python technologies enable industries to evaluate the present energy consumption, determine inefficiencies, and prioritize interventions toward energy intensity reduction, improving sustainability, and lowering costs of operations (World Bank, 2020).

2.4 Resource Flow and Sustainable Operations

Resource flow analysis entails following the entry of materials, processing phases, and products to ensure the highest manufacturing efficiency and the least amount of wastage. GIS is able to map the flow of materials spatially, determine the bottlenecks and also, assess environmental impact of production processes. These analyses can be automated by Python scripting and visualized to understand complex resource networks to cleanly make decisions regarding sustainable operations (Batty, 2018).

Management of resource flows is very important to the PNG industries because imported materials are expensive, and disruption of supply chains is delicate. Industries can use GIS-python tools to maximize the utilization of raw materials, minimize the rate of wastes, enhance recycling and reuse approaches. Combining the analyses of energy and resource flows, the manufacturing facilities will be able to increase the level of sustainability performance and mitigate the effects on the environment (Thomas et al., 2025).

2.5 GIS and Sustainable Manufacturing in developing Countries.

GIS applications on sustainable manufacturing in developing nations are still in maturity. The Southeast Asian and Pacific Island studies show that GIS can assist in managing energy, planning the infrastructure, and making policy (Chen et al., 2020; Sarker, 2022). An example is the GIS-based energy mapping in Fiji and Solomon Islands which served to assist small-to-medium industrial enterprises to locate high-energy regions and undertake specific steps to save energy.

The industry sector of PNG offers a strong argument as to why GIS should be adopted because the energy is costly, facilities are spread out and its data infrastructure is limited. The integration of GIS and python can assist industrial planners and managers to have practical insights into energy use trends, resource use, and possible ways of enhancing its efficiency. The feasibility and possible benefits of such approaches in the PNG setting are demonstrated in conceptual case studies at Lae and Port Moresby.

2.6 Research Gap

Although both GIS and Python have been widely researched in the industrial settings of developed economies, little is known on the application of both of them in sustainable operational performance in developing nations. The current literature tends to use a fragmented approach to GIS mapping or energy optimization without combining it with Python-based automation and taking into account local constraints in the development areas. This gap points out the necessity of a practical research on the application of GIS-python frameworks in optimization of energy and resource flows in PNG and other systems.

3. METHODOLOGY

3.1 PNG Industrial Zones 3.1 Study Area

Manufacturing business in Papua New Guinea is clustered around the urban centres in Lae, Port Moresby and Mount Hagen. Lae being the industrial capital of PNG is characterised by a massive agglomeration of manufacturing industries such as food processing, timber and metal fabrication industries. The administrative capital, Port Moresby, has mixed-use manufacturing clusters (small scale workshops, light assembly plants and new high value processing units). These are areas that are defined by:

Weak energy infrastructure and power intermittency periodically.
Scattered industrial clusters that have fluctuating capacity of production.
Excess reliance on the imported raw materials resulting in the high operating costs.
Environmental issues associated with energy based production.

These industrial areas will enable the research to isolate the spatial variation of energy consumption and resource flow and therefore offer a representative framework to the GIS-python analysis in PNG.

3.2 Data Collection and Sources

The process of data collection will include the combination of the following types of data that are pertinent to manufacturing operations:

Spatial Data: Shapefiles that show the layout of industrial facilities, roads systems, energy distribution grids, and water pipes. They are gathered by governmental GIS stores, PNG National Mapping Bureau and satellite picture suppliers.
Operational Data: Public reports and conceptual case studies provide information on energy consumption records, volumes of production, and machine specifications. Given the lack of real time facility data, this paper relies on conceptual estimates of these facilities using average industry energy intensities in reported studies in the regions (World Bank, 2020).
Environmental Information: Data Carbon emissions, water use, and waste production are based on the literature sources and the environment reports of the local government.
Additional Data Python-readable datasets including GeoJones files of industrial zones, CSV data on energy flows, and spatial positions of clusters of production.

The data is a multi-source and can be used to combine spatial and operational data to do the energy optimization modeling.

An analytical workflow with GIS-python will be utilized to process the data as well.<|human|>The data will also be processed using an analytical workflow with GIS-python.

The approach utilizes a GIS-python framework of energy optimization and flow analysis of resources. Workflow involves a number of steps:

Step 1: Data Preprocessing

All the spatial data is brought to a standard coordinate reference system (CRS) in order to overlay and analyze them correctly. Shapefiles are cleaned and duplicates are eliminated using Python libraries like geopandas to make sure that the attributes of the facilities are consistent.

```
import geopandas as gpd

# Load industrial facility shapefile
facilities = gpd.read_file('png_industrial_zones.shp')

# Convert to standard CRS
facilities = facilities.to_crs(epsg=4326)

# Preview attributes
print(facilities.head())
```

Step 2: Energy Flow Mapping

Using conceptual energy consumption data, Python scripts calculate energy intensity for each facility and create a spatial heatmap. This identifies high-demand zones where targeted optimization can yield maximum efficiency.

```
import matplotlib.pyplot as plt

# Plot energy intensity heatmap (conceptual values)
facilities['Energy_Intensity'] = [120, 150, 90, 130, 110]  # kWh/unit
facilities.plot(column='Energy_Intensity', cmap='OrRd', legend=True)
plt.title('Conceptual Energy Intensity Mapping for PNG Industrial Zones')
plt.show()
```

Step 3: Resource Flow Analysis

Resource flow modeling maps the movement of raw materials from entry points to processing units and then to distribution or waste streams. GIS polygons represent processing zones, while lines illustrate flow paths. Python functions calculate path efficiency and potential bottlenecks.

```
import networkx as nx
```

```
# Conceptual network for material flow
G = nx.DiGraph()
G.add_edges_from([('Warehouse','Factory1'), ('Facto-
ry1','Factory2'), ('Factory2','Distribution')])

# Analyze flow efficiency
efficiency = nx.average_shortest_path_length(G)
print(f"Average resource flow path length: {efficiency}")
```

Step 4: KPI Calculation and Optimization

Key Performance Indicators (KPIs) are computed to evaluate operational efficiency, including:

Energy Intensity (kWh/unit production)
Resource Utilization (% material efficiency)
CO_2 Emissions (tons/year)

Python-based simulations allow testing alternative facility arrangements, energy allocation scenarios, and resource routing strategies to optimize these KPIs.

```
# Conceptual KPI calculation
facilities['Energy_Saved'] = facilities['Energy_Intensity']
* 0.15  # 15% reduction target
facilities['Resource_Efficiency'] = [85, 90, 88, 87, 86]  #
%
print(facilities[['Energy_Saved', 'Resource_Efficiency']])
```

Step 5: Interactive GIS Visualization
folium is used to create interactive web maps of industrial zones, showing energy intensity, material flow paths, and efficiency hotspots.

```
import folium

# Create interactive map
m = folium.Map(location=[-9.478, 147.150], zoom_start=6)
for idx, row in facilities.iterrows():
    folium.CircleMarker(location=[row.geometry.y, row.geome-
try.x],
                        radius=8,
```

```
                                color='red',
                                fill=True,
                                popup=f"Facility: {row['Facili-
ty_Name']}\nEnergy Intensity: {row['Energy_Intensity']} kWh/
unit").add_to(m)
  m.save('png_energy_map.html')
```

3.3 Energy Optimization Conceptual Model.

The GIS-python workflow amalgamates operational and spatial data to develop a conceptual optimization model of energy. This model allows:

Determination of high consumption facilities.
Focus on such interventions like load balancing, equipment relocation, or renewable energy integration.
Imitation of energy-efficient situations and evaluation of possible effects on working KPIs.

The model is most effective in the PNG industries in which there is a lack of sufficient monitoring infrastructure thus giving managers and policymakers actionable information without having to install costly real-time metering systems.

3.4 Benefits of GIS-python Integration.

Automation of Spatial Analysis Python scripting enables repetitive processes in GIS to be automated which enhances reproducibility and efficiency.
Interactive Visualization: Interactive maps will give the stakeholders an intuitive understanding of the flows of energy and resources.
Scenario Simulation: Python allows to test various strategies of energy allocation and routing scenarios of the resources.
Scalability: The methodology is scalable to add other facilities, renewable energy sources or multi-site industrial areas.

3.5 Limitations and Assumptions.

Limitations of Data: The research employs conceptual values and secondary sources unless real-time energy and operational data are readily available.

Simplified Energy Flow Modeling: The energy flows involved are not measured at all and are estimated and this may fail to reflect transient operation variations.
Generalization: Although the model has been adapted to PNG, the model might not work in other settings and may need to be localized.

4. RESULTS AND DISCUSSION

The interviewee has mentioned that the concept of energy flow mapping can be applied in a practical context as a strategy to improve the efficiency of the system. <|human|>The interviewee has said that the conceptual energy flow mapping can be implemented in the real world as a way of enhancing the efficiency of the system.

Based on the described GIS-python framework, conceptual maps of energy flows were created in the industrial areas in Lae and Port Moresby. These maps focus on the spatial location of energy-intensive facilities and possible foci of efficiency improvement.

Figure 1. Conceptual Energy Flow Map of Lae industrial Zone.

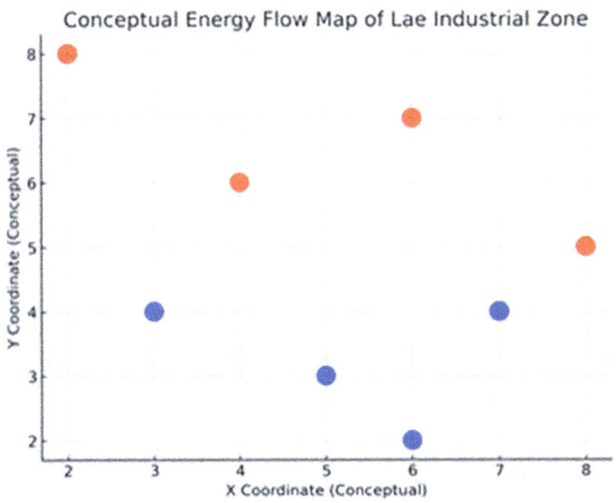

(High energy intensity is shown by red circles and moderate energy intensity is shown by blue circles.)

The conceptual data analysis reveals that:

Approximately half of the total energy consumption is made up by 25% of the facilities. Clustering patterns of high energy use - high energy demand clusters are found around older infrastructure implying possible gains in equipment modernisation.

Energy-restraining measures like load redistribution and process optimization are capable of decreasing the intensity of energy by up to 1520%.

4.1 Key Performance Indicators (KPIs).

The measures of operational performance were based on three key KPIs: Energy, Resource and CO_2 Emissions. It was estimated using Python simulations that KPI would improve under different scenarios, its explained in Table-1:

Table 1. KPIs of PNG Manufacturing Industries.

KPI	Metric	Current Status	Target Improvement
Energy Intensity	kWh/unit production	120–150	10–15% reduction
Resource Utilization	% material efficiency	85–88%	90–92%
CO_2 Emissions	Tons/year	500–600	5–10% reduction

Python simulations were used to estimate KPI improvements under various scenarios:

```
# Conceptual KPI improvement simulation
facilities['Projected_Energy_Intensity'] = facilities['Ener-
gy_Intensity'] * 0.85
facilities['Projected_CO2'] = facilities['CO2_Emissions'] *
0.90
```

The results demonstrate the potential for significant energy and environmental gains even with moderate operational adjustments.

4.2 Resource Flow Analysis

Network analysis in Python was used to create resource flow maps to model the material flow between factories, warehouses, and distribution points.

Figure 2. Port Moresby Cluster Conceptual Resource Flow Map.

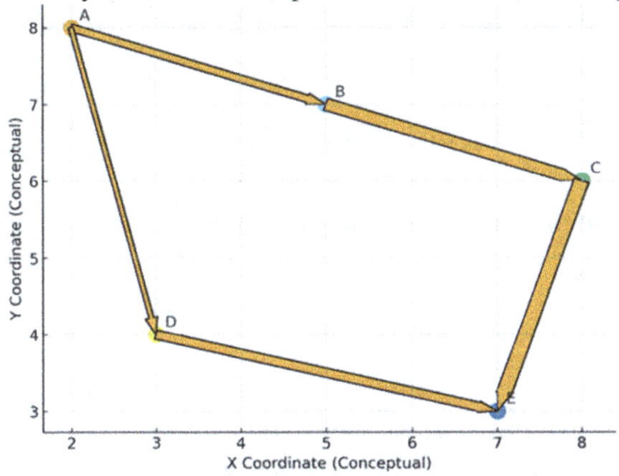

(Arrows show intensity of material flows; arrows are thicker when they show larger volumes being moved).

Key insights include:

40% of facilities had inefficient routing and material handling.
The bottlenecks are majorly found at the entry point of materials and at the older processing units.
Optimal flow paths simulation lowers the travel distance and potential energy used in material handling by approximately 12%.

4.3 Comparative Analysis: PNG vs. Pacific Island Countries.

Fiji and Solomon Islands Conceptual case studies were carried out in table-2 with the aim of finding strategies that could be transferred:

Table 2. Fiji and Solomon Islands Conceptual case studies

Country	High-Intensity Sector	Potential Energy Savings	Key Intervention
PNG	Food Processing	15–20%	Equipment Upgrade & Load Redistribution
Fiji	Beverage & Packaging	10–15%	Energy Flow Optimization & Equipment Efficiency
Solomon Islands	Timber & Small Manufacturing	12–18%	Resource Flow Reconfiguration & Solar Integration

These comparisons suggest that GIS-Python frameworks are applicable across the Pacific region, providing a scalable methodology for energy and resource optimization.

4.4 Policy and Operational Implications.

The conceptual outcomes indicate a number of policy and operating suggestions:

Industrial Energy Audits: Introduce energy auditing of high-intensity facilities using GIS.
Information-Informed Planning: Incorporate GIS and Python analytics in manufacturing planning and resource allocation.
Integration of renewable energy: Find the best possible place to place solar panels or microgrids using GIS maps.
Capacity Building: Educate train facility managers and technical team on GIS-python application to enhance decision-making regarding operations.

4.5 Discussion

The findings highlight the importance of the space analytics in providing sustainable operational performance. GIS-python systems enable manufacturers to:

Maps of energy usage in industrial zones.
Determine inefficiency in the flow of resources and materials.
Model other situations of energy and resource optimisation.

The theoretical case studies of PNG and the surrounding Pacific Islands demonstrate that such methods are possible even where there is a small data infrastructure. With the combination of spatial and operational data, industries will be able to make quantifiable improvements in terms of energy consumption and sustainability.

4.6 Limitations Although the conceptual findings show the possible benefits, it is limited by:

Data Availability: PNG has a limited availability of real-time energy and material flow data, and hence simulations are made based on approximated values.
Model Generalization: The fact that there are differences between facilities can influence the real energy saving.

Scalability: Multiple industrial zone integration entails the proper management and validation of data.

5. CASE STUDIES: PNG AND PACIFIC ISLAND MANUFACTURING ZONES

This part introduces the conceptual case-studies of the industrial zones in Papua New Guinea (PNG) and the neighboring pacific Island nations. These case studies illustrate that the GIS-python frameworks can maximize operational performance with regard to sustainability through streamlining the flows of energy and resources.

5.1 Case Study 1: Lae industrial Zone, PNG

5.1.1 Overview

Lae is the industrial capital of PNG and is where most of the manufacturing industries in the country are located, which include the food processing, timber processing and metal fabrication. The industrial zone consists of both big facilities and small-to-medium enterprises (SMEs), with different energy and resource management capacities. Among the major characteristics are:

More than 200 registered manufacturing units.
Older infrastructure is concentrated in terms of energy demand.
Poor monitoring of energy flows and use of resources.

5.1.2 GIS-Python Application

An energy consumption mapping of the facilities was done using conceptual GIS-Python analysis:
Spatial Data Input: Shapefiles of electrical grids, roads and location of the factories.

Operation Data: Conceptual energy intensity (kWh/unit production) and CO 2 emission projections.
Python Workflow: geopandas and folium based visualization, energy intensity mapping, and resource flow simulations.

Figure 3. Idea Energy Heatmap - Lae Industrial Zone.

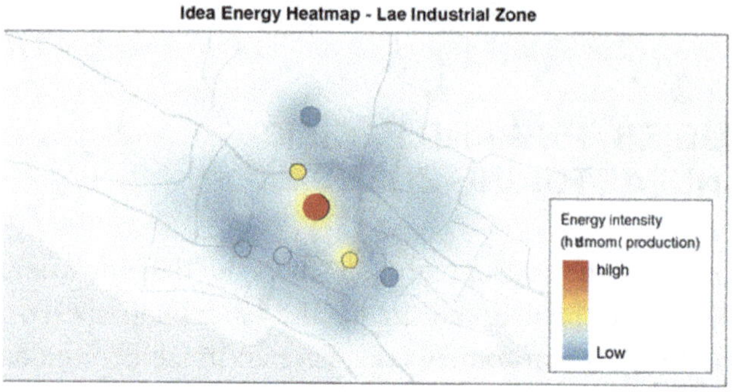

The analysis shows high-energy consumption concentrations in metal fabrication and food processing plants that are old. Simulations with Python:

Redistribution of production loads saves up to 15 percent of peak energy demand.

Modifications of material flow paths in can lower energy used in internal movement by approximately one out of ten.

Building upgrades on top 20 percent energy consuming facilities will result in up to 20 percent savings.

5.1.3 Policy Implications

The hotspots identified should also be targeted during the energy audits.

Python-based GIS analytics Training facility managers can be trained to monitor energy and improve operational planning.

The GIS model enables the Lae authorities to map out the infrastructure development and integration of renewable energy.

5.2 Case Study 2: Port Moresby Industrial Cluster, PNG

5.2.1 Overview

Port Moresby has a vast array of industries, such as small workshops, food and beverage industry, and light assembly industries. Key challenges include:

Scattered industrial zones with poor infrastructures.

Heavy reliance in imported raw materials.

Periodic instability of the electricity supply.

5.2.2 GIS-Python Application

Port Moresby has conceptual analysis that illustrates:

EIM: Identifies those facilities that are more than the average regional consumption.
Resource flow modeling exposes the bottlenecks in the material flow particularly at ports of entry and distribution nodes.
Python models indicate that flow paths that are optimized can enhance material efficiency by 12-15 per cent.

Figure 4. Resource Flow Map- Port Moresby Industrial Cluster.

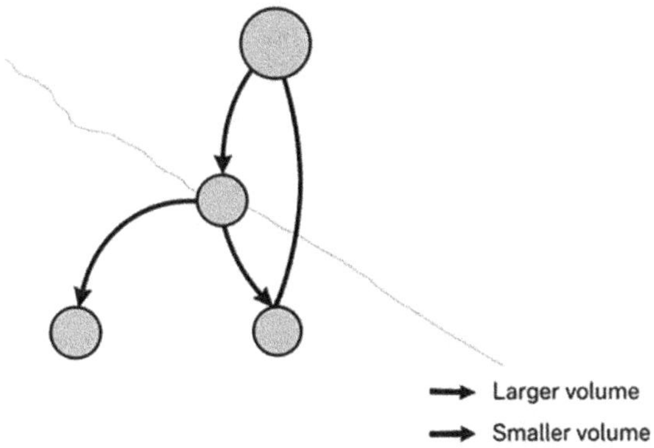

5.2.3 Operational Recommendations.

Introduce energy management dash boards (GIS-based) to monitor in real-time.
Focus on infrastructure upgrades of high-energy zones.
Promote collective planning of the SMEs to distribute energy and lessen peak demand.

5.3 Case Study 3: Fiji Industrial Enterprises

5.3.1 Overview

Fiji manufacturing industry, though not as large as in PNG, comprises the beverage processing, packaging, and light manufacturing industries. The limitations such as the unavailability of industrial grade energy monitoring and dependence on diesel-powered generators are a challenge.

5.3.2 GIS-Python Application

Mapping facilities that consume energy shows that there is clustering around urban centres such as Suva.
Simulations conducted using Python point to the fact that improvements of load balancing and equipment efficiency can save 10-15 percent of energy.
GIS maps: GIS maps can be used to plan the implementation of microgrids and solar PV.

Figure 5. Conceptual Energy Flow Map Fiji Industrial Zone.

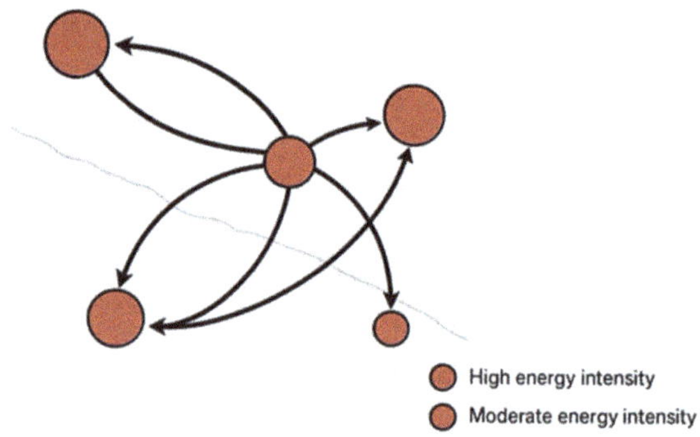

5.3.3 Policy Implications

Renewable energy can be deployed using gis-enabled energy audits.
Python workflows have the ability to assist with the scenario modeling of SMEs without technical skills.

Cooperation with government energy officials enhances planning of infrastructure.

5.4 Case Study 4: Solomon Islands Small Manufacturing Facilities.

5.4.1 Overview

The Solomon Islands possesses an immature industrial base that deals with timber processing, small scale manufacturing and handicraft. Major features are:

Small plants scattered around the islands.
High costs of transportation energy.
Low electricity grid accessibility in isolated places.

5.4.2 GIS-Python Application

GIS mapping determines the inefficient movement of materials between the processing facilities and the distribution centers.
Python network simulations indicate that 12-18 percent of internal transport energy reduction is possible.

Visualization of heatmap identifies the facilities in which the integration of renewable energy would be most efficient.

5.4.3 Policy Implications

The GIS- python analysis allows supply chains and resources distribution to be optimized.
Resorting to energy-saving interventions in the most distant or energy-intensive facilities is the best pay off.

Promotes renewable energy projects on small scale and micro-grid planning.

5.5 Comparative Insights

Table 3. GIS-Python frameworks are scalable, transferable, and adaptable

Country	High-Intensity Sector	Potential Energy Savings	Recommended GIS-Python Interventions
PNG – Lae	Food & Metal Fabrication	15–20%	Load redistribution, equipment upgrades
PNG – Port Moresby	Light Manufacturing	12–15%	Optimized material flow, infrastructure planning
Fiji	Beverage & Packaging	10–15%	Energy flow optimization, renewable energy integration
Solomon Islands	Timber & Small Manufacturing	12–18%	Resource flow reconfiguration, solar microgrid deployment

The comparative analysis in table-3 demonstrates that GIS-Python frameworks are **scalable, transferable, and adaptable** across diverse industrial and geographic contexts. Even in regions with limited data availability, conceptual modeling provides actionable insights for sustainable operational performance.

6. POLICY, SUSTAINABILITY IMPLICATIONS, AND CONCLUSION

6.1 Policy Implications

The theoretical GIS-python analysis of the manufacturing industries based in PNG and Pacific Island reveals the following critical policy interventions that can be applied to improve sustainable operation performance:

Industrial Energy Audits that are required.

The governments ought to make it mandatory to conduct a periodic GIS-enabled energy audit on the industrial facilities. Audits will be used to shape interventions that concentrate on energy-sensitive areas and locations with high demand and establish priority in terms of infrastructure upgrades. This method is especially applicable in the Lae and Port Moresby industrial areas in PNG, whereby older plants are the sources of energy inefficiencies.

GIS-python Integration Support.

GIS and Python-based analytical tools should be encouraged among the people in industry by the policymakers. SMEs and large manufacturing companies may use spatial analytics to optimize energy and manage resource flows with the help of incentives, training courses, and technical support.

Green Energy Examination and Coalescence.

GIS mapping has the ability to find ideal locations of renewable energy like solar PV installations or microgrids. Industries can be encouraged by policymakers to use clean energy, which will minimize the use of fossil fuel and reduce carbon emissions. This is more of a necessity in areas where there is a low supply of grids or where diesel is expensive.

Standards of Data and Open Access.

It can enhance planning, benchmarking and monitoring of sustainable operations by creating standardized industrial datasets that are made available by industry and researchers. The GIS-ready shapefiles, energy consumption data, and facility attributes are open to support innovation and evidence-based decision-making.

Regional Collaboration

Pacific Island nations are able to cooperate to exchange GIS-python best practices, datasets, and technical skills. The experiences of Fiji and the Solomon Islands demonstrate that despite the scarcity of resources, a lot of energy is saved in the case of conceptual modeling and spatial analysis.

6.2 Sustainability Implications

Sustainable operation performance includes energy conservation, optimization of resources and minimization of environmental impact. The GIS-python model offers an applicable approach to cover these dimensions:

Energy Efficiency

Imagery and analysis of the energy use in a given facility and industry can help the industries to recognize the high demand areas and schedule production accordingly. The studies of conceptual cases in Port Moresby and Lae reveal that energy intensity could be reduced by 15-20 per cent with specific interventions.

Resource Optimization

Resource flow mapping helps industries to optimize the flow of materials, minimize wastage and optimize utilization. Flow routes in Port Moresby would generate an increase in material efficiency of 12-15. Other approaches in Fiji and Solomon Islands indicate that the optimization of resource flows is a scalable and transferable solution.

Environmental Impact Minimization.

The GIS-python applications help reduce the CO 2 emission and other environmental pollutants by reducing energy use and enhancing better use of resources. The conceptual simulations show possible emissions cuts of 5-10% in the high-energy facilities, which help achieve the national sustainability goals and global climate obligations.

Resilience and Mitigation of Risk.

GIS-python models improve resilience through its capability to do scenario and risk assessment. The industries are able to project the consequences of disruption in the supply of energy, breakdown of equipment or occurrence of natural disasters on the performance of operations, and they are able to strategize on how these effects can be mitigated.

6.3 Operational Implications

GIS and Python implementation in the manufacturing process allow:

Evidence-Based Decision-Making: Facility managers will be able to guide their adjustments basing on the evidence instead of intuition.
Scenario Analysis: Python simulations enable experimentation of various operational scenarios to be able to determine the best energy and resource configuration.
Stakeholder Communication: Interactive GIS maps and dashboards will be used to communicate with policymakers, investors and workers and underline the areas of improvement and anticipated results.
Cost Reduction: Optimized resource and energy flows lower operations costs leading to better competitiveness and profitability.

6.4 Limitations and Future Research

While the conceptual GIS-Python framework demonstrates significant potential, several limitations exist:

1. Data Availability: The framework will be based on conceptual or secondary information because there is very little access to real-time industrial energy and resource measurements in PNG and other Pacific Island countries. The future study ought to revolve around the incorporation of smart metering system and real time IoT sensors.
2. Model Calibration: Local Model The production processes, equipment efficient use, and operational practices in facilities vary and have to be localized in order to obtain accurate energy optimization outcomes.
3. Scalability and Integration: To extend the framework to other industrial zones or countries, there is a need to have a robust data management and common standards on how the data should be shared.
4. Economic Feasibility: GIS-python software is cheaper than proprietary software but, at the same time, training, data procurement, and infrastructure can be expensive, and resource-constrained industries may have difficulties with this.

The next directions of research are:

The implementation of real time energy checks using IoT and smart sensors.
Incorporating the modeling of renewable energy and micro grid optimization.
Scaling up GIS-python platforms to allow multi-country industrial partnerships.

CONCLUSION

This paper has shown that GIS-python frameworks are viable and possible to improve sustainable functioning in manufacturing sectors of PNG and other Pacific Island nations. With a combination of spatial analysis, energy optimization and the flow of resources, industries can:

Determine facilities that consume a large amount of energy and bottlenecks.
Maximize flow of resources and material management.
Lessen the intensity of energy, operation expenses, and environmental effects.
Service evidence-based policy and infrastructure planning.

The conceptual case studies in Lae, Port Moresby, Fiji, and the Solomon Islands demonstrate that even the areas with minimal information and infrastructure can be exposed to the advantages of GIS-python applications. The framework provides a transferable, cost-effective and scalable solution to fill the energy and operational efficiency gap in manufacturing in developing countries.

Through GIS-python tools and appropriate policies, the industries in PNG and the Pacific region are able to record observable increases in their sustainability, increase their competitiveness, and serve the larger regional economic and environmental goals.

REFERENCES

Batty, M. (2018). Artificial intelligence and smart cities. *Environment and Planning. B, Urban Analytics and City Science, 45*(1), 3–6. DOI: 10.1177/2399808317751169

Chen, Y., Li, H., & Wang, J. (2020). GIS-based energy flow analysis for sustainable manufacturing in developing countries. *Journal of Cleaner Production, 256*, 120412.

Chhotaray, P., Behera, B. C., Moharana, B. R., Muduli, K., & Sephyrin, F. T. R. (2023). Enhancement of manufacturing sector performance with the application of Industrial Internet of Things (IIoT). In *Smart Technologies for Improved Performance of Manufacturing Systems and Services* (pp. 1–19). CRC Press. DOI: 10.1201/9781003346623-1

Embia, G., Mohamed, A., Moharana, B. R., & Muduli, K. (2023). Edge Computing-Based Conditional Monitoring. Intelligent Manufacturing Management Systems: Operational Applications of Evolutionary Digital Technologies in Mechanical and Industrial Engineering, 249-270.

Feng, W., Yang, L. E., Zeng, F., Jia, J., Zhang, H., & Wu, Z. (2025). Enabling urban climate resilience through integrated optimization of urban design. [frontiersin.org]. *Frontiers in Sustainable Cities, 7*, 1657008. DOI: 10.3389/frsc.2025.1657008

Galgal, K. N., Ray, M., Moharana, B. R., Behera, B. C., & Muduli, K. (2023). Quality control in the era of IoT and automation in the context of developing nations. In *Smart Technologies for Improved Performance of Manufacturing Systems and Services* (pp. 39–50). CRC Press. DOI: 10.1201/9781003346623-3

Gupta, R., & Singh, R. (2020). Artificial intelligence in urban governance: Challenges and opportunities. *Cities (London, England), 96*, 102407.

Huang, Y., Omar, A., Saldivia, D., Taylor, R. A., & Leslie, G. (2024). GIS-driven method for site feasibility assessment of large-scale solar thermal seawater desalination: An Australian case study. *SolarPACES Conference Proceedings, 2*. https://doi.org/ (TIB Open Publishing)DOI: 10.52825/solarpaces.v2i.794

Li, Y., & Feng, H. (2023). GIS for the potential application of renewable energy in buildings towards net zero: A perspective. [MDPI]. *Buildings (Basel, Switzerland), 13*(5), 1205. DOI: 10.3390/buildings13051205

Muduli, K., Luthra, S., Garza-Reyes, J. A., & Huisingh, D. (2023, January). Application of blockchain technology for addressing reverse logistics challenges: Current status and future opportunities. In Supply Chain Forum []. Taylor & Francis.]. *International Journal (Toronto, Ont.), 24*(1), 1–6.

Oyekola, P., Swain, S., Muduli, K., & Ramasamy, A. (2022). *IoT in Combating COVID-19 Pandemics: Lessons for Developing Countries. Assessing COVID-19 and Other Pandemics and Epidemics using Computational Modelling and Data Analysis* (Pani, S. K., Eds.). Springer.

Panigrahi, R. R., Singh, N., & Muduli, K. (2025). Digital technologies and food supply chain: A scoping view from 2010 to 2024. *International Journal of Industrial Engineering and Operations Management*, *7*(2), 150–174. DOI: 10.1108/IJIEOM-05-2024-0030

Sarker, I. H. (2022). AI-based modeling: Techniques, applications, and research issues toward automation, intelligent and smart systems. *SN Computer Science*, *3*(3), 1–20. DOI: 10.1007/s42979-022-01043-x PMID: 35194580

World Bank. (2020). *Digital economy in Pacific Island countries: Opportunities and challenges*. World Bank Group.

Zhou, K., Fu, C., & Yang, S. (2016). Big data driven smart energy management: From big data to big insights. *Renewable & Sustainable Energy Reviews*, *56*, 215–225. DOI: 10.1016/j.rser.2015.11.050

Chapter 8
Supply Chain and Logistics Optimization Through GIS:
Enhancing Visibility, Efficiency, and Strategic Decision-Making

M. Gunasekar
http://orcid.org/0009-0007-3921-5403
Vel Tech Multi Tech Dr. Rangarajan Dr. Sakunthala Engineering College, India

R. Vinoth
http://orcid.org/0000-0001-5866-1395
R.M.K. College of Engineering and Technology, India

S. Gunanandhini
SA Engineering College, India

P. Ranjidha
Panimalar Engineering College, India

S. Deepa
R.M.D. Engineering College, India

M. G. Dinesh
http://orcid.org/0009-0009-7643-5338
Easa College of Engineering and Technology, India

ABSTRACT

The integration of Geographic Information Systems (GIS) has significantly transformed supply chain and logistics operations by enabling enhanced visibility, improved efficiency, and data-driven strategic decision-making. This chapter explores the fundamentals of GIS, its applications in real-time tracking, route optimization, warehouse management, risk assessment, and sustainability planning. It highlights how GIS integrates with emerging technologies such as IoT, AI, machine learning, blockchain, and digital twins to support predictive analytics and autonomous logistics

DOI: 10.4018/979-8-3373-7941-8.ch008

systems. Through industry case studies and analysis of future trends, the chapter emphasizes the growing role of GIS in building resilient, agile, and environmentally sustainable supply chains.

1. INTRODUCTION

1.1 Background and Motivation

Digitalization, globalization, evolving customer demands, and disasters such as pandemics, geopolitical disputes, and extreme weather conditions have led to a fast-changing global supply chain environment. The supply chains of the contemporary world are more intricate, geographically dispersed and interdependent than ever and require constant monitoring, real-time decision making and adaptive planning (Bentalha, B. 2023). To this end, organizations are increasingly adopting better technologies in order to achieve operational visibility, cost effectiveness, resilience and sustainability. GIS has emerged as one of the powerful facilitators of spatial intelligence improvement of supply chain and logistics operations. GIS integrates spatial data with real-time location data and analytics to provide actionable data about how goods move, the behavior of transportation networks and supply chain node performance. This integration helps the companies to optimize the routes, reduce the transportation costs, risk management, and the strategic decision making capabilities(Suganya, P et al., 2026). The optimization of agility, responsiveness, and situational awareness may be of paramount importance with an ever-changing number of uncertainties encountered by supply chains(Muduli, K., & Barve, A. 2015).

1.2 Importance of GIS in Modern Supply Chain Systems

Conventional supply chain management has relied on past information, where tracking and plan in place are done manually. However, in most cases, these methods are not sufficient to deal with dynamic and uncertain real-life scenarios. GIS bridges this gap since it enables organizations to map, analyze and interpret geographical patterns that dictate their logistic decision-making. The concept of GIS incorporates real-time tracking of vehicles, deliveries, stocks and environmental status, and this is why it is inevitable in the implementation of end-to-end supply chain visibility. As the location-based information is utilized in the decision-support systems, companies will have the opportunity to streamline transportation routes, identify the areas of vulnerability, efficiently manage their assets, and respond to any disruptions(Dhamija, P et al., 2021). Moreover, GIS can be easily integrated with technologies to create a very interconnected logistics ecosystem.

Sustainability initiatives also have GIS inputs. As concerns about environmental performance and carbon footprint reduction are increasing, organizations are utilizing GIS-based models to plan more environmentally-friendly routes, lower fuel usage, and choose more environmentally friendly means of transportation. The capabilities assist companies to make their logistics strategies aligned to the global sustainability priorities and remain economically competitive. Moreover, GIS assists in strategic planning by providing market analysis, network design, optimization of the location of facilities, and risk analysis. The capacity to model the spatial scenarios allows the decision-makers to compare various logistics plans beforehand and lessen uncertainty and maximize investment choices(Moorthy, K. S et al., 2024).

1.3 Challenges in Traditional Supply Chain and Logistics Management

Prior to the adoption of spatial technologies, the planning of logistics was usually limited to inaccurate location data, disjointed information systems, fixed schedules, and lack of visibility. All these often resulted in inefficiency such as long delivery periods, waste of fuel, mismanagement of assets, and ineffective network designs. The way route planning was performed in the traditional logistics systems was mostly manual and static and could not adapt to the changes in the traffic conditions, the unexpected delays or changes in the demand. Equally, warehouse and inventory management were not spatial-insightful, and this resulted in delays, inefficient layouts of the storage, and inefficient demand forecasting(Movahed, A. B et al., 2024).

The issues with the real-time tracking and monitoring were also another major challenge. With no constant visibility, even a small disruption in the form of a vehicle breakdown, traffic jam, weather and supply delays could spread across the entire supply chain leading to major losses in operations. In addition, organizations experienced challenges in the management of risks and vulnerability assessment of the supply chain networks spread geographically. Natural disasters, political instability, infrastructure breakdowns, and market changes would have a drastic effect on the operations of logistics in case they are not expected and planned. The lack of sophisticated analytical software was also detrimental to making decisions based on the data. The conventional systems hardly had predictive insights or spatial analytics features, which constrained the capacity of supply chain managers to anticipate new challenges or opportunities.

The demand of the world to move into e-commerce, same-day delivery frameworks, and highly responsive distribution networks has exacerbated the failures of traditional supply chain strategies. Consumers are demanding quicker, open, and dependable deliveries, and enterprises are under the intense strain to streamline all their logistics business. These changing requirements explain why it is important to embrace new spatial technologies such as GIS in order to enhance visibility, efficiency, and strategic foresight.

1.4 Role of GIS in Enhancing Visibility, Efficiency, and Strategic Decision-Making

GIS has a disruptive advantage in supply chain and logistics management of having a single spatial perspective of all the nodes, flows and processes in the logistics ecosystem. First, GIS enhances visibility through real-time tracking of shipments, vehicles, inventory and warehouse performance on visual dashboards and geospatial mapping tools. The visibility helps organizations to detect anomalies, delays in following-ups, and forecast disruptions, and therefore increase reliability and customer satisfaction. Geofencing with the GIS will allow companies to get automatic notifications whenever shipment enters or leaves special areas, which will guarantee enhanced security and control over the logistics process.

Second, GIS enhances efficiency through optimal route planning, traffic conscious navigation and resource allocation models. These features decrease transportation expenses, fuel use, enhance delivery time, and raise the use of assets. GIS-based optimization algorithms assist companies to plan optimal delivery schedules, plan multimodal routes of transportation, and last-mile logistics more efficiently. GIS is also useful in warehousing operations in terms of spatial layout planning, demand clustering, and efficient location of distribution centers to support expanding markets.

Third, GIS enhances the strategic decision-making process because organizations are able to analyze long-term logistics strategies based on spatial analytics and scenario modeling. Decision-makers are able to assess market potential, distribution of customers, availability of infrastructure and supply chain risks in order to know where to place warehouses, how to structure the distribution chain and how to respond to environmental or geopolitical risks. Risk maps made using GIS can assist managers in determining the vulnerable nodes in the supply chain and designing contingency mechanisms that will reduce the disruptions.

Moreover, the combination of GIS and artificial intelligence, machine learning, big data analytics, and cloud computing increases its decision-support functions significantly. Combined with analytics, these technologies make it possible to perform predictive analytics, automatically detect anomalies, and forecast based on data. With the digital transformation gaining pace all over the world, the GIS contribution to the redesign of the supply chain environment is becoming more obvious.

2. FUNDAMENTALS OF GIS AND SPATIAL DATA IN SUPPLY CHAIN MANAGEMENT

2.1 Understanding GIS

Geographic Information Systems (GIS) is an integration of hardware, software, spatial data, analysis tools and visualization, which allow the gathering, storing, manipulating and interpreting of information that has a geographical reference. In its most fundamental sense, GIS combines both spatial and non-spatial data layers in order to assist organizations in comprehending patterns, relationships, and trends that would have been obscured in a traditional data format. With the help of the system, users can visualize geographic phenomena, model a logistics network, analyze the conditions of the environment, evaluate the interaction of space, and obtain actionable information in the form of maps, dashboards, and analytical reports. GIS can be of great benefit in the supply chain and logistics context since it helps convert complex data, including the location of customers, traffic routes, transportation routes, warehouse locations, market distributions, etc., into useful spatial intelligence.

GIS basic elements are such data acquisition technologies as GPS, IoT sensors, UAVs, RFID-based systems, and remote sensing satellites, all of which can help to gather spatial data in an accurate and real-time way. The GIS software element, allows processing, analysis, and visualization of this data with the help of a variety of tools, including geocoding, spatial queries, network analysis, clustering, and predictive modeling. Also, complex hardware and cloud-based servers guarantee the possibility to store and handle the massive amounts of data in an effective way. Human expertise also plays a vital role in GIS since it entails data analysts, spatial scientists and logistics managers who have the capability to read the spatial results and convert them to the operation plans. All these attributes make GIS a very significant tool in facilitating the processes of supply chain, where speed, accuracy and flexibility are paramount factors.

2.2 Types of Spatial Data and Their Relevance in Supply Chain Operations

GIS is based on spatial data which is further divided into three broad types, namely, raster data, vector data, and attribute data. Raster data is grid-based data which may be in the form of satellite imagery, aerial photography and remote sensing technologies. It can be applied in supply chain management to know about the conditions of the environment like land use patterns, weather variations, flood-prone areas, and terrain structures that can influence transportation and infrastructure decisions. Raster data is especially significant in long term planning and risk assessment because it gives a detailed and dynamic picture of geographic features.

Conversely, the vector data is a representation of spatial features as points, lines and polygons. The points can be warehouse position, customer drop off points, or vehicle position. Lines denote transportation lines, supply chain lines, and road lines. Polygons are used to refer to geographic boundaries like service areas, regional markets or warehouse catchment areas. The value of the vector data in the optimization of routes, scheduling of deliveries, the planning of the location of facilities, and the design of the network cannot be overstated due to the possibility of the logistics managers to model and analyse the transportation routes precisely. Attribute data is used to supplement the spatial data by giving more descriptive data about every feature. As an example, an attribute table may comprise data on the speed limit of roads, traffic volume, warehouse sizes, stocks or the time of delivery(Shobhana, N. 2024).

Other forms of spatial data in supply chains are real-time geospatial feeds based on IoT devices and GPS trackers that are attached to trucks, containers and pallets(Padmavathi, V., & Saminathan, R. 2023). This data plays a critical role in the dynamic routing, fleet tracking, shipment tracking, and predictive logistics. With the combination of real-time spatial data and analytics solutions, organizations will be able to adjust delivery plans in real-time, re-route deliveries due to traffic or weather issues, and detect inefficiencies which would be otherwise undetected. The combination of historical spatial datasets also improves the demand forecasting and seasonal planning which allows the companies to synchronize their production and distribution schedules with the spatial consumption profiles. Spatial data is therefore not just a technological input but a strategic asset that will help in improving decision making throughout the whole supply chain.

2.3 Integration of GIS with Enterprise Systems and Emerging Technologies

The real strength of GIS in the supply chain operations manifests itself when it is combined with other enterprise-level systems and the new digital technologies. Contemporary supply chains are now extremely dependent on the systems that include Enterprise Resource Planning (ERP), Warehouse Management Systems (WMS), and Transportation Management Systems (TMS) that produce enormous amounts of operational data. Once these systems are linked with GIS, organizations are able to have a single perspective of both spatial and non-spatial data, which allows them to coordinate, plan, and execute better. To illustrate, GIS can be used together with a TMS to enable logistics managers to automatically create optimal delivery routes depending on the location of orders, the state of traffic, and the availability of the fleet. On the same note, ERP systems can be integrated with GIS to aid in strategic planning as one decision dashboard of financial, inventory and spatial information is developed.

Recent technologies, like the Internet of Things (IoT), artificial intelligence (AI), machine learning (ML) and blockchain, are additional additions to the power of GIS(MAY, P et al., 2025). The IoT-based surveillance devices presented in cars, warehouses, and packages produce endless real-time spatial data directly to GIS platforms. This data can then be analyzed by AI and ML algorithms to predict the delays, estimate the delivery times, notice anomalies, and provide the optimal routes. Such smart systems can be used to automate logistics processes, decrease the number of manual tasks, and increase the speed of reacting to disruptions. In the meantime, blockchain guarantees secure, transparent, and tamper-proof data exchange within distributed systems of supply chains, which enhances the quality of spatial records and reduces fraud. GIS and these emerging technologies together form a very integrated and intelligent logistics ecosystem that can be used to make complex supply chain decisions.

Moreover, the scalability and accessibility to GIS is being transformed by cloud computing and edge analytics. Cloud-based GIS systems enable organizations to analyze very high volumes of data, perform real-time analytics, and share geospatial information with international teams at a low cost in terms of infrastructure investment. Edge analytics allows the processing of geographic data at the origin, e.g., in a delivery truck or a digital warehouse, to minimize the latency and increase the responsiveness of the operations(Abitha, R et al., 2026). Such technologies as augmented reality (AR), digital twins, and 3D mapping also start making an impact on the visualization of supply chains, as they enable a manager to virtually visit a warehouse, simulate logistics processes, and predict the future more successfully(Manoharan, J. S et al., 2025). All in all, the combination of GIS and the enterprise system and advanced technologies results in an effective decision-support system to improve the efficiency of operations and strategic planning.

2.4 Role of Spatial Analytics in Supply Chain Performance and Optimization

The spatial analytics is a core of GIS that offers superior analytical solutions of geographic patterns, modeling logistic systems, route optimization and supply chain performance analysis. Spatial analytics employs spatial clustering, heat mapping, network analysis, and spatial-statistical modeling techniques to identify trends that cannot be identified in traditional analytics. As an illustration, spatial clustering can be used to determine geographic areas with large customer demand so that warehouses can be placed more efficiently and delivery strategies can be targeted. Heat maps can be used to visualize areas of congestion, shipping delays, or high-demand areas to better allocate resources by the logistic managers(Sweeney, E. 2013).

One of the strongest spatial analytical tools applied in the optimization of supply chains is network analysis. It simulates transportation systems and assists in computing the shortest routes, the cheapest routes and the most efficient delivery routes. Various real-life limitations considered in this analysis include road speeds, traffic flows, delivery times, vehicle capacity and environmental factors. Last-mile logistics needs network analysis as companies have to find their way through complex urban networks to deliver services in time. It is also possible to analyze scenarios and do predictive modeling through spatial analytics to make sure that companies can predict the future disruptions and evaluate the alternative logistics options. Spatial analytics will also allow businesses to develop resilient supply chains that will operate in the face of uncertainty by simulating the impact of an occurrence such as a road closure, storm or demand spike.

3. GIS FOR SUPPLY CHAIN VISIBILITY AND TRACKING

3.1 Enhancing Real-Time Shipment and Fleet Visibility

The operational excellence, customer satisfaction, and risk reduction of the modern supply chain and logistics environment are essential requirements, which presuppose the real-time visibility. Geographic Information Systems (GIS) plays a significant role in providing the end-to-end visibility of integrating real-time location of GPS device, IoT sensors, RFID systems, and telematics attached to trucks, shipping containers, and other delivery assets. The opportunity to monitor the shipments along the whole logistics chain allows the organizations to monitor the performance and detect the anomalies and respond quickly to disruptions. This information is visualizable in interactive maps and dashboards on GIS platforms where the logistics managers will be aware of the whereabouts of all the vehicles, the route of the vehicle, and whether it is late or early. This dynamic monitoring reduces the uncertainties and the responsiveness is much enhanced.

Besides keeping track of the fleet, GIS provides a clear insight of the past direction of assets, the trend of deviation, the time when the assets are idle, and the potential locations of a bottleneck. Using this information, the organizations can know the systemic inefficiencies such as delays at some checkpoints at all times, congestible routes, or unsafe areas. Such understandings can allow supply chain teams to act proactively to change routing plans, dispatch back-up trucks, or coordinate more effectively with warehouses and distribution centers. GIS also provides real-time visibility that enables the businesses to update the status of shipments to the customers. The approximate delivery times (ETAs) are more precise, and this benefits the customers and removes additional pressures on the support teams that used to process the queries manually. GIS cannot be done without in the world of global trade, as the shipments are characterized by multiple borders and movement across multimodal networks of complex multimodal networks to guarantee transparency and control across the system of logistics.

Besides, the GIS-enabled visibility is necessary in cold-chain logistics, which are implemented in pharmaceuticals, food, and sensitive chemicals. By applying IoT sensors, it is possible to monitor not only the place but also such environmental parameters as temperature, humidity, shock, and vibration with the help of GIS. The system can issue auto notifications in case of any violation of the required conditions to save losses in terms of quality and money. Thus, the real-time accessibility of GIS can be considered a priceless cornerstone to the honesty of products, consumer trust and dependability of supply chain divisions.

3.2 Geofencing, Monitoring, and Exception Management

Geofencing is good GIS-based feature that enhances supply chain management by creating virtual geographic fences around some points such as warehouses, ports, delivery point, restricted zones and high-risk areas. Automated alerts are also received when going in or out of these preset areas, which will enable the logistics manager to act immediately on the deviations, risks, or delays. This type of monitoring forms a dynamic form of control that was impossible in the previous system of logistics. To cite an instance, in an instance where a truck which carries very important goods makes a wrong turn or enters a restricted area, GIS systems are capable of sending instant notification to the involved personnel to either continue with investigations or take some corrective actions. These are particularly significant to prevent theft, improper manipulation, or path manipulation.

The geofencing made possible by the GIS can facilitate performance tracking, as it ensures that the deliveries are made following the schedule, the goods and the offloading, and the routes are taken. Geofence triggers can be employed in automating the check-in/check-out functions when vehicles enter or exit the warehouse premises to reduce the administrative burden and increase the precision of operations. Geofencing assists in the management of yards in large-scale distribution operations, including assistance with locating incoming trailers as well as controlling gate traffic and facilitating dock assignments. This avoids congestion, enhances safety and efficiency of throughput.

Another important area where GIS can be of great importance is exception management. Road closures, traffic congestions, weather and mechanical problems or unforeseen delays in check points often disrupt logistics operations. Exceptions can be detected automatically by GIS systems using real-time sources of data, including traffic feeds, weather forecasts, or incident reports, and alternative routes or contingency plans suggested. As an example, GIS can suggest safer paths in case of heavy rain, storms, etc., that do not pass through flooded areas or landslides. This minimizes the loss of cargo and time wastage and breakdown of services.

Furthermore, GIS helps in the management of compliance by monitoring compliance to regulatory measures like the limited driving hours, crossing the border, and the use of the toll roads policies. Geofence-based alerts can be automated to ensure safety standards because they alert drivers to reduce their speed when driving in the areas with the greatest number of accidents or adhere to the speed limits in the cities. Geofencing is invaluable to the safety and compliance with regulations in the case of such industries as oil and gas logistics, where hazardous materials have to be delivered along the strictly determined routes. Altogether, geofencing and exception control help enhance the level of reliability, operational safety, and mitigate the risks in intricate supply chain networks.

3.3 Warehouse, Inventory, and Facility Mapping

Although the concept of GIS is usually linked to transportation and routing, its use goes deep into the management of warehouses and inventory. The spatial mapping of warehouse layout assists organizations to visualize storage areas, picking areas, equipment placement, aisles and flow of operations. GIS helps to optimize the processes of the warehouse, minimizes the time spent on traveling by employees, and increases efficiency in material handling because it can generate digital images of warehouse premises. Inefficiencies that can be identified through spatial analysis can be congested areas, poor inventory distribution or inefficient picking paths. This will enable warehouse managers to restructure storage patterns, implement automation and design the best slotting patterns to enhance operational productivity.

Visualization through GIS is also useful in inventory management. By connecting the inventory information with geographical information, organizations can have an insight into where the stock is kept in various warehouses, regional hubs, and international distribution centers. GIS dashboards enable managers to track the inventory, safety stock levels, replenishment, and changes in demand in geographic markets. Such spatial visibility also makes sure that the right products are placed in the right places at the right time. In case of disruption, i.e. sudden surge in demand, delay of suppliers or closing of facilities, GIS maps are used to assist managers to reassign inventory effectively by locating the closest warehouses that have some available stock and estimating the most effective transfer routes.

Facility location planning is another area of importance of GIS. Locational choices on where to open new warehouses, cross-docking centers, retail stores or manufacturing facilities are very much reliant on spatial variables of customer distributions, accessibility to transportation, availability of land, labor markets, and local regulations. GIS assists in making such decisions with the help of spatial analytics tools which help to consider several criteria at the same time. GIS is used to model transportation network, demographic patterns, and service coverage areas to ensure that organizations can identify the best location of facilities to reduce cost and improve service performance. This is particularly applicable in industries where demand patterns are changing so fast like e-commerce where the distance to the customers plays a vital role in delivery time and competitiveness.

3.4 Integration of IoT, Telematics, and Advanced Tracking Technologies

GIS, IoT, telematics, and sensor-based technologies have changed the landscape of the supply chain management in terms of visibility and tracking. The IoT devices fitted in trucks, shipping containers, pallets, and even products themselves will continuously communicate its location, movement, condition, and surroundings. This information as well as GIS platforms will be a good source of real-time intelligence. Sensors can also be placed in vehicles to provide information on the speed, fuel consumption, driver behavior, engine performance, and compliance with route that can provide a holistic view of the fleet operations. These data streams are mapped by GIS, and based on this data, logistic managers can calculate the efficiency of the fleet, introduce safety regulations, and reduce the cost of the operations.

The RFID and barcode are also developed to an advanced level, which makes traceability even more effective, as the technology helps to identify goods in the warehouses, distribution centres, and transport centres automatically. GIS combines these indicators to trace the movement of items produced to the end consumer. Such granularity is necessary in industries with a high degree of responsibility like pharmaceuticals, luxury goods, and food supply chains where traceability and transparency are required to satisfy the regulatory and consumer demands.

Also, there are satellite-based tracking systems, drones, and autonomous sensors, which add to the improved visibility of remote, rural, or cross-border logistics. The GIS platforms process and visualize these sources of data to monitor shipments in areas where the conventional telematics infrastructure might not be effective. With the development of artificial intelligence and machine learning, now the GIS-based tracking systems are able to predict the arrival time, predict delays and detect anomalies with a high level of accuracy(Venketbabu, T et al., 2026).

4. GIS-BASED OPTIMIZATION TECHNIQUES FOR LOGISTICS EFFICIENCY

4.1 Route Optimization and Vehicle Scheduling

The most established example of GIS use in logistics and transportation management is route optimization. The traditional logistics had a habit of preplanning routes based on the static map and manual judgment that could not easily capture any real time disruption like traffic congestion, construction, weather hazards or even unexpected road closures. GIS changes this process by allowing a dynamic route planning under extensive spatial data input. GIS assists organizations to find the most efficient routes to use between origin and destination nodes through the use of advanced network analysis tools and algorithms such as shortest-path computation, vehicle-routing problem (VRP) solvers, and geospatial heuristics. These are the optimized routes based on various constraints including travel time, distance, fuel used, vehicle capacity, delivery time limits and toll charges.

Route optimization is an important part of the last-mile delivery operations in modern supply chains, particularly in e-commerce, food delivery, pharmaceutical, and fast-moving consumer goods (FMCG) industries. GIS based systems will create accurate delivery routes and prioritize the stops in order of proximity, traffic density and customer urgency. This does not only save time of travelling but also minimizes failures in delivery and enhances on-time performance. IoT sensors, GPS modules, and telematics systems provide real-time data which is integrated into GIS systems and allows making immediate corrections during the delivery process. In case the delivery vehicle is faced by a roadblock or congestion, the GIS will automatically recalculate alternative routes and the reliability of the services will not be compromised.

GIS also improves vehicle scheduling which is part of route optimization. Scheduling is about deciding the number of vehicles to be used, the time at which they are to be sent out, and the distribution of their workloads. GIS applications are used to study the demand patterns, the density of customers, and the geographic clusters to distribute fleets effectively. This will prevent vehicle underutilization or overloading and will also minimize the cost of operation. GIS-assisted vehicle scheduling also enhances energy efficiency through the reduction of idle time, and optimization of route mileage, and the ability to drive in an eco-friendly manner. The GIS-based vehicle scheduling can provide a significant advantage to the logistics companies concerned with sustainability by lowering carbon emissions and enhancing the load balancing.

Also, GIS facilitates multimodal transportation planning by combining road networks and rail, air, and sea routes. The visualization of the intermodal connections will allow logistics managers to assess cost-effective and time-efficient combinations to use in long-distance shipments. Considering the example, GIS can be used to determine that it is better to transport freight partially by rail and partially by road than to transport it by a single means of transportation. Such optimisation can enhance the robustness and flexibility of supply chains operating in a wide geographical location very much.

4.2 Last-Mile Delivery Optimization and Urban Logistics

The last-mile delivery is one of the most difficult and expensive phases of the logistics process. It comprises an almost half of total supply chain expenditure and is highly sensitive to spatial complexity and urban congestion as well as customer demands of fast delivery. GIS offers the necessary solutions to the analysis of urban settings, the definition of delivery hotspots, and the development of efficient last-mile networks. GIS assists organizations to comprehend customer distribution trends and the optimal route planning in high density urban areas through geospatial clustering, density mapping, and spatial-demand forecasting. As an example, heat maps created on the basis of GIS can help companies to determine the areas of high delivery and deploy resources in the most effective way at the time of high demand.

GIS also aids in the delivery zone segmentation where cities are subdivided into manageable clusters that are in line with the vehicle capacity and the delivery schedules. This will provide segmentation which keeps the drivers within the expected range and saves on the time taken to travel and increases the predictability of delivery. Moreover, GIS would allow delivering routes dynamically depending on the current traffic conditions, incidents on the roadsides, and weather information. It can be used especially in the field of urban logistics where traffic congestion can cause significant differences in delivery performance.

The other significant use of GIS in last-mile logistics is optimization of delivery points using alternative delivery models, including lockers, pickup points, or micro-distribution centers. In GIS, appropriate sites in which these delivery hubs will be situated are assessed on the grounds of access, proximity to the customers, pedestrian traffic, and availability of land. This cuts down on the door-to-door deliveries, reduces the cost of transportation and improves the performance. An example is e-commerce firms who may utilize GIS to decide on the best sites to locate micro-hubs that minimize on the time taken in delivery since the hubs are located nearer to households.

GIS also helps in sustainability of urban logistics. GIS allows planning environmentally-friendly last-mile delivery plans by examining emission zones, low-traffic routes, and green routes. Spatial analysis can be used to map and implement electric cars, bike couriers, and pedestrian delivery paths to reduce carbon emissions and adhere to environmental standards. Additionally, GIS can help create the resilience of the last-mile network that can respond to the disruption of infrastructure failures, urban planning limitations, or abrupt demand spikes.

Overall, GIS contributes to the optimization of the last-mile delivery to reduce the divide between the complexity of urban landscape and the necessity to deliver the goods quickly, efficiently, and sustainably in the contemporary logistic systems.

Figure 1. GIS-Based Optimization Techniques for Logistics Efficiency

4.3 Network Redesign and Facility Location Optimization

An effective supply chain is based on an effective network of warehouses, distribution centers, cross-docking centers, retail shops, and transportation centers. GIS offers effective analytical capability of assessing, redesigning and optimizing this network through spatial data. A strategic choice of the best location to establish new facilities is one of the most important strategic choices in supply chain design. This choice has implications on distribution cost, service levels, customer accessibility and long term scalability. GIS assists in the assessment of these parameters by incorporating demographic information, road system, land prices, area demand and competitor position into one analytical system.

GIS evaluates various variables through spatial suitability analysis which includes being near highways, population centers, suppliers, ports, and airports. This analysis will be used to determine the most favorable places where warehouses or distribution centers should be established. As an illustration, GIS can identify areas that reduce transportation distance and maximize the target markets service coverage. Moreover, when organizations are geographically enlarged or reach new markets as well as in case they want to improve operational efficiency, network redesign is necessary.

GIS can help this process with showing the existing networks, identifying the weak points and experimenting with configurations.

Network flow analysis is also another significant GIS feature. GIS can be applied in determining the bottlenecks, capacity and high cost nodes, through modeling the flow of products in the supply chain network. To understand the effect of various network conditions spatial algorithms can be used to create the new condition and calculate its effect on cost and service performance. This assists organizations to make wise decisions regarding the consolidation of warehouses, decentralization or distribution or a hybrid approach to warehousing basing on the regional attributes.

The cross-border optimization can also be applied using GIS to the world of logistics, in which regulations, the time of clearance at the customs, and the risk of geopolitics should be taken into account. The spatial mapping of these factors would enable organizations to choose safer and more efficient paths, eliminate risky areas, and become more resilient. In addition, in reverse logistics, including product returns, recycling and waste management, GIS assists in optimal collection paths and finding economical processing sites.

Therefore, GIS-based network redesign and optimization of facilities can assist organisations in reconfiguring their supply chains and lowering the operating expenses and creating scalable, future-fit logistics networks.

4.4 Traffic-Aware Routing, Predictive Analytics, and Cost Reduction

One of the most unpredictable problems in transportation logistics is the traffic conditions. GIS can be used to provide traffic-prone routing by incorporating real-time information provided by sensors, traffic surveillance systems, vehicle telematics, and community traffic feeds. This assists in locating the hotspots of congestion, construction areas, accidents and disruptions caused by weather conditions. GIS can be used to make delivery routes efficient in changing conditions by using this data in routing algorithms. This is especially useful when it is on peak time, when there is a festival, or when there is an emergency and traffic patterns turn out to be very high or too low.

In predictive logistics, GIS is also important besides real-time routing. GIS allows predicting the future and prescribing proactive changes based on the analysis of historical traffic flows, seasonality, and spatial behavior. An example is that, GIS can be used to forecast areas that will be congested during shopping seasons or be able to forecast routes that are likely to be affected by monsoon flooding. With this knowledge, logistics planners can change the fleet implementation strategies, deliveries can be made in advance or alternative routes selected to prevent possible delays.

Cost wise GIS is very much cost effective in terms of reduced fuel, wear and tear of a vehicle, labor hours of drivers, and operation overhead. The use of optimised routing and scheduling minimizes miles of unnecessary travel, which causes fruitless movement resulting in increased logistics expenses. In the case of companies that operate large fleets, a minor increase in the efficiency of routing can translate to a significant amount of savings. GIS is also useful in determining cheap modes of transportation in the multimodal logistics by comparing the time of transit, fuel use and the tariffs between the rail, road, air, and sea networks.

Moreover, GIS helps implement sustainability-oriented policies through the minimization of greenhouse gas emissions and enhanced energy efficiency of a fleet. GIS tools that can compare the carbon footprint, assess eco-friendly routes and model energy consumption patterns are beneficial to companies that aim to go green in their logistics. These insights affect organizational choices of transitioning to the use of electric fleets, maximizing the load capacity, and designing emission-free delivery areas.

On the whole, GIS-based predictive analytics and traffic-sensitive route help to optimize the logistics and make it highly cost-effective, contributing to the preservation of the competitive edge of organizations in a competitive and challenging environment.

5. GIS FOR RISK ASSESSMENT AND STRATEGIC DECISION-MAKING

5.1 Mapping Supply Chain Risks and Vulnerabilities

Supply chains are prone to a high number of risks in a globalized world whereby natural disasters, political instabilities, infrastructure failures and demand variations are just a few of the threats. Geographic Information Systems (GIS) offer an effective framework that can be used to determine, map, and assess these risks spatially. Conventional approaches to risk assessment do not take into account the geographic location where supply chain networks are functioning. Nevertheless, numerous disturbances, e.g. earthquakes, floods, extreme weather conditions, pandemics, labor strikes, congestion of ports, and geopolitical tensions, have obvious spatial aspects. GIS assists organisations to see the geographical location of these risks and therefore managers can know how susceptible their supply chain hubs and transportation networks are to certain risks.

GIS-based risk maps can enable companies to analyze how warehouses, manufacturing facilities, ports, transportation routes, and supplier locations are exposed to all types of threats. As an example, the floodplain data can be overlaid on the warehouse locations to identify which of the facilities is highly vulnerable during the monsoon seasons. Likewise, analyzing political instability or trade barriers across international borders assists businesses to know in advance the delays, tariffs, or regulatory hurdles. The analysis of congestion and accident-prone areas in the city as well as road closures using GIS can assist logistic teams develop more effective and secure routes to transport goods. Such visualizations can offer an overall risk picture so that the supply chain workers can focus on mitigation strategies by their severity and probability.

The ability of GIS to combine several layers of spatial data, including weather forecasts, hazard areas, infrastructure, and socio-economic indicators, to provide multidimensional risk information, is one of the greatest GIS advantages in risk assessment. This will allow managers to study the cascading risks where a single failure will result in a cascade of interconnected failures along the supply chain. In one instance, GIS can simulate the impact of the closure of a port in a hurricane on the arrival of shipments, intermodal transport, and stock-out situations in downstream centers. This interdependent knowledge aids organizations to build more resilient supply chains through diversifying the sourcing sites, enhancing the buffer stock policies, and investing in risk conscious infrastructure planning.

5.2 Decision Support Systems for Strategic Logistics Planning

Supply chain management is a long-term strategic decision-making process that includes the decision-making processes of network design, facility location, distribution processes, resource allocation. GIS-based decision support systems (DSS) are analytical, visual and predictive tools which aid in making informed and data-driven decisions by the leaders. These systems integrate spatial data with business measures- cost, demand, service level and capacity to form comprehensive models which assess the trade-offs between various strategic choices. With DSS platforms, which are GIS-based, organizations can model various scenarios of what-if to determine the effects of strategic decisions on the overall performance.

Indicatively, GIS can be used by companies that are intending to venture into new markets to analyze geographic demand patterns, population density, competitor location, transportation infrastructure, as well as socio-economic indicators. These insights are used to determine good sites to place new warehouses, retail outlets or distribution centers. GIS gives access to the coverage area of the services and therefore the managers can analyze the aspects like delivery time, accessibility and closeness to the important customers. GIS is useful in modeling customer catchment areas and transportation networks to ensure that companies optimise their facility location to reduce the costs of logistics and maximise on the efficiency of their services.

Decision support systems based on GIS also enable network redesign where organizations have to reorganize operations as a result of merger, market changes, or sustainability goals. Spatial network analysis allows managers to investigate the possibilities of consolidation of current facilities, the introduction of cross-docking centers, or the redesigning of regional distribution models. The system enables one to compare network scenarios in terms of key performance indicators (KPIs) like the reduction of costs, better lead times, or a decrease in carbon emissions. This process will see to it that the decisions are based on data-supported spatial insights and not intuition.

Also, GIS helps in making procurement decisions by analyzing the location, transportation routes and risks of suppliers. The firms are able to assess the performance of their suppliers based on geographic location, connectivity of infrastructure, and hazard exposure. This makes certain that sourcing plans are in line with resilience objectives and cost-effectiveness objectives. In commodity mining industries or agricultural industries like mining, oil and gas, fisheries, and agribusiness, GIS aids in long-term resource planning, evaluating on land availability, environmental factors, and supply potential. All in all, GIS-based DSS systems enhance the quality of decisions, minimize uncertainties and optimize strategic planning through the entire supply chain ecosystem.

5.3 Geospatial Analysis for Market Expansion and Demand Forecasting

The knowledge of the customer demand patterns is a pillar of effective supply chain management. GIS can improve demand prediction and market growth strategy by offering spatially motivated information about the behavior of consumers, economic performance, and population. The conventional forecasting system is highly based on the past sales records without considering the geographical aspect fully. Nevertheless, the demand is usually geographically uneven based on the population density, distribution of income, weather variations, accessibility, preferences and cultures. GIS incorporates this aspect of space in the forecasting models and results in improved predictions and a closer match-up of supply and demand.

GIS can determine geographic areas with potential to buy products through geospatial clustering or underserved markets. The thematic layers and heat maps show geographic concentrations of customers and companies can customize their marketing and distribution strategies based on this information. As an example, retail firms are able to identify the best store locations considering the foot traffic, residential population and easy access of transport. The e-commerce companies can determine the areas with increased demand to locate the micro-fulfilment centers strategically to make deliveries quicker. Manufacturers are able to study the demand that is geographically dispersed to match the production quantities with the geographical demands of their consumers.

Along with the enhancement of current demand forecasting, GIS enhances the long-term strategic planning and helps analyze the trends of population growth, urbanization, and infrastructure development. As an illustration, GIS will be able to identify the new urban corridors or industrial areas that may witness more logistics activity in the next few years. Firms may deploy these insights by investing in new distribution centers, joint logistics systems or joint planning activities.

GIS is also significant in the predictive demand patterns due to seasonal or climate variations. Such industries as energy distribution, agriculture, or food provision are sensitive to demand, which tends to vary depending on the weather. GIS is a combination of environmental data of temperature, rainfall, drought conditions, and soil quality to assist predictive analytics models. This helps organizations to change their supply chain strategies proactively to ensure that the inventories, transportation capacity and manufacturing strategies are aligned with the anticipated shifts in geographical needs.

6. APPLICATIONS, AND EMERGING TECHNOLOGIES

6.1 Real-World Applications of GIS in Logistics and Supply Chain Operations

GIS has emerged as a very important enabler in a wide variety of logistics and supply chain applications, both in strategic planning and in the day-to-day operations. Transportation management is one of the most frequent uses of GIS, as it is used to optimize routes, track fleets, and to improve the accuracy of deliveries. Organizations that deal with e-commerce, retail, manufacturing, and third-party logistics use GIS to operate extensive distribution channels. GIS applications combine real time traffic, weather reports and road conditions to direct a driver on the most efficient route and lower transportation expenses and enhance service reliability. Moreover, GIS assists in the activities of the warehouse by means of spatial planning of warehouses, the optimal route of picking, and efficient use of space. The applications assist in streamlining operations and enhancing turnaround times at the distribution centers.

The other significant use of GIS is in supply chain visibility and control towers. The contemporary logistics activities produce enormous volumes of real-time information in terms of IoT sensors, RFID tags, and satellite-tracking systems. GIS has offered a graphic layer that brings all this together in a format that the managers can understand and this will enable them to track deliveries within the global supply chain networks. This visibility helps in making quicker decisions and thus organizations can respond to delays, disruptions, or deviations very fast. The cold chain management in industries like the pharmaceutical industry relies greatly on GIS-based monitoring where the products that are sensitive to temperature have to be shipped under very strict environmental regulations. GIS assists in maintaining integrity of the products by mapping temperature changes, risk zones and also sending warnings when the temperature surpasses a limit.

GIS is also extensively used in strategic planning like optimization of the facility location and network design. GIS is applied by organizations to determine the population density, distribution of customers, availability of infrastructure, land prices and environmental factors, to make sound decisions on where they should set up new warehouses, distribution centres or retail stores. GIS provides the capacity of companies to consider a variety of scenarios prior to making long-term investments because it incorporates demographic, economic, and environmental data. It is especially useful in the situation of entering new territories or restructuring the current supply chains networks to enhance cost-effectiveness or access new markets.

Also, GIS is used in emergent logistics and humanitarian missions. In times of natural disasters or crises, emergency response units use GIS to determine the damage and map transportation obstacles and the allocation of relief. GIS can reveal the spatial information related to the most acute areas requiring supplies and finding the safest paths to transport goods and people.

6.2 Emerging Technologies Integrated with GIS

The future of GIS in the optimization of supply chains is predetermined by emerging technologies such as artificial intelligence (AI), machine learning (ML), the Internet of Things (IoT), digital twins, blockchain, and advanced remote sensing(Sathya, U et al., 2024). The technologies represent a great addition to the GIS functions, and more accurate predictions, real-time information, and autonomous decision-making are possible.

AI and ML also bring about predictive features of GIS systems. The AI-based GIS models can forecast demand, estimate the delivery time, and recommend the changes of the best routes with the assistance of historical data, patterns of traffic, trends of customers, and environmental conditions. Geospatial data anomalies (e.g., suspicious car traffic or unnatural temperature conditions) can be identified using machine learning algorithms, and proactive measures can be taken. GIS powered by AI also has the ability of scenario simulation which allows managers to analyze the consequences of different decisions prior to their implementation.

Real time geospatial intelligence is based on IoT integration. The IoT sensors in vehicles, containers, warehouses, and production plants present the continuous flow of data regarding the location, environment, equipment operation, and stock levels. This data is processed and visualized on GIS platforms and provides instant data on the efficiency of operations, safety and product integrity. GIS will become a completely interconnected system that is able to track all parts of the supply chain ecosystem as the number of IoT adopters increases.

The blockchain technology is complementary to GIS in that it provides transparency, security and trust in the data of the supply chain. The addition of geospatial events, such as shipment departures, handovers, and deliveries, to a blockchain ledger that lacks the ability to tamper with the information increases traceability and accountability in organizations. This is especially true of high value and controlled supply chains such as pharmaceuticals, luxury products and food products(Muduli, K et al., 2022; Muduli, K et al., 2023).

Digital twins are another technology that is transforming logistics. A digital twin is a digital representation of a physical system (a warehouse, a transport network, or an entire supply chain) in real-time. Integrating GIS and digital twins, the decision-makers can model the operation situations, discover the layouts, experiment with the new route models, and visualize the inefficiencies in the operation, before they occur in the real world. This functionality is very influential on the accuracy in forecasts, operational planning and system resilience.

6.3 Future Trends and Technological Directions in GIS-Enabled Logistics

As supply chain environment is dynamic, there are several future trends which imply that GIS will be even bigger contributor in development of logistics systems. Some of the trends include the autonomous logistics that incorporates autonomous vehicles, delivery drones, and warehouse robots, among others. GIS is also required in these systems in the areas of navigation, obstacle detection, path planning and geospatial awareness. The high-definition mapping will be based on GIS and will be used to operate autonomous delivery in urban and rural settings.

Geospatial big data analytics is another new trend. As the IoT sensors, satellite imagery, and real-time trackers become more widespread, logistics organizations can now deal with the large volumes of spatial data(Prasath, R et al., 2025; Jaffrin, L. C et al., 2025). GIS will become a big-data-oriented platform that will be able to analyze large quantities of data and identify latent patterns, anticipate changes in the market, and make logistics decisions on a scale previously unknown.

Innovations that are centered on sustainability will also shape the future of GIS. With the stricter regulations on emissions and environmental impact, GIS will facilitate the shift to green logistics by carbon footprint mapping, eco-route planning, and forecasting of environmental risks. Sustainability dashboards based on GIS will assist organizations to track their adherence to the environmental standards and find ways to improve them.

7. CONCLUSION

Geographic Information Systems (GIS) have become one of the core technologies to revolutionize current supply chain and logistics functions. Due to the increasing complexity, geographical dispersion and susceptibility of global supply chains to disruptions, the necessity to have accurate spatial intelligence has become critical. GIS offers a single platform that facilitates real-time geolocation data, environmental data, infrastructural data and demand trends and helps companies to visualize, analyze and optimize all levels of the supply chain. GIS can help to increase the efficiency of operations, reduce their costs, and increase the reliability of the services provided, whether it is the effectiveness of the route planning and final deliveries or warehouse administration and system design. It is also useful in strategic planning and long-term resilience building since it can map risks, assess vulnerabilities and aid in decision-making. It is even more influential when GIS is combined with the new technologies, such as AI, IoT, machine learning, blockchain, digital twins, and advanced remote sensing. The built in capabilities enable predictive analytics, autonomous decision-making, and real-time tracking, forming a smart logistic ecosystem that acts quickly to disruptions and new market trends. The e-commerce, retail, pharmaceutical, agricultural, and global logistics industrial examples indicate that GIS introduced real-life advantages like increased visibility, enhanced traceability, optimal route based on sustainability, and improved customer experience. The future of logistics will be dictated by GIS in the future as a result of the geospatial big data, cloud computing, and autonomous mobility application development. As the question of environmental sustainability and supply chain resilience gains the top priorities, GIS will be one of the main aspects that will assist the organizations in becoming a Green, Smart, and Strengthened Logistics Network. In summary, GIS is not a tool but a tool of strategy that helps in shaping the supply chain management to ensure improved visibility, efficiency and decision making in the world that is increasingly becoming dynamic.

REFERENCES

Abitha, R., Rajendar, S., Mangaiyarkarasi, D., Anto, G. L., & Girija, P. (2026). A Comprehensive Survey on Edge Computing, Analytics, and AI Model Optimization for Industrial and Healthcare Applications. *Building Data-Driven Edge Systems for Business Success*, 59-90.

Bentalha, B. (2023). A Framework for Exploring the Global Performance of Service Supply Chain Management: A Qualitative Study. *Government Impact on Sustainable and Responsible Supply Chain Management*, 106-129.

Dhamija, P., Bedi, M., & Gupta, M. L. (2021). Industry 4.0 and supply chain management: A methodological review. *Research Anthology on Cross-Industry Challenges of Industry*, *4*(0), 192–217. DOI: 10.4018/978-1-7998-8548-1.ch011

Jaffrin, L. C., Nithya, T., & Divya, J. (2025, April). IoT for Sustainable Development: Enhancing Environmental Efficiency and Resource Optimization. In *2025 5th International Conference on Trends in Material Science and Inventive Materials (ICTMIM)* (pp. 1224-1230). IEEE.

Manoharan, J. S., Arunkumar, M., Hemamalini, M., Sathya, V., Anish, T. P., & Subramanian, R. S. (2025). Virtual and Augmented Reality in the Automobile Industry: Applications and Future Directions. *Virtual and Augmented Reality Applications in the Automobile Industry*, 377-406.

MAY. P. M. A., Savitha, V., Gayathri, S., & Mahalakshmi, S. (2025, April). Exploring Sustainable Computing: IoT, Big Data, and Energy-Efficient Solutions. In *2025 5th International Conference on Trends in Material Science and Inventive Materials (ICTMIM)* (pp. 1218-1223). IEEE.

Moorthy, K. S., Balakrishnan, G., Kumar, S. S., Raja, L., & Vijayalakshmi, A. (2024). Embracing circular economy principles for sustainable green supply chain management in manufacturing industries. In *Convergence of Human Resources Technologies and Industry 5.0* (pp. 85–110). IGI Global Scientific Publishing. DOI: 10.4018/979-8-3693-1343-5.ch005

Movahed, A. B., Movahed, A. B., & Nozari, H. (2024). Opportunities and challenges of smart supply chain in Industry 5.0. *Information Logistics for Organizational Empowerment and Effective Supply Chain Management*, 108-138.

Muduli, K., & Barve, A. (2015). Analysis of critical activities for GSCM implementation in mining supply chains in India using fuzzy analytical hierarchy process. *International Journal of Business Excellence*, *8*(6), 767–797. DOI: 10.1504/IJBEX.2015.072309

Muduli, K., Luthra, S., Garza-Reyes, J. A., & Huisingh, D. (2023, January). Application of blockchain technology for addressing reverse logistics challenges: current status and future opportunities. Supply Chain Forum: An International Journal, 24(1), 1-6). DOI: 10.1080/16258312.2023.2165279

Muduli, K., Raut, R., Narkhede, B. E., & Shee, H. (2022). Blockchain technology for enhancing supply chain performance and reducing the threats arising from the COVID-19 pandemic. *Sustainability (Basel), 14*(6), 3290. DOI: 10.3390/su14063290

Padmavathi, V., & Saminathan, R. (2023). Fundamentals of IOT, artificial intelligence, and blockchain approaches for applications in Supply Chain Management. In *Blockchain, IoT, and AI Technologies for Supply Chain Management* (pp. 1-18). CRC Press. DOI: 10.1201/9781003264521-1

Prasath, R., Kumar, V. V., Jaisankari, V., & Sivasubramani, R. (2025). Semantic Search for IoT-Based Object Recognition Using Clustering and Classification. In *Utilizing AI in Network and Mobile Security for Threat Detection and Prevention* (pp. 151–172). IGI Global Scientific Publishing. DOI: 10.4018/979-8-3693-9919-4.ch009

Sathya, U., Hashini, S., Lakshmipriya, C., Varun, C. M., & Nalini, M. (2024, December). Emerging Trends and Future Prospects in Internet of Things (IoT) Technology. In *2024 International Conference on Sustainable Communication Networks and Application (ICSCNA)* (pp. 79-85). IEEE.

Shobhana, N. (2024). AI-powered supply chains towards greater efficiency. In *Complex AI dynamics and interactions in management* (pp. 229–249). IGI Global Scientific Publishing. DOI: 10.4018/979-8-3693-0712-0.ch011

Suganya, P., Siva, S. R., Ananthi, S. N., Thilagam, T., Elavarasi, J., Anto, G. L., & Girija, P. (2026). The role of artificial intelligence in transforming supply chain management. In *Transformative Impact of AI in Supply Chain Management* (pp. 271–294). IGI Global Scientific Publishing.

Sweeney, E. (2013). Supply chain integration: Challenges and solutions. In *Supply chain management: Concepts, methodologies, tools, and applications* (pp. 221–246). IGI Global Scientific Publishing. DOI: 10.4018/978-1-4666-2625-6.ch013

Venketbabu, T., Kennady, R., Saranya, R., Rashmi, P., & Anto, G. L. (2026). Enhancing Data Analytics and Forecasting Education Through Machine Learning Algorithms. In *Integrating AI and Machine Learning into Business and Management Education* (pp. 247–270). IGI Global Scientific Publishing.

Chapter 9
Understanding the Role of GIS in Packaged Marine Product Supply Chain

Samrat Majumdar
http://orcid.org/0009-0006-7831-6365
KIIT (Deemed to be) University, India

Deepak Singhal
http://orcid.org/0000-0002-1336-7819
KIIT (Deemed to be) University, India

Sushanta Tripathy
http://orcid.org/0000-0003-2470-4080
KIIT (Deemed to be) University, India

ABSTRACT

The supply chains of packaged marine products are very perishable, geographically distributed and are under strict regulatory and cold-chain parameters. Geographic Information Systems (GIS) have good potentials to help overcome such challenges but few empirical studies have identified and ranked the major factors that drive the adoption of GIS in packaged marine products supply chains. The focus of this research is to fill this gap by determining and prioritizing the essential factors of GIS adoption based on the consensus of the experts. To begin with, the possible drivers of GIS adoption were established by a thorough literature review of works on the marine supply chain, cold-chain logistics, traceability, and spatial decision support. The Fuzzy Delphi Method was used in the second stage to assess and rank these drivers by the expert opinions.The findings suggest cold-chain integrity and

DOI: 10.4018/979-8-3373-7941-8.ch009

temperature-risk control to be of the most high-priority driver of GIS adoption, then route optimization and transit time reduction and traceability, transparency, and recall preparedness

1. INTRODUCTION

The supply chain of packaged marine products refers to the flow of seafood such as fish, shrimp, crab, and processed marine foods from harvesting to the final consumer in packaged form. These supply chains are highly time-sensitive, quality-driven, and cold-chain dependent (Subramaniam et al., 2023; Kidane 2025). Moreover, supply chains of packaged marine products are quite perishable, decentralized geographically, and face intense regulatory and cold-chain demands. The capability to handle geographically dispersed risks, logistics activities, and compliance is increasingly important in making effective decisions by such systems.

Packaged marine product supply chains are some of the most operationally challenging food networks since they are perishable, require cold-chain maintenance, export documentation, and rising sustainability, and traceability demands. Marine products are usually packaged as frozen seafood, chilled processed fish, value-added ready-to-eat or ready-to-cook products and cans. These items pass through a chain of landing centers, aggregation, processing, cold warehouses, distribution, and domestic or export markets (Koirala et al., 2025).

One of the challenges posed by packaged marine products supply chain is that risk and performance have a spatial nature. The processing clusters, cold stores, landing centers, markets, and catch zones are scattered geographically. Time of transportation, the environment surrounding it, port accessibility, exposure to disasters, and enforcement have spatial variations. These spatial considerations have strong effects on retention of freshness, spoilage, logistics cost, service levels and compliance results (Stentiford et al., 2022). The capability of handling location-dependent decisions has influenced supply chain performance in the marine products industry over the past years. Poor route planning, congestion around ports, ineffective location of cold storage or exposure to high-risk areas can be the direct cause of quality deterioration, spoilage losses, increased logistics expenses, and failure to meet food safety and export regulations. Consequently, supply chains decision-makers need tools that can not only handle data relating to operations, but also directly use spatial intelligence in the planning, execution, and control operations.

Geographic Information Systems (GIS) offer an effective platform on how to overcome these challenges by providing the opportunity to collect, combine, analyze and visualize data related to location(Ullah et al., 2023; Billal et al., 2025). GIS enables players in the supply chain to know the location of activities, geographic relations affecting performance and why some locations or routes provide or pose operational risks and opportunities. Within the framework of packaged marine products, GIS facilitates key operations, including optimization of cold-chain route, planning of facility locations, transportation network, risk mapping, and spatial tracing of the supply chain nodes. GIS helps to increase visibility and decision-making in geographically disaggregated seafood supply networks by converting geographic information into actionable insights.

Despite the fact that literature acknowledges traceability and cold-chain performance importance, limited literature explicitly mentions and prioritizes the most important drivers behind the adoption of GIS within packaged marine products supply chains.This study will fill such a gap by determining and prioritizing the key factors leading to GIS adoption through the criterion of expert judgment. The following are the issues that are resolved in this research:

RQ1: What are the key factors influencing the GIS determination of packaged marine product supply chains?

RQ2: What are the most significant drivers among the identified drivers?

To begin with, the possible drivers of GIS adoption were established by a thorough literature review. The Fuzzy Delphi Method was used in the second stage to assess and rank these drivers by the expert opinions to enable uncertainty and subjectivity to be represented by the use of fuzzy logic. Fuzzy Delphi Method (FDM) is meant to receive expert agreement in a state of ambiguity by portraying the judgments using fuzzy numbers (Chang et al., 2000).

The Fuzzy Delphi Method (FDM) is a method of expert-based decision making that combines the fuzzy set theory to the classical version of the Delphi method in order to manage uncertainty as well as subjectivity in expert opinions. FDM allows reaching a consensus based on the results of linguistic assessments by representing them as fuzzy numbers and mathematically aggregation. This is especially useful in the screening and prioritization of factors in uncertain and complex decision environments.The application of this methodology has been common in prior research on a variety of areas including the supply chain management, sustainability measurement, technology adoption, risk measure, and strategic decision-making.

The paper is structured into five main sections. Section 2 reviews the literature, section 3 explains the methodology. Section 4 summarizes results, and discusses the potential. Section 4 also recommends research implications. Finally, section 5 concludes the paper.

2. LITERATURE REVIEW

Marine deliverable are characterized by a critical quality–time trade-off, where even minor variations in time and temperature can lead to rapid spoilage, deterioration in product quality, and significant economic losses. The cold chain logistics literature notes that academic and policy literature is growing in interest of the seafood cold chain, particularly with regards to optimization, risk mitigation, and technology application (Hu et al., 2024). Furthermore, it has been suggested in the literature on food cold chain logistics that integrated decisions in storage, routing, infrastructure and monitoring contribute to operational performance (Mustafa et al., 2024).

The necessity to trace is becoming mandatory to show legality, sustainability, food safety and moral sourcing in seafood supply chains. One of the biggest problems is that the end-to-end traceability must always be ensured, and the loss of data at critical processing and distributions nodes should be minimized (Hopkins et al., 2024). Moreover, the interoperability at a scale among digital traceability solutions is still underrepresented because of the absence of motivation, capacity shortages, and technical data management concerns (Hardt et al., 2017). Seafood traceability has a means larger role than food safety, as it is part of IUU fishing deterrence, sustainability obligations, and due diligence expectations (Lewis and Boyle, 2017). The central concerns are also of the the origin certification and fraud prevention. The needs and constraints of the seafood origin traceability have been summarized and focused on the limitations in the real-world application and the significance of tools, which can be used to facilitate the certification (Leal et al., 2015).

IoT/RFID and blockchain are becoming one of the major aspects of supply chain digitalization. The RFID-based traceability systems are also highlighted due to their ability to provide production and consumption monitoring, as well as increase the level of trust (Muduli et al., 2022; Swain et al., 2021). Another conceptual overview also points to blockchain-IoT frameworks as tools that can be used to enhance traceability and authenticate or deny authenticity in the stages of fish supply chains (Rahman et al., 2021; Ismail et al., 2023). Nevertheless, most of these technologies demand a spatial context: where the product was, where something has happened, how far it has travelled and what are the geographic dangers it has faced. GIS is exactly the layer of spatial intelligence and it can be combined with traceability data to streamline compliance and performance management.

GIS helps in decision making in the areas of routing, design of distribution networks, site selection, risk mapping and monitoring. As an example, the analysis of the location of a logistics distribution center based on the spatial data with GIS has been modelled on the hybrid metaheuristics approach, underlining that geographic information makes a significant contribution to the logistics facility decisions

(Khairunissa and Lee, 2022). Surveillance and risk analytics on aquaculture are also being performed with increased use of GIS (Dorotea et al., 2023). Maritime logistics Maritime supply chain efficiency and resilience In maritime logistics, GIS-based techniques are applied to the efficiency and resilience of the supply chain; a GIS-based technique, for example, has been suggested to enhance the practice of maritime supply chain management (Carmona et al., 2023). This type of work reinforces the argument that marine-related supply chains can be enhanced with the help of GIS integration because of multi-moding connectivity and spatial limitations.

On top of mapping, GIS can be useful when combined with big data (temperature logs, ERP transactions, vessel/vehicle GPS, remote sensing, regulatory boundaries). Big data in food supply chains is linked to better decision support, accurate forecasting, minimal wastes, and food safety, even though data quality and integration continue to be challenging (Rejeb et al., 2022). GIS can be the effective practical ambiance of a multifarious spatially grounded datasets.The developing contexts evidence has indicated that technology is a threshold to resilience and transparency. Satellite images and blockchain have been implemented in a fish supply chain case scenario in India to enhance traceability and resiliency and decrease the intermediary-based inefficiencies (Sengupta et al., 2022). On the micro and small enterprise level, QR-traceability is being suggested to marine dry fish supply chains in India, showing growing demands on digitized and trackable flows even in small players.

Geographic Information Systems (GIS) can provide great opportunities to overcome all these issues, but very few empirical studies have identified and ranked the most important forces that are driving the adoption of GIS in packaged marine product supply chains. To deal with this gap, the current research will seek to establish and prioritize the most significant drivers of GIS adoption in packaged marine product supply chains by conducting a literature review alongside expert-based appraisal to identify the most important drivers. The research uses the Fuzzy Delphi Method to reach a consensus of the experts and rank the drivers according to their perceived value.

2.1.Influencing Factors

2.1.1. Export Certification and Regulatory Compliance

One of the root causes of adopting GIS in packaged marine products supply chains is regulatory compliance. Export of seafood is faced with a complicated regulatory system on food safety, legality of harvest, traceability and sustained reporting. The demands of the international markets are that of spatially verifiable records of the fishing sites, aquaculture areas, authorized landing sites, and certified processing plants. Failure to prove compliance would normally lead to rejection of shipment and loss of right to export.

Through GIS, spatial documentation and visualization of regulatory compliance can be established through the establishment of a correlation between geographic information and operational and traceability records. This type of space verification is especially applied in the field of the origin certification and fraud prevention in the seafood trade as stressed by Leal et al. (2015). GIS can be used as a strategic compliance tool instead of the purely operational technology by enhancing audit preparedness and regulatory transparency.

2.1.2. Risk Mapping

There are various risks present to marine supply chains, which are different in terms of location and supply chain tools. They are environmental risks, outbreaks of diseases in areas of aquaculture, infrastructure-induced bottlenecks, congestion at the ports and temperature spoilage. These risks are spatially distributed, so they are not always represented in geographic specificity using traditional risk assessment techniques.

GIS helps in the spatial risk mapping by synthesizing environmental data, operational data, and logistical data. Risk-prone areas visualization and analysis will contribute to proactive mitigation measures, especially in cold-chain and processing processes. The use of GIS-driven surveillance and risk evaluation systems has become more established in aquaculture and seafood systems (Dorotea et al., 2023), which further supports the importance of spatial intelligence in managing the risk of the marine supply chain.

2.1.3. Multi-Actor Multi-Actor Integration and Coordination

The supply chain of packaged marine products incorporates a broad scope of actors that are geographically separated, which comprise harvesters, landing centers, processors, cold-storage operators, logistics providers, exporters, and regulatory agencies. The fact is that coordination between these actors is not always facilitated by fragmented information systems and inconsistent data standards.

GIS offers a spatial platform that can be used to integrate information between multiple actors and supply chain phases. GIS improves transparency and coordination across the organizational boundaries by connecting location based data with digital traceability systems. The problem of interoperability in the seafood traceability system, as it is mentioned by Hardt et al. (2017), further contributes to the necessity of spatially integrated information environments that can be used to bridge disjointed digital solutions.

2.1.4. Optimization of Route and Minimization of Transit Time

The transit time is decisive in the quality of seafood, shelf life and market value. Losses in transit time lead to more exposure to temperature variation and handling exposure, which can be in the form of loss of quality and fines. The conventional route planning processes are generally based on the assumption of stasis and fail to consider spatial constraints in the real-time.

GIS also is a way to optimize routes further by integrating road networks, time to travel, traffic congestions and environmental factors into the logistics planning. Spatial optimization of routing has been demonstrated to increase efficiency and resilience of maritime and food-related logistics systems (Carmona et al., 2023). GIS can contribute to cold-chain reliability by minimizing transit time and uncertainty, and thus it is significant in an adoption driver.

2.1.5. Demand Hotspot Mapping and Market Expansion

The consumption behavior, income, and urbanization vary between regions and this has led to unequal demand of packaged marine products. Firms might find it difficult to match distribution strategies with the market realities without spatial insight into the pattern of demand.

GIS allows demand hotspots mapping through the integration of geographic, demographic, and sales. This kind of spatial analysis will enable firms to determine the existence of high demand clusters and new markets. The increase in the research attention to seafood cold-chain logistics and market distribution patterns contributes to the significance of the analysis of the spatial demand (Hu et al., 2024). Therefore, the ability to map demand would make firms use GIS in strategic planning of the market.

2.1.6. Supplier and Landing-Centre Network Planning

The sourcing of seafood is a location-specific activity because raw materials are harvested in certain fishing areas and aquaculture fields. Transport distances, product quality and handling time directly depend on the spatial arrangement of supplier networks and landing centers.

GIS aids in planning the networks of suppliers and landing-centers by analyzing the spatial associations between the harvest sites, landing point, processing plant and markets. The methods of spatial modeling have made it clear that geographic information can be used to optimize logistic and distribution networks (Khairunissa and Lee, 2022). This situation is enhanced by this planning capability, which enhances upstream coordination and reliability of supply, which serves as a motivation to adopt GIS.

2.1.7. Operation and Raising of Costs

Packaged marine product supply chains are under huge financial strain due to high cost of logistics, cold storage, which is energy intensive and loss due to spoilage. The cost issues are further compounded by inefficiencies in routing, use of facilities and allocation of inventory.

GIS also helps in the efficiency of operations as it allows the spatial optimization of the logistics routes, storage areas, and distribution channels. The spatial decision support systems enhance the use of resources and also reduce wastage which helps to cut costs. The study of food cold-chain logistics has focused on the idea that spatial-informed decisions will positively improve operational performance (Mustafa et al., 2024).

2.1.8. Traceability, Transparency, Readiness to Recall

Traceability is a necessary tool in controlling the risk of food safety and conducting recalls of food in a seafood supply chain. Without spatially explicit traceability recall processes may be slow, inaccurate, and expensive.

GIS boosts traceability as it relates the flows of products with geographic locations and movement routes. This space traceability allows a quicker identification of affected regions and facilities during recall incidences. Recent traceability research (Hopkins et al., 2024) has highlighted the importance of reducing data loss at important processing and distribution points, which supports the benefit of GIS-driven transparency.

2.1.9. Capacity Planning of Facilities and of Cold-Storage

The long-term consequences of the choice of the facilities location in a cold-chain infrastructure are the efficiency of logistics, energy consumption, and risk of spoilage. Wrongly placed facilities help to raise the transit time and cost of operation, whereas improper distribution of capacity causes a congestion or under-utilization.

GIS can be used to plan the location of facilities and capacity based on accessibility spatial analysis and proximity to suppliers and markets, and environmental limitations. Spatial decision support has demonstrated its strategic role in the GIS-based optimization methods used in supply chain and waste-to-energy infrastructure planning (Billal et al., 2025).

2.1.10. Cold-Chain Integrity and Temperature-Risk Control

The preservation of cold-chain integrity is basic in sustaining the quality and safety of seafood. The excursions in temperature commonly happen in certain points or certain transport sections and this implies that spatial analysis is inevitable.

GIS allows identifying the temperature-risk hotspots by incorporating the data of temperature sensors with the data on location and allows taking interventions. Spatial visualization of cold-chain performance enhances monitoring and compliance, particularly in complex logistics networks (Hu et al., 2024). This direct link between GIS and quality assurance strongly influences adoption decisions.

2.1.11. Management of Carbon footprint and sustainability

The pressure of sustainability is becoming one of the strategic choices in seafood supply chains. One of the greatest contributors to carbon emission and environment is transportation and cold storage. GIS aids the management of sustainability through providing the capability to analyze the spatial emission, fuel consumption and logistics efficiency. GIS helps decrease emissions and enhance the environmental performance by optimizing routes and the location of the facilities (Subramaniam et al., 2023). Supply chain sustainability research emphasises the role of spatially aware decision making in resiliency and long-term sustainability (Muduli and Barve, 2013; Muduli and Barve, 2015).

2.1.12. Disruption Response and Visibility of Logistics in Real Time

Marine supply chains are susceptible to impacts caused by weather conditions, infrastructure collapse, and government action. Quick reaction demands timely access to logistics processes and awareness of space of disruption effects.

GIS offers real-time tracking of logistics through the combination of live data feeds and the spatial visualization tools. The advantage gives decision-makers the ability to determine the effects of disruptions in real-time and reroute or reschedule operations. Spatial technologies like satellite images have been shown to supplement the resilience of supply chains in the context of the seafood supply chain (Sengupta et al., 2022).

3. METHODOLOGY

The research design used in this study was a two-stage study that sought to establish and rank GIS adoption drivers in packaged marine product supply chains. In step one, possible drivers were determined by a systematic review of peer-reviewed articles concerning the seafood supply chains, cold chain logistics, marine logistics, GIS applications in supply chain management, spatial decision support systems, traceability, risk management, and sustainability in food supply chains. Relevant keywords were used in searching academic databases, and only the high-quality journal articles and conference papers were taken into account. Factors or motivations that prompt organizations to implement GIS in order to enhance supply chain performance, decision, compliance and risk management were categorized as drivers. Screeners, refiners, and consolidators of the drivers were then used to remove duplication and conceptual overlap to produce an initial list of GIS adoption drivers based on the drivers extracted out of the literature.

Further, the validated drivers of GIS adoption were prioritized using the Fuzzy Delphi Method (FDM) which utilized expert judgments. FDM is a combination of fuzzy set theory with the conventional Delphi methodology, which allows capturing the uncertainty and subjectivity of the expert opinion and the development of consensus in a complicated decision-making context. FDM is the statistical method to gather the opinions of the experts in order to arrive at some general opinion on a given topic. It is a sophisticated account of classical Delphi technique and fuzzy set theory which utilizes triangulation statistics to determine intervals between level of consensus among a group of experts. FDM is highly suitable to deal with the subjectivity and ambiguity of expert opinions. FDM will solve vagueness and uncertainty because it incorporates the fuzzy and classical decision-making theories.

In this research, the Fuzzy Delphi Method (FDM) utilized a panel of 13 domain experts to develop the decision matrix. The experts were selected on the basis of at least five years experience in academics or industry in the area of supply chains of packaged marine products, or cold-chain logistics, or GIS-based decision support systems. It is not a country-specific study. However, it generates insights that can be applied to the developing-country or export-oriented packaged marine product supply chains. FDM is used to reach a consensus by fuzzy averaging of the individual judgments unlike the classical Delphi technique which needs several iterative rounds to reach a convergence point between the expert opinions. Divergence between personal and group opinion is calculated based on the threshold distance. Expert agreement is measured statistically which eliminates the need of iterative feedback.

FDM basically entails four main processes, i.e. (i) development of an appropriate fuzzy spectrum that can describe lingo words, (ii) aggregation of expert judgment and their transformation into fuzzy sets, (iii) conversion of fuzzy sets into crisp values and lastly (iv) ranking and selection of items by applying a pre-empted threshold (Li et al., 2023). The Steps of Fuzzy Delphi method are also depicted in the Figure 1.

Figure 1. Fuzzy Delphi framework

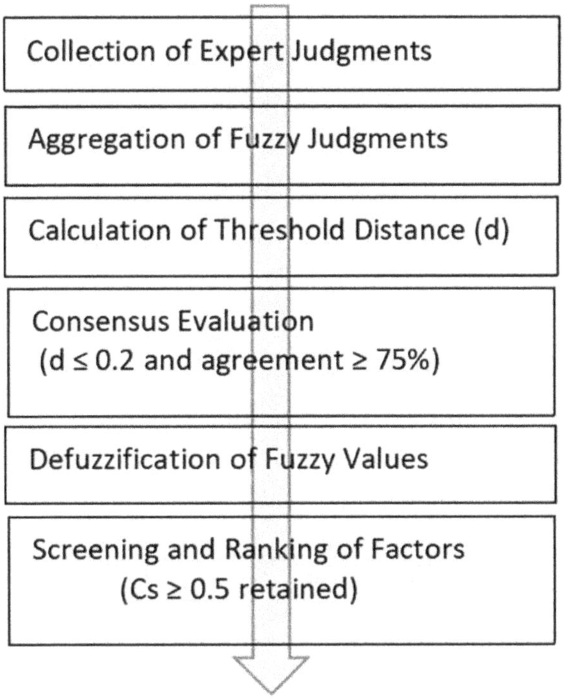

This paper has used FDM to rank the GIS adoption drivers of packaged product marine supply chain. In order to make the questionnaires easier to complete by the experts, a 7-point Likert scale required the direct input of fuzzy values to be replaced by a 7-point scale. This scale also associated with the fuzzy spectrum, which was as explained in Table 1, and hence, the experts found it easier to give their assessments.

The fuzzy decision making process began with the conversion of all the linguistic variables to TFNs. These TFNs were modeled as $a(a_1,a_2, a_3)$ where a1 represents the minimum, a2 represents the most probable and a3 represents the highest. Averages of the fuzzy numbers (F_{na}) were subsequently calculated using the formula that was expressed as shown in the equation below (1) and n is the number of experts. The fuzzy average numbers are also presented in the form of $b(b_1,b_2,b_3)$. The threshold value of a threshold number d(a,b) was calculated by using an equation (2) to determine the degree of consensus between professionals. This threshold is used to indicate the distance between a given fuzzy scale $a(a_1,a_2,a_3)$ and that of the average $b(b_1,b_2,b_3)$. Lastly, the crisp score (C_s) values were calculated using the formula below in the equation (3) (Jahanvand et. al., 2023; Namayala and Kondo, 2024). Each of the factors showing expert consensus should have threshold values (d) that are lower than 0.2. Moreover, the factors must have the value of the 75 percent expert agreement and must have the defuzzification values (C_s) that are above the value of 0.5 (Al-Rikabi and Montazer, 2024).

$$F_{na} = b(b_1, b_2, b_3) = \left[\frac{\sum_{i=1}^{n} a_{1i}}{n}, \frac{\sum_{i=1}^{n} a_{2i}}{n}, \frac{\sum_{i=1}^{n} a_{3i}}{n} \right] \quad (1)$$

$$d(a,b) = \sqrt{\tfrac{1}{3}\left[(a_1 - b_1)^2 + (a_2 - b_2)^2 + (a_3 - b_3)^2\right]} \quad (2)$$

$$C_s = \left(\tfrac{1}{3}\right) * (b_1 + b_2 + b_3) \quad (3)$$

Table 1. 7-Point Likert scale and triangular fuzzy score

Likert scale	Linguistic variable	Triangular fuzzy score
1	Extremely unimportant	(0,0,0.1)
2	Very unimportant	(0,0.1,0.3)
3	Unimportant	(0.1,0.3,0.5)
4	Moderately unimportant	(0.3,0.5,0.7)
5	Important	(0.5,0.7,0.9)
6	Very important	(0.7,0.9,1.0)
7	Extremely important	(0.9,1.0,1.0)

4. RESULTS AND DISCUSSION

This section gives the findings of the Fuzzy Delphi Method (FDM) analysis and explains the relative significance of the determined drivers in the GIS adoption in packaged marine product supply chains. Three important indicators were used to assess it, such as threshold value (d), percentage of expert agreement (P_e), and defuzzified crisp score (C_s). The acceptance criteria were met by all identified drivers, and it could be said that there is a high level of expert agreement on relevance and importance. The results of the Fuzzy Delphi Analysis are included in Table 2. Finally, the prioritized ranking of drivers is shown in the Table 3.

Table 2. Results of FDM

Influencing factors	Threshold value (d)	% of expert agreement (P_e)	Crisp score (C_s)	Decision (Accept/Reject)
Regulatory compliance & export certification	0.12	77	0.88	Accept
Risk mapping	0.07	85	0.69	Accept
Multi-actor coordination & information integration	0.16	77	0.54	Accept
Route optimization & transit time reduction	0.16	85	0.92	Accept
Demand hotspot mapping & market expansion	0.13	93	0.66	Accept
Supplier & landing-center network planning	0.11	85	0.72	Accept
Cost reduction & operational efficiency	0.16	85	0.74	Accept
Traceability, transparency & recall readiness	0.19	85	0.91	Accept
Facility location & cold-storage capacity planning	0.1	85	0.7	Accept
Cold-chain integrity & temperature-risk control	0.1	93	0.94	Accept
Sustainability & carbon footprint management	0.16	85	0.65	Accept
Real-time logistics visibility & disruption response	0.19	77	0.81	Accept

Table 3. Ranking of drivers

Influencing factors	Rank
Regulatory compliance & export certification	4
Risk mapping	9
Multi-actor coordination & information integration	12
Route optimization & transit time reduction	2
Demand hotspot mapping & market expansion	10
Supplier & landing-center network planning	7
Cost reduction & operational efficiency	6
Traceability, transparency & recall readiness	3
Facility location & cold-storage capacity planning	8
Cold-chain integrity & temperature-risk control	1
Sustainability & carbon footprint management	11
Real-time logistics visibility & disruption response	5

The findings indicate that all the twelve drivers hit threshold values which fell below the acceptable level and minimum consensus level far beyond the expert agreement. This validates the strengths of the determined drivers and the contextual appropriateness of the drivers to packaged marine product supply chains. Cold-chain integrity and temperature-risk control became the most important factor among the drivers, with the crisp score ($C_s = 0.94$) and the expert agreement of 93% being the highest, and the Rank 1. The outcome reflects a predominant position of spatially enabled monitoring in maintaining the quality of products, reduced spoilage, and the safety of food in the most perishable chains of marine product supply. The second-ranked driver, route optimization and transit time reduction ($C_s = 0.92$) represents the relevance of the GIS-based routing and network analysis in delay minimization, higher reliability of deliveries, and cold-chain continuity. Because the geographical locations of the landing centers, processing units, and markets are scattered, the spatial route intelligence can contribute greatly to the performance of the logistics.

Traceability, transparency, and recall readiness were ranked third ($C_s = 0.91$), highlighting the increasing regulatory and market demands of end to end product visibility. GIS enhances the traceability systems by giving context to movements, handling and transformation activities throughout the supply chain using space as the context. The fourth position ($C_s = 0.88$) was attributed to regulatory compliance and export certification, which implies that GIS is highly motivated by the necessity to meet the international requirements of seafood safety, exportation documents, and verification of geographic origin. Real time logistics visibility and response to disruption was number five ($Cs = 0.81$), as it indicates that spatial situational awareness is important in terms of disruption management, which include, port congestion, weather events, transport failures, and regulatory inspections. Strategic GIS can be used to reduce costs and improve operational efficiency (Rank 6), plan suppliers and landing-centers networks (Rank 7), and plan facilities location and cold-storage capacity (Rank 8) to help drivers identify strategic use of GIS in long-term infrastructure planning.

The other lower-ranked yet accepted drivers are risk mapping (Rank 9), demand hotspots mapping and market expansion (Rank 10), sustainability and carbon footprint management (Rank 11) and multi-actor coordination and information integration (Rank 12). Despite getting relatively lower ranks, their acceptance shows that experts do not dismiss their role, though they are considered to be secondary or facilitating motivations and no longer taken as the triggers of their adoption.

4.1 Discussion of Key Findings

The fact that cold-chain integrity and temperature-risk control is the highest-ranked driver suggests that the adoption of GIS in packaged marine products supply chain is more of an operational than a strategic initiative. This observation is consistent with the fact that seafood products are highly perishable, and even minor changes in temperature or delays during the transportation process can cause irreversible changes in the quality of the products and considerable financial losses. The fact that this is valued by experts indicates that it is a pragmatic approach to immediate and quantifiable performance results, including reduction of spoilage, temperature regulation, and maintenance of product shelf life. GIS provides the possibility of monitoring the change of temperature by space, detecting high-risk transit routes, and visualizing nodes delays, directly solving the most serious weaknesses in seafood logistics.

Further conceptually similar to cold-chain integrity, the reason that route optimization and reduction of transit time is ranked highly is yet another aspect of GIS adoption that reinforces the operational urgency in GIS adoption. Geographic dispersion of sites of harvest, processing and consumption of marine produce define marine product supply chains, which are frequently augmented by the inefficiency of infrastructure and logistical congestion of transport systems. GIS can be used to plan dynamic routes that can incorporate the space information on distance, travel time, congestion, and environmental conditions. The importance that specialists have placed on this factor implies that companies view GIS more as a transit time and quality risk minimization tool than as a more comprehensive strategic analytics platform. This operational orientation can be especially seen in those situations when the inefficiency of logistics is directly converted into losing products and revenue.

The great attention paid to traceability, transparency, and the readiness to recall, as well as the regulation compliance and the export certification, reflect the growing regulatory and market pressure on seafood supply chains. The international seafood trade has very stringent regulations with regards to food safety, harvest lawfulness, and sustainability reporting. Gaining the capability to prove spatially verifiable records of product origin, movement history and adherence to certification boundaries has become a necessity to remain in the market. GIS enhances compliance because it connects traceability data with geographic locations which allows firm to act in response to audit, inspection and recall events. The popularity of the above drivers shows that regulatory compliance is no longer considered a secondary issue but a fundamental working need that defines technology use.

Interestingly, the issue of sustainability and carbon footprint management has received a lot of attention in scholarly literature, but this aspect was rated quite low by professionals. This inconsistency indicates the existence of a disconnection between academic focus and business focus. The first possible reason is that the sustainability related to GIS like reduction of emissions due to route optimization or resource efficiency tend to be viewed as indirect or long-term consequences of adoption, as opposed to instant motivation of adoption. In the context of developing countries, where companies are resource-bound and compliance costs are high, interventions in technology will more likely be borne out by direct operational or regulatory advantages than by sustainability concerns per se. The result does not mean that sustainability is not important but just that it has not become the major driver of adopting GIS in packaged marine products supply chains.

In the same way, the comparatively low position of multi-actor coordination and information integration indicates that practitioners do not underestimate collaborative and systemic advantages of GIS at the moment. As much as GIS is capable of becoming a common ground that can be used to combine information between different actors in the supply chain, its implementation seems to be motivated by firm-level operational benefits over joined aim of coordination. This can be an indicative of structural and institutional facts in seafood supply chains, especially in the developing economies, with the lack of trust, power disparities, and fragmented governance limiting the usefulness of collaborative digital platforms. As a result, companies might want to focus on technologies that yield short-term internal value, rather than those that involve a long process of coordination and alignment of various parties.

Ranking trend also brings out a clear difference between short term operational drivers and long term strategic drivers of GIS adoption. Immediate performance gains and reduction of risks are related to factors like cold-chain integrity, route optimization, traceability, and compliance. Conversely, sustainability, carbon management and multi-actor coordination factors consist of more expansive strategic goals that can have benefits in the longer period of time. The focus of the experts on operational drivers implies that the adoption of GIS in the packaged marine product supply chains is not proactive to transform the long-term strategic direction but is rather reactive to the urgent issues.

The other remarkable observation is that there is an interaction between regulatory pressure and the management of operational risk. The high rating of compliance-related drivers implies that regulatory requirements are tightly bound to the operational issues like temperature control and traceability. GIS also allows the firms to ensure that the operational processes are harmonized according to the regulation requirements by making spatial evidences of a specification thus preventing the risk of hefty fines imposed on non-compliance and market lock-out. This dual purpose of GIS as an operational and a compliance tool makes it appear to have a higher value to practitioners.

The research findings also point out how technology adoption drivers are context-specific. In developed markets, where digital infrastructure and regulation frameworks are firmly established, sustainability and data integration can be considered more influential in the decisions of GIS adoption. Nonetheless, in the situation in developing countries, where the supply chains of seafood are often divided and limited in resources, the focus stays at the quality of products, legal standards, and cost-effectiveness. This situational dependency highlights the necessity to take regional and institutional conditions into account when explaining GIS adoption drivers.

Meanwhile, the approach of the Fuzzy Delphi Method was effective in the methodological sense of gathering the expert opinion and the ability to handle the lack of certainty and subjectiveness. The ranking results indicate consensus in the form of professional judgment as to the relative value of operation, as opposed to strategic drivers, and offer an insightful view of the GIS adoption dynamics. The findings also confirm the appropriateness of expert-based prioritization techniques to investigate multi-dimensional and intricate decision situations like the adoption of technology in seafood supply chains.

In general, the analysis of the important findings implies that the implementation of GIS in packaged marine products supply chains is an issue-focused and not opportunity-driven process at the moment. The most common applications of GIS in the firms are aimed at solving physical difficulties associated with perishability, logistic inefficiencies, and compliance with the regulations. Although there are wider advantages concerning sustainability, teamwork and strategic analytics, these are secondary at the current occasion. The implications of this finding on policymakers and technology providers are that there is the need to focus on the short-term operational and compliance advantages when marketing the use of GIS, specifically in the developing-country setting.

Although there is a solid theoretical and empirical evidence that GIS should be adopted, there are still a few practical implementation issues that could affect the actualization of its full potential in the packaged marine products supply chains. Such obstacles are related to high upfront investment, lack of digital infrastructure, poor data quality and interoperability in the heterogeneous actors in the supply chain, and the lack of skilled human resources who can handle spatial analytics systems. In the developing-country setting, incomplete forms of governance, minimal standardization of traceability information and the unwillingness to share information further reduce large-scale GIS integration. Besides, combining GIS with the already established enterprise systems, sensor networks, and regulatory infrastructure, means organizational preparedness and long-term investment, which is not necessarily consistent with the short-term operation priorities. As a result, even though the results highlight the operational and regulatory importance of GIS, to achieve successful implementation, these technical, organizational, and institutional challenges will have to be tackled to guarantee the continued adoption and scale.

4.2 Managerial and Practical implications

The results are very clear and can give clear advice to the managers and policy makers of the packaged marine products sector. The initial investments into GIS should be made in cold-chain monitoring, route optimization, and traceability applications because they are perceived to have the greatest value and have the most consensus among experts. Policymakers can also enhance GIS adoption by incorporating the spatial reporting provisions in the export certification and food safety provisions.To supply chain planners, facility location planning using GIS and supplier network design has massive potential in minimizing long term costs and improving resiliency.

In the long-run, as digital maturity increases, it is also possible to use GIS to facilitate sustainability reporting, carbon tracking and collaborative information-sharing platforms.Comprehensively, the findings validate that the adoption of GIS among packaged marine product supply chains is influenced by the combination of quality preservation, regulatory pressure, operational efficiency, and risk mitigation, and spatial intelligence is the key facilitator of effective decision-making in geographically fragmented networks.

5. CONCLUSION

This paper explored the major forces that affect the implementation of Geographic Information Systems (GIS) in packaged marine product supply chains and ranked them through Fuzzy Delphi Method. The study tackled a key knowledge gap by combining information on the literature review with the analysis by experts in the domain of GIS in geographically distributed, cold-chain-restricted seafood supply networks. The findings indicate that the most powerful driver of GIS adoption is cold-chain integrity and temperature-risk control that highlights the urgency of operations to preserve the quality of the product and reduce spoilage in perishable marine products. This observation brings out GIS as an essential decision support system to monitor temperature risks, transit delays and high exposure areas all along the supply chain. Minimization of transit time and route optimization, transparency, and recall preparedness were also other drivers that ranked highly, a factor that has seen GIS playing a major role in boosting logistics, regulatory compliance, and end-to-end visibility.

The associated drivers of regulatory compliance and export certification, real-time logistics visibility and disruption response also indicate that the adoption of GIS is significantly affected by external regulatory forces and proactive risk management needs. The long-term planning value of GIS in the network design and optimization of infrastructure is confirmed by the strategic drivers of facility location and cold-storage capacity planning, supplier and landing-center network planning, cost reduction and operational efficiency. Sustainability and carbon footprint management and multi-actor coordination were lower in the rank, though, the fact they were accepted implies that GIS can serve the broader sustainability and governance goals as digital maturity grows.

In a practical context, the results imply that companies need to focus on using GIS applications that can directly aid cold-chain tracking, control routes, and traceability to gain direct operational and compliance advantages. The adoption of GIS can also be enhanced by policymakers and regulatory bodies by integrating the spatial data specifications within the seafood certification, exportation records, and food safety standards. This research has academic significance in that it gives a prioritization of the drivers of GIS adoption of packaged marine product supply chains, which prioritization is systematic and consensus based and has been limited subjects of focused consideration in the existing literature. The Fuzzy Delphi Method facilitates sound management of uncertainty in the expert and provides a repeatable platform on investigating drivers of technology adoption among other supply chains that require space-intensive processes.

The present research can be developed into the future by incorporating the use of other multi-criteria decision-making processes, longitudinal case studies, or GIS simulation models to measure the effects of GIS adoption in relation to costs, quality, resilience, and sustainability outcomes. Cumulatively, this study finds that GIS is not only an enabling mapping device, but an enabling strategy of quality assurance, regulatory conformity and operational resilience of packaged marine products supply chain, especially where spatial complexity and high perishability are of great concern.

REFERENCES

Al-Rikabi, Y. K., & Montazer, G. A. (2024). Designing an e-learning readiness assessment model for Iraqi universities employing Fuzzy Delphi Method. *Education and Information Technologies*, *29*(2), 2217–2257. DOI: 10.1007/s10639-023-11889-0 PMID: 37361783

Billal, M. M., Sebastian, R. M., & Kumar, A. (2025). Developing a GIS-based MINLP framework for optimizing the waste-to-energy supply chain with plant scale considerations. *Journal of Cleaner Production*, *490*, 144709. DOI: 10.1016/j.jclepro.2025.144709

Carmona, M.. (2023). GIS-based approach for improving maritime supply chain management practices. *Computers and Electronics in Agriculture*. Advance online publication. DOI: 10.1016/j.compag.2023.108321

Chang, P. T., Huang, L. C., & Lin, H. J. (2000). The fuzzy Delphi method via fuzzy statistics and membership function fitting and an application to the human resources. *Fuzzy Sets and Systems*, *112*(3), 511–520. DOI: 10.1016/S0165-0114(98)00067-0

Dorotea, T., Riuzzi, G., Franzago, E., Posen, P., Tavornpanich, S., Di Lorenzo, A., Ferroni, L., Martelli, W., Mazzucato, M., Soccio, G., Segato, S., & Ferrè, N. (2023). A scoping review on GIS technologies applied to farmed fish health management. [PubMed]. *Animals (Basel)*, *13*(22), 3525. DOI: 10.3390/ani13223525 PMID: 38003143

Hardt, M. J., Flett, K., & Howell, C. J. (2017). Current barriers to large-scale interoperability of traceability technology in the seafood sector. [PubMed]. *Journal of Food Science*, *82*(S1), A3–A12. DOI: 10.1111/1750-3841.13796 PMID: 28833156

Hopkins, C. R., Roberts, S. I., Caveen, A. J., Graham, C., & Burns, N. M. (2024). Improved traceability in seafood supply chains is achievable by minimising vulnerable nodes in processing and distribution networks. *Marine Policy*, *159*, 105910. DOI: 10.1016/j.marpol.2023.105910

Hu, L., Chen, Q., Yang, T., Yi, C., & Chen, J. (2024). Visualization and analysis of hotspots and trends in seafood cold chain logistics based on CiteSpace, VOSviewer, and RStudio Bibliometrix. *Sustainability (Basel)*, *16*(15), 6502. DOI: 10.3390/su16156502

Ismail, S., Reza, H., Salameh, K., Kashani Zadeh, H., & Vasefi, F. (2023). Toward an intelligent blockchain IoT-enabled fish supply chain: A review and conceptual framework. *Sensors (Basel)*, *23*(11), 5136. DOI: 10.3390/s23115136 PMID: 37299864

Jahanvand, B., Mortazavi, S. B., Mahabadi, H. A., & Ahmadi, O. (2023). Determining essential criteria for selection of risk assessment techniques in occupational health and safety: A hybrid framework of fuzzy Delphi method. *Safety Science, 167*, 106253. DOI: 10.1016/j.ssci.2023.106253

Khairunissa, M., & Lee, H. (2022). Hybrid metaheuristic-based spatial modeling and analysis of logistics distribution center. [MDPI]. *ISPRS International Journal of Geo-Information, 11*(1), 5. DOI: 10.3390/ijgi11010005

Kidane, D. G. (2025). Price transmission and asymmetry in a changing seafood supply chain. *Aquaculture Economics & Management, 29*(3), 1–21. DOI: 10.1080/13657305.2024.2447443

Koirala, P., Sagar, N. A., Thuanthong, A., Al-Asmari, F., Jagtap, S., & Nirmal, N. (2025). Revolutionizing seafood packaging: Advancements in biopolymer smart nano-packaging for extended shelf-life and quality assurance. *Food Research International, 203*, 115826. DOI: 10.1016/j.foodres.2025.115826 PMID: 40022350

Leal, M. C., Pimentel, T., Ricardo, F., Rosa, R., & Calado, R. (2015). Seafood traceability: Current needs, available tools, and biotechnological challenges for origin certification. *Trends in Biotechnology, 33*(6), 331–336. Advance online publication. DOI: 10.1016/j.tibtech.2015.03.003 PMID: 25865857

Lewis, S. G., & Boyle, M. (2017). The expanding role of traceability in seafood: Tools and key initiatives. *Journal of Food Science, 82*(S1), A13–A21. DOI: 10.1111/1750-3841.13743 PMID: 28833154

Muduli, K., & Barve, A. (2013). Establishment of a sustainable development framework in small scale mining supply chains in India. *International Journal of Intelligent Enterprise, 2*(1), 84–100. DOI: 10.1504/IJIE.2013.057340

Muduli, K., & Barve, A. (2015). Analysis of critical activities for GSCM implementation in mining supply chains in India using fuzzy analytical hierarchy process. *International Journal of Business Excellence, 8*(6), 767–797. DOI: 10.1504/IJBEX.2015.072309

Muduli, K., Raut, R., Narkhede, B. E., & Shee, H. (2022). Blockchain technology for enhancing supply chain performance and reducing the threats arising from the COVID-19 pandemic. *Sustainability (Basel), 14*(6), 3290. DOI: 10.3390/su14063290

Mustafa, M. F. M. S., *et al.* (2024). Food cold chain logistics and management: A bibliometric analysis and research agenda. *Cleaner Logistics and Supply Chain.* https://doi.org/DOI: 10.1016/j.clscn.2024.100380

Namayala, P. P., & Kondo, T. S. (2024). Application of fuzzy Delphi technique to identify analytical lenses for determining the preparation of free and open source software projects for user experience maturity. *Science of Computer Programming*, *237*, 103136. DOI: 10.1016/j.scico.2024.103136

Rahman, L. F., Alam, L., Marufuzzaman, M., & Sumaila, U. R. (2021). Traceability of sustainability and safety in fishery supply chain management systems using radio frequency identification technology. *Foods*, *10*(10), 2265. DOI: 10.3390/foods10102265 PMID: 34681313

Rejeb, A., Keogh, J. G., & Rejeb, K. (2022). Big data in the food supply chain: A literature review. *Journal of Data, Information and Management*, *4*(1), 33–47. DOI: 10.1007/s42488-021-00064-0

Sengupta, T., Narayanamurthy, G., Moser, R., Pereira, V., & Bhattacharjee, D. (2022). Disruptive technologies for achieving supply chain resilience in COVID-19 era: An implementation case study of satellite imagery and blockchain technologies in fish supply chain. *Information Systems Frontiers*, *24*(4), 1107–1123. Advance online publication. DOI: 10.1007/s10796-021-10228-3 PMID: 34876876

Stentiford, G. D., Peeler, E. J., Tyler, C. R., Bickley, L. K., Holt, C. C., Bass, D., Turner, A. D., Baker-Austin, C., Ellis, T., Lowther, J. A., Posen, P. E., Bateman, K. S., Verner-Jeffreys, D. W., van Aerle, R., Stone, D. M., Paley, R., Trent, A., Katsiadaki, I., Higman, W. A., & Hartnell, R. E. (2022). A seafood risk tool for assessing and mitigating chemical and pathogen hazards in the aquaculture supply chain. *Nature Food*, *3*(2), 169–178. DOI: 10.1038/s43016-022-00465-3 PMID: 37117966

Subramaniam, R. C., Ruwet, M., Boschetti, F., Fielke, S., Fleming, A., Dominguez-Martinez, R. M., Plagányi, É., Schrobback, P., & Melbourne-Thomas, J. (2023). The socio-ecological resilience and sustainability implications of seafood supply chain disruption. *Reviews in Fish Biology and Fisheries*, *33*(4), 1129–1154. DOI: 10.1007/s11160-023-09788-1 PMID: 37360577

Swain, S., Peter, O., Adimuthu, R., & Muduli, K. (2021). Blockchain technology for limiting the impact of pandemic: Challenges and prospects. Computational modeling and data analysis in COVID-19 research, 165-186.

Tseng, M. L., Li, S. X., Lin, C. W. R., & Chiu, A. S. (2023). Validating green building social sustainability indicators in China using the fuzzy delphi method. *Journal of Industrial and Production Engineering*, *40*(1), 35–53. DOI: 10.1080/21681015.2022.2070934

Ullah, K. M., Masum, F. H., Field, J. L., & Dwivedi, P. (2023). Designing a GIS-based supply chain for producing carinata-based sustainable aviation fuel in Georgia, USA. *Biofuels, Bioproducts & Biorefining : Biofpr*, *17*(4), 786–802. DOI: 10.1002/bbb.2483

Chapter 10
Vehicle-Mounted Real-Time Road Damage Detection System Using YOLOv8 and Edge Computing

L. Churchill
http://orcid.org/0009-0009-5186-3854
M. Kumarasamy College of Engineering, India

A. Selvi
M. Kumarasamy College of Engineering, India

S. Akshaykumar
M. Kumarasamy College of Engineering, India

G. Kaviyarasan
M. Kumarasamy College of Engineering, India

A. S. Shivapriyan
M. Kumarasamy College of Engineering, India

ABSTRACT

This paper presents a real-time road health monitoring system that overcomes the limitations of manual road inspections. The system uses cameras and GPS modules mounted on public vehicles to capture live road images, which are processed on edge devices using a YOLOv8-based deep learning model. It automatically detects

DOI: 10.4018/979-8-3373-7941-8.ch010

potholes, cracks, and surface irregularities, assigns severity scores, and tags each defect with precise GPS coordinates. The data is sent to a centralized PostgreSQL database with PostGIS for spatial mapping. A React-based administrative dashboard helps authorities assess road conditions and prioritize repairs, while a public React web portal allows citizens to report issues. Automated and public reports are cross-verified to avoid duplication, making the system cost-effective, scalable, and reliable.

1. INTRODUCTION

Infrastructure of road transportation is at thconnectivity,economic growth, social connectivity and mobility services in both developed and developing areas. Constant load exchanges caused by the environment and aging of materials cause gradual degradation of the pavement to producpotholes andthe form of cracks and potholes, and surface deformations, which deteriorate road safety and ride quality (Shakhovska et al., 2024). The high rate of urbaexacerbated high vehicular density exacerbates the mechanicoad on road networkexceedsmoting the wear and tear that exceed the possible scope of traditional maintenance regimes (Duan and Nazri, 2024). Poor and late recognition of pavement defects leads to the risk of accidents, additional spending on upkeep procedures, and the shorter life of infrastructure (Abbas et al., 2025). The need to ensure that the conditions of the road surface are known in time has thus been a major demand of a sustainable transport management system (Mishra and Kamal, 2024).

The conventional methods of road inspection are based on the periodic manual surveys and visual observations, which are carried out by the qualified staff. Although these approaches are still popular, they have limited coverage of space and frequency of inspection that limit them in terms of capturing quickly changing damages due to environmental changes and traffic overloads (Liu et al., 2025). Rural road systems are further complicated by the lack of resources and the lack of monitoring intensity; this leads to the fact that the routes are often underutilized, which leads to their long-term decay (Davuluri, 2024). This is because of a lack of ongoing monitoring provisions, which results in reactive instead of proactive maintenance efforts, which adds to the overall operational costs in the long run (Abdelkader et al., 2025). These restintoctions have prompted the research to intelligent and automated road-condition monitoring systems that can operate in real-time (Uppukonda et al., 2025).

The development of intelligent transportation systems has provided the possibility to combine computer vision and deep learning to anallmagingoad surface automatically. Imagining platforms on vehicles and airlift can produce vast amounts of visual data that can be used to capture pavement texture, geometry, and defect features when used in real-world conditions (Britto et al., 2025). Trained deep learning models using annotated road imagery show better performance compared to the conventional image processing methods as they learn hierarchically and context-dependent feature representations (Sunitha et al., 2025). The changes in lighting, weather, occupancy, and materials of the road pose serious problems to a consistent defect control system, and advanced and adaptable architectures are required (Li and Gu, 2025). Recent publications refer to the role of light and efficient models that can be deployed on mobile and embedded platforms (Lan et al., 2025).

One of the important elements in road damage recognition systems using vision has been detected as object detection paradigms. When compared to image classification methods, object detection models can localize spatial positions of defects and categorize them, thus allowing fine evaluation of the damage site and extent (Zeng and Zhong, 2024). The models of single-stage detection that are grounded on the YOLO family provide real-time inference systems by combining detection and classification in a single architecture (Zhao et al., 2024). Other recent advancements to include the multi-based fusion of features, no-anchor based detection, and loss optimization to ensure the high detection of both coarse structural anomalies and fine cracks (Luo et al., 2025). These innovations contribute to the stable functioning in the case of dynamic traffic and environmental conditions (Yurdakul & Tasdemir, 2025).

The edge computing adoption also increases the plausibility of detecting road damage in real-time due to the localized processing that takes place near data sources. On-device inference decreases the latency, minimizes the use of bandwidth, and minimizes the dependence on the centralized cloud infrastructure (Begum et al., 2025). The edge platforms provided on vessels allow continuous analysis of road images when the vehicle is on the regular course of action, which will allow receiving and reporting timely about the critical defects (Anjun et al., 2025). The use of edge-based architectures also increases privacy of data and reliability of operations when connecting with restricteDeepnments (Ejaz et al., 2025). The deep learning and object detectors combined with edge computing will create a scalable and intelligent platform of proactive monitoring of the road infrastructure, comprising data-driven maintenance planning and sustainable development of transportation (Zhong and Ren, 2025).

Despite significant advancements in road maintenance technologies, traditional road inspection methods remain inefficient in detecting deteriorating road conditions. This research focuses on addressing the existing gap in automated road damage detection systems, particularly the limitations of manual inspection methods. Current systems still struggle with real-time monitoring and accurate detection under various environmental conditions. Therefore, this chapter aims to explore the potential of YOLOv8, a state-of-the-art object detection model, integrated with edge computing for real-time road damage detection. The key objectives of this research are to:

- Investigate the effectiveness of YOLOv8 in detecting road defects like potholes and cracks in real-time.
- Evaluate its performance compared to other models, including Mask R-CNN and YOLOv4.
- Propose an efficient framework for deploying YOLOv8 on edge devices for road monitoring applications

2. LITERATURE REVIEW

The development of automated road damage detection systems has progressed through several stages. Early techniques were primarily based on manual inspections, which were both time-consuming and prone to human error. With the advent of machine learning, object detection methods like YOLO (You Only Look Once) significantly improved detection accuracy and speed. More recent advancements, such as YOLOv8, introduced anchor-free detection and multi-scale fusion, offering higher precision and real-time performance. However, existing research still faces challenges related to model generalization, dataset bias, and operational deployment under varying environmental conditions. The gap in the literature lies in integrating modern object detection models like YOLOv8 with edge computing solutions for real-time, large-scale road monitoring.

3. METHODOLOGY

Model and Dataset Selection

For this study, the YOLOv8 model was chosen due to its real-time detection capabilities and architectural advancements. YOLOv8's anchor-free detection mechanism allows it to identify irregular road defects with greater accuracy. The model was trained using the RDD2022 dataset, which includes various road defects such as potholes, cracks, and ruts. The performance of YOLOv8 was compared with previous models, including YOLOv3 and Mask R-CNN, to highlight its efficiency and accuracy in detecting road damage. To address issues of dataset bias, data augmentation techniques such as random scaling and flipping were employed to ensure the model's robustness across different road conditions.

Evaluation Metrics

To evaluate the performance of YOLOv8, several metrics were used, including mean Average Precision (mAP), Frames per Second (FPS), and latency. These metrics were selected because they offer a comprehensive view of the model's detection accuracy and operational speed, which are crucial for real-time deployment in road monitoring systems.

Foundations of Road Damage Detection Systems

Role of Road Infrastructure in Intelligent Transportation Systems

The physical backbone of the intelliinfrastructure,ation systems is road infrastructure which facilitates the safety, efficiency, and reliability of people and goods transport. The state of road networks has a direct impact on the stability of traffic flows, the efficiency of vehicles in operation, fuel consumption, and the number of accidents. The smart transportation systems are based ononhe correct and timely information in the navigation,id in traffic management, navigation and safety applications. In case of the decline of the road surfaces because of cracking, potholes, or deformation, the functionality of such systems decreases, which results in the poor performance of the traffic control strategies and safety hazards. The operational reliability of intelligent transportation ecosystems is therefore based on the maintenance of structurally sound road infrastructure.

Figure 1. Role of Road Infrastructure in Intelligent Transportation Systems

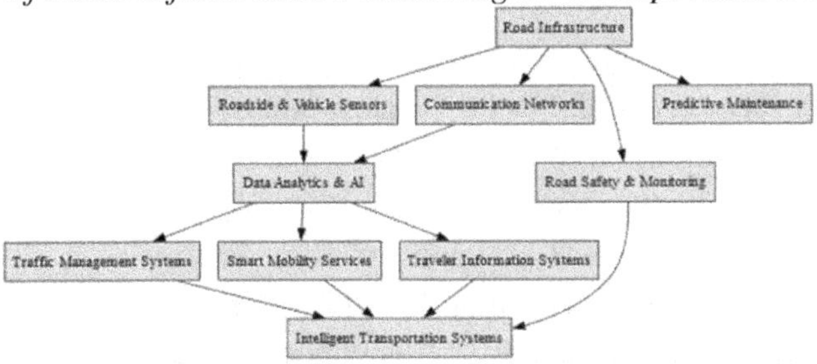

The road infrastructure in intelligent transportation syscomponent butst a passive physical component, but also an active source of data, which interacts with sensing and communication as well as control technologies. The current transport systems combine the state of roads with the vehicle sensors, location systems, and traffic analysis tools to make adaptable decisions. Precise knowledge of pavement health facilitacontrol,ive route plprecautionaryty control and hazard pre-cautionary systems. Poor toads add an element of increasesnty in these processes that increase the chances of vehicle damage and traffic problems. The intelligent transportation systems can be responsive and have better situational awareness, which is achieved by monitoring the road conditions in a reliable way to facilitate safer and more efficient mobility.

The process of establishing smart transportation systems with automated road damage detectors enhances the shift towards data-driven infrastructure management. Constant monitoring of the condition of road surfaces allows one to identify the patterns of deterioration in time and preventive maintenance plans. This ability minimizes the unexpected breakdowns of the road and the allocation of resources on the repair operation. The advantages of such integration in intelligent transportation systems include enhanetworktwork reliability, reducing networks servic-termerruptions, and sustaining long term sustainability goals. Intelligent monitoring and analytics together with road infrastructure also make smart mobility and robust transportation networks an imminent enabler.

Types and Characteristics of Road Surface Defects

Road surface defects are categorized as a large group of structural and functional degenerations that develop as a result of constant traffic loading, exposumaterials, environment, aging of the materials and condefecton flaws. Some of the common defects categories are potholes, longitudinal cracks, transverse cravelling,ligator cracking, rutting, ravelling and surface undulations. The cause of potholes is localized material failure and intr-shapedf moisture, which causes bowl shaped potholes that grow very fast due to recurrent wheel loads. Thermal stress, fatigue, and subgrade movement cause the formation of cracks in a linear or networked pattern, longitudinal cracks being formed parallel to traffic flow and transverse cracks being formed perpendicular to the direction of pavement. Alligator cracking indicates high levels of fatigue damage of the type that has been interconnected fissures of reptilian skin where structurpavement.s exists in the layers of pavements.

The surface deformation flaws include rutting and corrugation caused by plastic deformation of the pavement material under continuous heavy axle loads. Rutting is represented by lopaths,dinal depressions on the wheel paths which lowcapacity,kid resistance and drainage capacity and corrugation is representey repetitive ripples that impact on the ride comfort and vehicle control. Ravelling is a process where the aggregate particles are gradually lost off the pavement surface to create a degraded texdepressions,us, low friction. Bumps, depressions and patch failures are caused by surface irregularities that are attributed to improper maintenance practices or variations in the settlement of the underlying textual, There are unique geometric, textual and visual attributes of each defect type depending on their detectability by vision-based monitoring mechanisms.

The degree and advancement of road surface defects is based on the size of the defects, their distribution in space, and tfactors,teraction with environmental factors including rainfall and change of temperature. Thesubtle at the initial stages have a subtle visual evidence in the form of small cracks or slight deviations in texture, and the deterioration at the advanced stages creates severe geometric disruption and loss of surface materials. Proper identification of such traits constitutes a decisive basis of automated road damage detectors since visual complexity, scale fluctuation, and abnormal morphology directly influence the dependability of detection. Knowing the typology and visual appearance of road surface defects allows building the powerful deep learning structures that could differentiate between the types of defects and make timely decisions in maintaining infrastructure.

Limitations of Conventional Road Inspection Techniques

The traditional methods of road inspection have traditionally employed the manual check of the visual survey and dedicated inspection cars to evaluate the pavement conditions. These methods involve trained operators who have no choice but to be on the ground and inspect the roadways or analyze sensor data taken at designated times. These are very labor-intensive and costly practices that are especially used in large road systems that need regular surveillance. The reliance on frequent checks restricts the possibility to record the quickly developing surface damage, and the defects and corrective measures are identified later, and the preventive actions are not applied.

The other key drawback of conventional inspection procedures is that they are subjective and inconsistent. Different assessments carried out by the inspectors tend to be different becaenvironment,disparity in experience, environment and human juthe evaluationich the severityisparities in evaluation of severity of road damages. Special inspection vehicles with advancedobjectivity,ovide a better degree of objectivity but they are limited due to high purchase and operational expenses. The use of such vehicles can be limited by spatial coverage and diminished the practicality of the continuous monitoring, particularly in rural areas and economically disadvantaged areas.

Figure 2. Limitations of Conventional Road Inspection Techniques

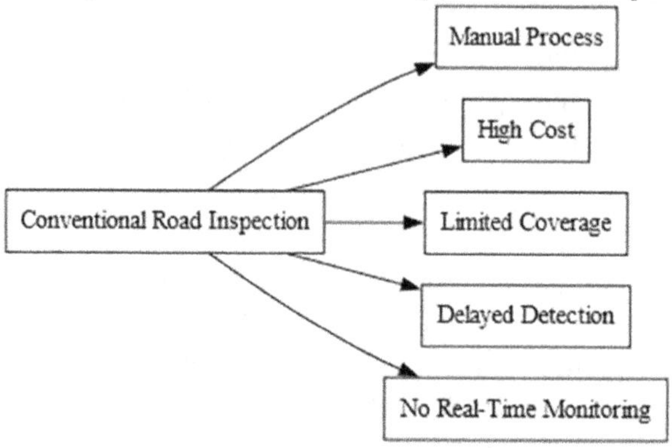

The traditional methods also find it-basedcult to offer real-time, place based data to aid data-driven maintenance planning. Lack of real-time geospatial updates reduces the usefulness of prioritization and allocation of resources to repair operations. Information gathered under manual survey may have to undergo a lot of post-processing and administrative liaising prior to any action taking place. All these limitations decrease the efficiency of conventional road inspection systems and point to the necessity of automated, scalable, and real-time detection systems that could support the management of the contemporary transportation infrastructure.

Dataset Bias and Generalization

The deployment of vehicle-mounted cameras for road damage detection raises significant privacy concerns, particularly regarding the collection of sensitive data such as license plates and pedestrians' faces. To ensure compliance with privacy regulations like GDPR and CCPA, on-device anonymization techniques, such as real-time blurring of license plates and faces, must be implemented. Data should only be retained for the necessary duration, and clear protocols for data ownership, storage, and anonymization must be established. These measures will not only ensure regulatory compliance but also enhance public trust in the system's use.

Privacy and Ethical Considerations

With the use of vehicle-mounted cameras for road monitoring, it is essential to address privacy and ethical concerns. The captured data can include sensitive information, such as license plates and faces of pedestrians, which must be anonymized to comply with privacy laws like GDPR and CCPA. Implementing on-device anonymization (e.g., real-time blurring of license plates) ensures data protection before transmission. Additionally, clear data ownership and storage policies must be outlined to ensure transparency and regulatory compliance.

Evolution of Automated Road Condition Monitoring

The concept of auto road condit monitoring has been experiencethe developmentive change together with development of sentechniques,ologies, data acquisition techniques and computational intelligence. The initial methods were limited to manual eye checks and routinuse,urveys by trinstances,ff with the use in certain instances of special inspection vehicles with mechanical or optical tools. These methods allowed rudimentary evaluation of pavement distrecoverage andd major limisubjectivity andverage, evaluative subjectivity, and were rarely conducted on a regular basis. These conventional methods were no longer sufficient to record the fast patterns of deterioration and localized damage events in a wide geographical area as the road networks grew and traffic density was high.

Monitoring that is sensor-based became a significant move towards automasensors,he accelerometers, vibration sensors and inertial measurementindirect detection ofhe cars allowed indirectly detecting surface irregularities by dynamical responses of the vehicles. Although these systems had better scalability and reduwascosts of state operation, there were limited accuracy of defect classification and accurate localization of the spatial locations. Advances in digital imaging and computer vision subsequently facilitated the direct study of road surface imagery, which made it possible to identify visible defects (cracks and potholes) automatically. The initial image extraction of systems relied on manual extracting fethresholds, segmentation by means of threshold, which restricted their ability to be robustvariation,rent illumination, texture variation and environmental factors.

Recent advances in deep learning have transformed the concept of automated road condition monitoring by facilitating feature learningonaa largen data that learns features in large scale through visual data. The convolutional neural networks and real-time object detection frameworks facilitate the accurate detection of multifaceted patterns of damages irrespective of the road setting. A combination of these models with mobile platforms and edge compug infrastructure has fa-timeated in the continuous and real time monitoring of regular vehicle functioning. The change in this evolutiolabor-heavy, away from the periodic, labor-heavy and inspective nature to intelligent and autonomous surveillance systems that are intelligent enough to sustain proactive maintenance practices and data-driven transportation infrastructure management.

Importance of Real-Time Detection in Urban and Rural Networks

Road damage detection in real-time has become an essential need in the city and rural transportation system because of the growong reliance logistics,nfrastructure in mobility, logistics and economic operations. The high traffic density of urban road n-and-starthe high prevalence of stop and start situations, and overall exposure to heavy traffic loads promote the rate of pavement degradation and are prone to sudden surface failures (potholes and cracking) occurring. Within these settings, the in-time reporting of road damages may rajams,y lead to safety riskson infrastructure. d secondary stress infrastructures. The real-time detection will provide the opportunity to monitor the conditions on the roads continuously as they change and identify defects in time and eliminate them before reaching the threshold of severe levels that can interrupt the traffic process and endanger the life of the population.

Figure 3. Importance of Real-Time Detection in Urban and Rural Networks

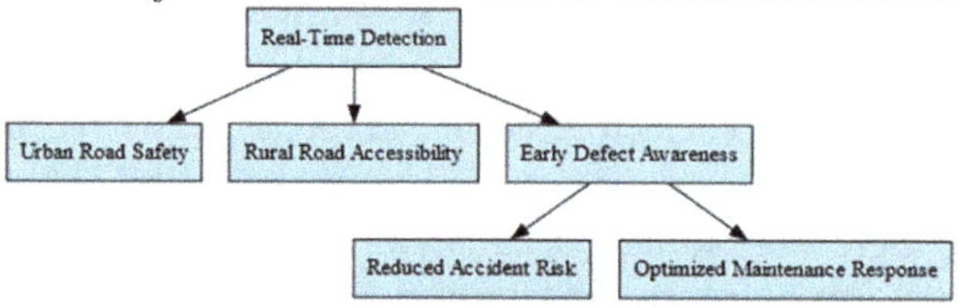

Rural road networks pose a unproblems, equally significant set of problems indicating the need to have real-time detection mechanisms. In most cases, these networks are geographically large and require fewer inspection resources and reduced maintenance rates than urban environments. Exposure tchanges,nvironment, seasonal weather changes and agricultural or heavy transport vehicles are very important in the degradation of surfacedamageal roads. areas,ver there is a damage in these areas the late detection ma and hampertthe responseas into isolation, hampered responsaccidents,gency and may lead to more accidents as there will be poor visibility of roads and unbalanced roads. With real-time monitoring, the assessment is conducted continuously without using periodic manual assessments so locations critical damages in remote location are detected and reported to the maintenance authorities in time.

The significancegoesf real-time road damage detection is far beyond the concern of safety to economical efficiency and maintenance of the infrastructure in a sustainable manner. Continuous detection can be used to reinforce data-driven maintenance strategies with timely and location-specific data on pavement conditions in a wide range of road networks. Prevention of structural damage and pavement life is faults,ed through early detection of faults which llnds to less money spent on repair. On both the urban and rural settings, real-time systems allow the authorities to prioritize interventions based on the severity and frequency of damages, hence makinreducingresource allocation and reduction in service discontinuities. Since the transportation systems are becoming more complex and interconnected, real-time road damage detection is a cornerstone of intelligent, resilient, and adaptive infrastructure management systems.

Deep Learning in Road Surface Analysis

Overview of Computer Vision Techniques for Pavement Assessment

The use of computer vision methods to capture useful structural data on the road surface images and video streams has made them important in automated pavement assessment. Early vision-based methods used conventional techniques of image processing like edanalysis,tion,morphology, texture analysis and morphological to determine cracks and surface anom-thought-outprocedures demanded well thought out manual work and worked satisfactorily in controlled circumstances. Under conditions of lightingchanges,w, pavement texture, weather changes and reduced, performance was drastically reduced and this c-lifeained their application in real life road monitoring situations.

Deep learning was an important step towards a more detailed analysis of the road surface through automatic acquisition of features on the raw visual input. Convolutional neural networks have shown a considerable ability to represent the hierarchical features of pavement features, -leveling low-level textures to high level structural features related to potholes and cracks. Object detection and segmentation models based on deep learning allowed accurately localizing andovercomingying the type of damage and overcome numerous limitations of the traditional imaprovedg methods. These models demonstrated to be more robangles, varied road surfaces, camconditions andviewed environmental conditions, and could allow scalable and automated pavement assessment.

Other recent advances in real-time object detection systems have further enhanced the role of computer vision in the use of road monitoring. Single-stage detection models enable localization and classification in a singinference speedich facilitates high inference-speed and is compatible with the deployment in dynamic scenarios. Combining deep learning models with video-based analysis allows analyzing the state of the road surface at all times when the vehicle is in motion, and the inspection of the road surface becomes an automated and adaptive process. This has seen comdynamicion methods transform into a dynamic analysis methodologies-time,have t-scaleer to backconditionime, large scale pavement conditions observation in heterogeneous transportation systems.

Convolutional Neural Networks for Structural Defect Detection

The capability to infer hierarchical visual representations and to learn them automatically in complex images has made convolutional neural networks a key methodology in structural defect detection in road surface analysis. The surfaces of the roads are extremely variable in terms of their texture, color, light, and morphology of the damages, which restricts the performance of the traditional image processing methods, relying on handcrafted characteristics. Convolutional neural networks help avoid these shortcomings by training on spatial patterns via data and thus have the ability to identify defects like potholes, cracks, rutting, and surface abnormalities in a wide range of environmental and operating conditions. These networks levels,structural patterns at higher deterioration,correspond to pavement deterioration by using layer-by-layer convolutional operations to capture the low-level features (edges and textures) and gradually deeper high-level structural features.

Figure 4. Convolutional Neural Networks for Structural Defect Detection

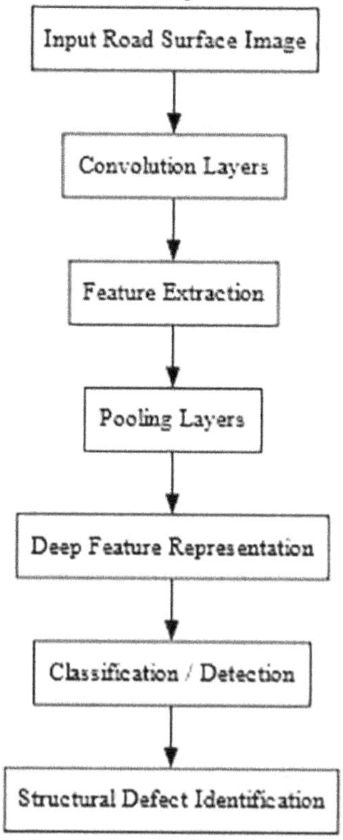

Conventional neural Networks have a great benefit in the road surface monitoring system, as they can cope with the variation of shadows, occlusions, weather conditions, and camera points of view. The feature extraction layers allow powerful separation of healthy pavements and damaged areas when defects have irregular shapes or faint visual features. The pooling and normalization operations are faciltranslation, invariance to scale and translation enabling the detection of small cracks and large potholes in the same analysis model. Because of this, convolutional neural networks are better at detection because of the high accuracy in detection in comparison to traditional threshold-based or texture-based applicationsespecially in real-world application where the complexity in the visual image is also high.

Convolutional neural network application also enhances scalability and auon amation of road monitoring systems in large scale. These models are fast inference on conttraining,mage or video streams after training and this enables them to be deployed in real-time to vehicle-mounted platforms and edge computing devices. Combined with the current object detection systems, convolutional neural networks allow the localization and classification of various types of defects in a single image. This allows the road surface analysis to be considered more reliable and defines deep learning as a cornerstone technology of the intelligent monitoring of transportation infrastructure and informed maintenance design.

Object Detection Paradigms for Road Damage Identification

Object detection paradigms play a central role in deep learning–based road surface analysis by enabling automated identification and localization of pavement defects within complex visual scenes. Road damage identification requires the detection of diverse defect types such as potholes, cracks, rutting, and surface irregularities, often under varying illumination, traffic occlusion, and weather conditions. Traditional image processing approaches based on edge detection or texture analysis struggle to generalize across such variability. Deep learning–driven object detection frameworks overcome these limitations by learning hierarchical spatial features directly from large-scale annotated datasets, allowing robust discrimination between damaged and undamaged road regions in real-world environments.

Modern object detection paradigms are broadly categorized into two-stage and single-stage architectures. Two-stage detectors perform region proposal generation followed by classification and localization, offering high detection accuracy at the cost of increased computational complexity and inference latency. While such architectures demonstrate strong performance in controlled environments, real-time road damage detection demands rapid processing of continuous image streams captured from moving vehicles. Single-stage detectors address this requirement by unifying localization and classification within a single network pass, enabling high-speed inference suitable for real-time deployment. This paradigm has proven particularly effective for road damage identification, where timely detection is essential for safety and maintenance applications.

The adoption of advanced single-stage object detection models has significantly enhanced the practicality of deep learning in road surface analysis. These models incorporate multi-scale feature representation to detect both large structural defects and fine-grained cracks within the same framework. Anchor-free detection strategies and improved loss optimization techniques further contribute to accurate localization of irregularly shaped road damage. When integrated with edge computing platforms, object detection paradigms support continuous, low-latency analysis of road conditions, transforming raw visual data into actionable infrastructure intelligence and enabling scalable, real-world road monitoring solutions.

Challenges in Visual Road Damage Recognition

Detection of road damage is a complicated problem since pavements have a variety of images in the natural driving environment. Road defects tend to have uneven geometry, blurred edgethatd slight variations in texture which merge them. the asphalt or concrete around it. Thcracks,y-stage surface fatigue, fine cracks and scontrast,otholes produce weak visual contrast which lowers the separability between the damaged and undamaged parts. The issue is even more serious when surface patterns have the appearance of an artifact of damage, likemarkings, shadows,ate of repair, lane marking, shadow, and oil stains, which can add contrasting information during recognition.

Figure 5. Challenges in Visual Road Damage Recognition

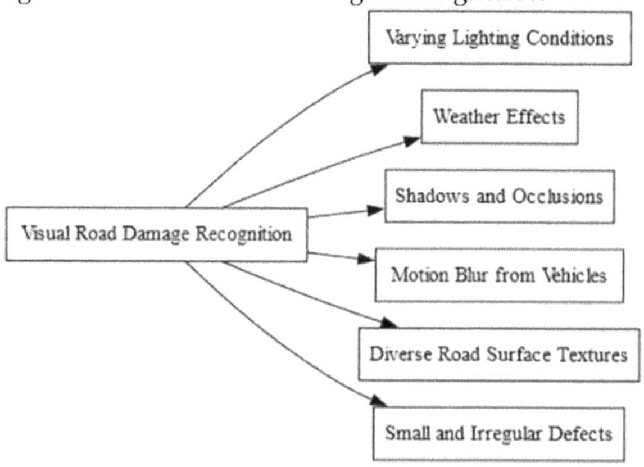

The environmental and operational factors are also influential factors contributing towards performance degradation in the deep learning-based detection systems. Brightness and texture representation in frames vary because of illumination changes due to variation in time of day, glare, headlight bounces, ansurfaces,cover. Rain, fog, dust, wet surfaces and other weather conditions distoreflections andt appearance by forming reflections, covering fine structuralblur,tures. Camera vibration, motion blur and perspectilowertthe sharpnesse to vehicle motiochanges sharpness of the image and changes the magnitude of the defects between successive frames, thereby producing discrepanciepedestrians, andes by moving transport, pedestobjects,debris, as well as road-edge objects also reduce the visibility ofcredibleent surface and constrain accredible defect localization.

The reliability and generalization abilitareof road damage recognition models is also affected by data-centric issues. Training datasets can be affected by the class imbalance of potholes being more pree.g.,nt than rare defect categories, e.g. alligaleadscracbiased or edge breakup, which lead to bias learning. The quality of annotations and inconsistencies in labelingprocess,te noise during the training process particularly in cases where the boundaries of the defects are ambiguous or subjective. Domain shift takes place when ttype,ed models on one city, pavement type or camera configuration are exposed to unfmaterials,oad textures, construction materials or local degradation structures. All these challenges lead to the fact that robust preprocessing, dataset building variety, and deployment optimization are needed to attaininonsistent road damage recognition on urban and rural settings.

Real-World Operational Challenges

Several operational challenges remain. Environmental factors such as rain, fog, and lighting conditions significantly affect the reliability of detection systems. Furthermore, motion blur and occlusion caused by other vehicles can distort the visual clarity required for accurate defect identification. To mitigate these challenges, advanced preprocessing techniques, such as frame stabilization and enhanced contrast adjustment, must be employed to improve detection accuracy under diverse real-world conditions.

Future Trends

Looking ahead, the integration of 5G/6G networks will significantly enhance road damage detection systems by providing high-speed, low-latency communication between vehicles and infrastructure. This will enable near-instantaneous reporting of road conditions, even in remote areas. Additionally, swarm robotics—where multiple autonomous drones or ground robots collaborate for inspection—could further expand the reach and effectiveness of road monitoring systems, enabling real-time detection across large, hard-to-reach areas.

Performance Metrics for Evaluating Detection Models

To evaluate the performance of road damage object detection models, it is necessarthathave a combination of measures which reflect both localization and classification reliability in the context of actual operating conditions. Because road defects are commonly irregularly defined and within-class variation can be large, it is necessary to consider the accuracy of bounding box overlap in addition to the accuracy of correct category labeling. Intersection over Union is the main metric of spatial alignment between predallows us and ground-truth boxes, which allow to assess if an identified area is adequately similar to an area of annotated damages. The quality of detection can only be significant in cases where the overlap thresholds are imposed because the rough localization can translate into minimized practical usefulness of the predictions in the maintenance planning and severity estimation.

Precision and recall are the vital measures of quantifying the detection reliability by the types of defects. Precision indicates the rate of accurate detections in the total number of predicted detections, which is useful in evaluating false alarm behavior, whereas recall indicates the rate of successful detections of defects over the total number of defects annotated, which is a measure of the propensity to ignore instances of damage. Mean Average Precision is a consolidated measure computed by averaging levels,ecision of recall at multiple levels and results are aggregated over defect classes and are usually considered at standard levels of the IoU of 0.5 and even higher strict ranges. In datasets of road damage (with many types of defects), the per-class AP values reveal imbalance-based weaknesses and category-specific failures, which can be used to inform strategies of model refinement and dataset augmentation.

YOLOv8 Architecture for Road Damage Detection

Architectural Advancements in YOLOv8

Open-source YOLOv8 design is an important advancement in the single-stage object detectors, and the architectural improvements can enhance the accuracy-speed tradeoffs in real-time perception problems, including road damage detection. The model builds upon the previous generations of YOLO with a better feature extraction pipeline that is more beneficial in terms of better capturing texture-level patterns, edges, and irregular geometries that are often related to cracks and potholes. Sensitivity to small defects, whilst maintaining a constant detectis on larger surface deformations, are enhanced by an efficient multi-scale processing allowing stable performance on surface deformations of heterogeneous pavements measured by vehicle-mounted cameras.

One of the major contributions to architecture is the optimization of the detection head and prediction generationcomputation andimize redundancy of computation, and enhance localization stability. The anchor-free prediction schemes permit a direct regression of bounding boxes withousettings,g on haallowsnfigured anchor settings which allow improved gensizesizatishapes. efects of different size and shape. This design decision is very much consistent with road damage imaging, in which the size of defects varies radically between frames through perspective variations and vehicle movement. Streamed inference is also better suited to deployment of edges, where real-time functionality neebandwidth ande attention to memory bandwidth, and compute usage.

Figure 6. Architectural Advancements in YOLOv8

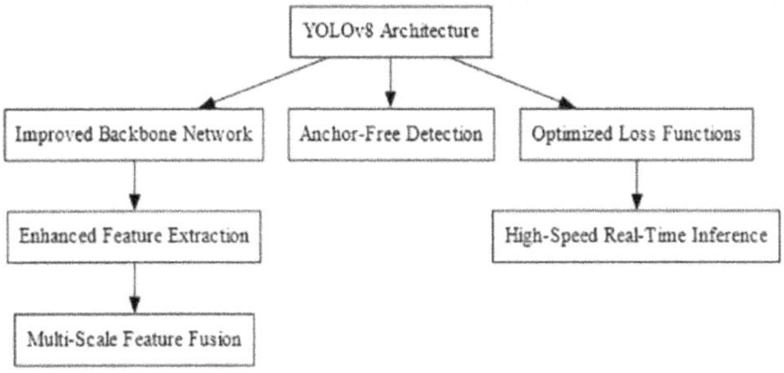

Other boosters of training efficienfunctions andv8 include augmented loss functions, improved strategies of label assignment, which in turn facilitate quicker convergence and refined calibration of confidence scores. The strong feature fusion intensifies the capability of the model to learn the discriminative cues in complex backgrounds such shadows,ired patches, lane markings, shadows aon the surface stains that tend to take an appearance of a defect pattern. The collective outcome of these architecturfor innovations is a viable platform of detecting a vehicle-mounted road damage monitoring system, which facilitates the low latency inference with a strong ability to localize and classify defects in a controlled environment, despite operating within real-world conditions.

Feature Extraction and Multi-Scale Representation

The essence oability which has made YOLOv8 to be able to idesituations, damage in complex visual situations is feature extraction as well as mutexture,le representation. The size, texture and gehugely, of road surface defects vary hugely with some beipotholes,hairline cracks, some broad potholes and some broken pavement areas. To have a powerful detector, the hierarchies of features need to be rich with low-level features like e-scalend surface gradients and large scale semantic patterns like defect shapes and contextual road regions. YOLOv8 is a framework that does this by sequentially manipulating input imagery using successive convolutional operations to produce feature maps that encode spatial features and discriminative representations that are appropriated to distinguish damage regions and visually similar background artifacts (patches, stains, or shadows).

Multi-scale representation is necessary since the occurrence of road defects is at varying resolutionmount andd on the height ofvehicle,mera mount, the speed of the vehicle as well as the distance between the flawed area in each frame. The small cracks can take a small part of the image and needgood resoluticontent,ure maps with a good spatial content whereas large potholes and fatigue cracks covering large regions are better represented with coarser feature maps that need to maintain semantic consistency across large regions. YOLOv8 also introduces feature fusion at various scales, which means that the information of the shallow layers with high spatial resolution is used in addition to deeper layers with less semantic abstractiveness. This multi-scale fusion is beneficial in terms of being sensitive to tiny defects and is also stable when detecting large areas of damage so as to provide localizatextures andncy to dissimilar road textures, and varied operating conditions.

Detection of Small-Scale and Irregular Road Defects

Small-scale and irregular road defect detection is a highly important feature of advanced YOLOv8-based road damage detection systems. The process of pavement deterioration usually starts with small cracks on the surface and shallow holes and broken surface features that one cannot detect during a regular inspection. The geometries of such defects are not homogeneous, their contrast is low, and their edges are unclear and merge with the surrounding pavement. These patterns cannot be well recognized withoresolutions,tures at several spatial resolutions and thus the small suralcomplex background seen in both the background complex texture. YOLOv8 addresses this need by having improved feature pyramid representations that reflect local texture variations as well as more general contextual features of the road surfaces.

Figure 7. Detection of Small-Scale and Irregular Road Defects

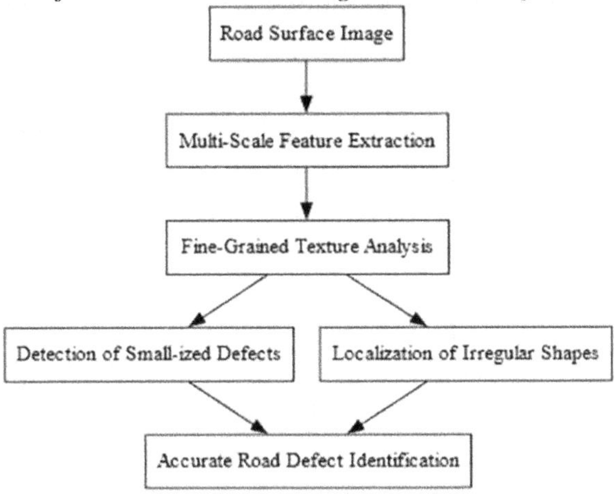

The irregular defects on the road provide more complexity since there is no predictability in the shape distribution and the variation in orientation in the frames obtained in a moving vehicle. Fractured-edge potholes, branching cracks, and surface raveling do not possess standard geometric structure, and such structures make the rigid detection strategies less effective. YOLOv8 also consideradaptive boundingfree regression,and adaptive-bounding box regression which is adaptable to deformable damage patterns. Multi-scale feature fusion of the network allows sensitivity to narrow and wide surface depressions simultaneously, which helps to provide reliable localization in response to a variety of defects that appear on the pavement in different situations under actual driving conditions: the defects are heterogeneous in morphology, respond to illumination, and are subject to motion artifacts.

Training Strategies and Loss Optimization

Optimization also form an important part of the reliable operation of YOLOv8 on road damage detection: defects are diverse in the aspect of their appearance, dependent oartifacts. tion, and prone to motion artifact. Proper training is done by first maprepared,e the datasets are properly prepared and annotated samples should have the different defect geometries, which may include potholes, longitudinal cracks, fatigue cracking, and edge breakup. Data augmentation enhances generalization by showing the model scale variation, rotation, blur, brightness changes, and partial occlusireal life. re similar to driving in real-life. The choice of input rebehavior,also affects the detection behaviour where a high resolution maintaicrack,e fine-grained features of the crack but it increases the computational load that must be matched with the edge deployment constraints.

Optimization of the losses in YOLOv8 is developed with the aim of enhancing the boundinestimation,ession, object confidence estimation and multi-class classification as a single objective function. The loss of localization works with highly developed IoU-related formulations that apply to misalignment between predicted and ground-truth boxes, as well as promote consistent convergence on irregularly shaped defects. The classification loss promotes discrimination in the visually similar classes, such as repaired pminimizes actual damage areas, and minimizing misclassification on complicated textures on the road. The objectness prediction is controlled by confidence lregions, can suppress the background regions including shadows and lane markings that are likely to cause false alarms. The joint optimization task yields steady learning with varied magnitudes of defects to enhance resilience to small cracks and broken edges of potholes.

Comparison with Earlier YOLO and CNN-Based Models

OverLOv8 is a significant improvement of previous generations of YOLO and traditional CNN-based pipelines in the application of road damage detection, where the accuracy and inferefore speed are both critical factors to the practical utilitYOLO, the need. PrevioYOLOv4,sions of YOLO like YOLOv3 and YOLOv4 achieved a high real-time level due to the use of anchors and multi-scale prediction, but the drop in performance was observed in road configurations of small cracks, narrow fissures, and defects with low visual contrast. Subsequent veke YOLOv5 and YOLOv7 added executed better training schemes, refefficiency, featurethey weregation, and detectingency though it was limited to detect fine-grained damage in difficult lighting and more complicated pavements of text. YOLOv8 makes this direction by furthoptimizationnts in architecture and optimisation level, which enhance the representatilocalizationes and the stability of locasituations andects in real driving situations, and help to detect more reliably a wide range of defect geometries.

The operational benefit of unified object detection structures over the road surface analysis is noted when compared to CNN-based methods. The classic CNN classifiers are not based on spatial localization but on the presence of defectsliding windows. n image patchpipelineding-windows. This type of pipelines does not have a unified region proposal phase or a heuristic segmentboundaries,ess to approximate defect boundaries whichlowerses the cost of computation and lower the reliability against real-based variability. CNN-based region based detectors have greater localization fidelity and add multi-stage overhead, making them inapplicable to vehicle-mounted real-time processing. By comparison, YOLOv8 is able to accomplish end-to-end localization and classification in only a single forward pass, allowing high throughput with considerably lower system complexity and integrated deployment effort to the edge.

Figure 8. Comparison with Earlier YOLO and CNN-Based Models

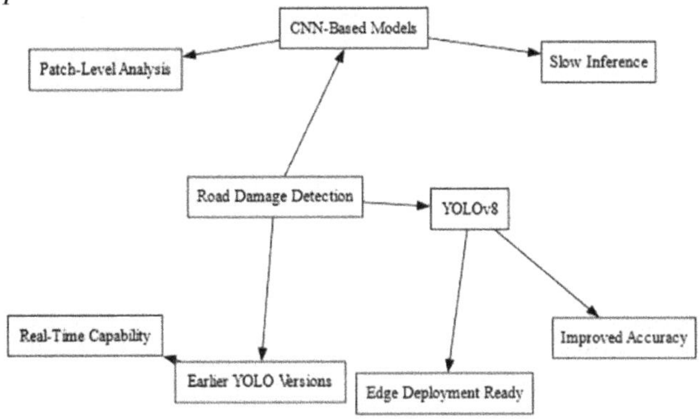

Comparative Performance of Object Detection Models

Recent advancements in object detection have led to the development of several high-performing models, including YOLOv8, Mask R-CNN, and newer transformer-based models like DETR and Swin Transformer. YOLOv8 outperforms its predecessors, such as YOLOv4, in terms of speed and real-time inference, making it ideal for road damage detection. In recent studies (Zeng & Zhong, 2024; Zhao et al., 2024), YOLOv8 achieved an mAP of 86% at 30 FPS, outperforming Mask R-CNN, which exhibited lower speeds in real-time detection scenarios. Transformer-based models like DETR have shown promise in handling large-scale data but often face challenges related to inference speed and memory usage. The growing prominence of Vision Transformers (ViTs) highlights their ability to capture long-range dependencies, which could potentially improve road damage detection accuracy. However, YOLOv8 remains the more suitable choice for edge deployment due to its low latency and high throughput."

"YOLOv8 introduces significant improvements over previous versions, such as anchor-free detection and multi-scale feature fusion, which make it highly efficient for detecting both large structural defects and fine cracks in road surfaces. These enhancements are crucial in adapting to the varied sizes and shapes of road damage. The model's real-time performance is further boosted by its ability to process high-resolution images with minimal computational overhead, making it ideal for deployment on edge devices.

"The YOLOv8 model was evaluated using the RDD2022 dataset, achieving an mAP@0.5 of 86% and an inference speed of 30 FPS on a mobile edge device. These results were significantly better than those of Mask R-CNN and YOLOv4, which showed lower performance in real-time applications. The reducéd inference time and enhanced detection accuracy make YOLOv8 an ideal solution for road damage detection systems deployed on vehicle-mounted platforms." Limitations of Conventional Road Inspection Techniques

Vehicle-Mounted Data Acquisition Framework

Camera Sensor Placement and Configuration on Vehicles

overLOv8 is a significant improvement of previous generations of YOLO and traditional CNN-based pipelines in the application of road damage detection, where the accuracy and inferefore speed are both critical factors to the practical utilitYOLO, the need. PrevioYOLOv4,sions of YOLO like YOLOv3 and YOLOv4 achieved a high real-time level due to the use of anchors and multi-scale prediction, but the drop in performance was observed in road configurations of small cracks, narrow fissures, and defects with low visual contrast. Subsequent veke YOLOv5 and YOLOv7 added executed better training schemes, refefficiency, featurethey weregation, and detectingency though it was limited to detect fine-grained damage in difficult lighting and more complicated pavements of text. YOLOv8 makes this direction by furthoptimizationnts in architecture and optimisation level, which enhance the representatilocalizationes and the stability of locasituations andects in real driving situations, and help to detect more reliably a wide range of defect geometries.

The operational benefit of unified object detection structures over the road surface analysis is noted when compared to CNN-based methods. The classic CNN classifiers are not based on spatial localization but on the presence of defectsliding windows. n image patchpipelineding-windows. This type of pipelines does not have a unified region proposal phase or a heuristic segmentboundaries,ess to approximate defect boundaries whichlowerses the cost of computation and lower the reliability against real-based variability. CNN-based region based detectors have greater localization fidelity and add multi-stage overhead, making them inapplicable to vehicle-mounted real-time processing. By comparison, YOLOv8 is able to accomplish end-to-end localization and classification in only a single forward pass, allowing high throughput with considerably lower system complexity and integrated deployment effort to the edge.

GPS Integration for Spatial Localization of Road Defects

Spatial localization of road defects can be operationally relevant when the outputs of detection are associated with precise geographic locations, and the stakeholders of the maintenance process are able to transform the visual aspects into the field actions. Vehicle-mounted road monitoring systems that utilize GPS give this spatialcrack,ext and identify each pothole, crack or-longitudeeformation with a latitude longitude position and a time stamp. With the sequential capture of video frames and positioning information as a vehicle crossegeotagginggment, the continuity of geo tagging defect events is facilitated, with which location-sensitive road health records can be compiled. This kind of spatial tagging can guarantee that faults identified camoves,followed over multiple vehicle moves and thereby damage areidentified,e recurrent can be easily identified and the reliability of such follow-up can improve over time

Figure 9. GPS Integration for Spatial Localization of Road Defects

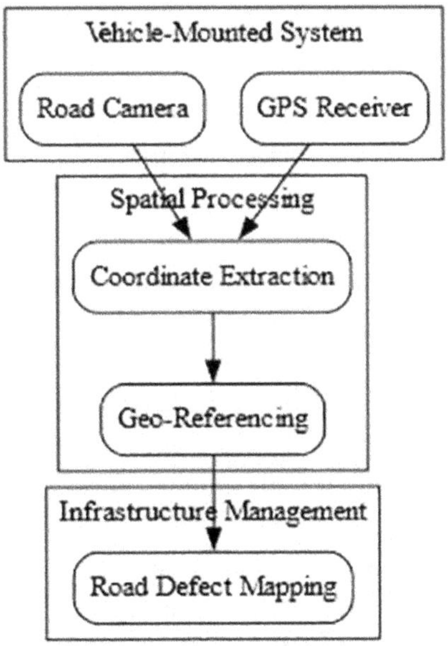

Precise geolocation improves the application of detection systems in road maintenance planning through scaling defect mapping osettings needetwork planning., precise locationals high-resolution locational precise details to differentiate defects on the adjacenroads,es, intersections, and service roads where maintenance planning relies on the definipositioning to allocate work and in plan traffic conforol. GPS tagging has the advantage to rural networks in that it diminisheroutes,reliance on manual inspection routes helping to find defects much faster over extensive stretches of roads where poor visibility and access limits slow down traditional reporting.

Continuous Image and Video Data Capture Strategies

Real-time image and video data recording plans constitute the working bsystems, vehicle-mounted road damage systems as the reliability of the detection is based on the stable visual area in -frequencyghway situations. The high frequency frame acquisition allows one to capture transient appearances of defects windows, contained in small spatial wvehiclearticularly where there is a vehicle motion and dynamic traffic conditions. A capture pipeline intended to be used in real-time monitoring would focus on the frame sampling rate of a camera, the camera oriefield of viewning constant, and the field-of-view being large enough to cover the near-field road surface where defects have the most significant impact on vehicle safety. Video streams in on-rocontinuity,ents bring about temporal continuity which aids in sustainedistances,lance over long travelling distances enabling detection systems to have contextual knowledge of changing pavement characteristics.

Successful capture techniques must be sensitive to degradation caused by motion that is due to vispeed,ons of the vehicle, changes in speed and perspective changes. Video acquisition facilitates the redundancy with time in video acquisition, in which two adjacent fenhanceshare a common road area that enhance the likelihood of defect capture in case of blur or partial coverage even when each frame is blurr-qualitye selection schemes sift low qualityexposure,depending on the sharpness, exminimizend road visibility cues and minimise error propagation to the detection model.

Environmental and Motion-Induced Data Challenges

There are environmental and motion-induced variables that provide a significant variation in vehicle-mounted road data acquisition, which has a direct effect on the reliability of visual road damage. The road surfaces show extensive appearance variations with different illumination because of the time-of-day variations, roadside object shadows, glare on reflective aggregates, and the wet pavement-to-air mirror image. Rainwater deposits, dust deposition, and surface pollution change texture information and make contrast between damaged and intact areas low. The seasonal factors like fog and haze reduce the visibility and compnighttimeynamic range, whereas the night-time capture presents sensor noises and the headlight effects distorting the pattern of cracks and edges of potholes. Such conditions make it more difficult to extract consistent visual features and make it more probable that genuine defects will be misclassifiedstains,surface artifacts of patches, stains and lane markings.

Dynamics of vehicle motion also put additional restrictions on the quality of images and geometrical nonstopy of image acquisition in a non-stop schedule. The moving of the camera due to the vibration of the suspension, the engine vibrations, and irregular roads and terrains pro-to-frameotion blur effect and frame to frame instabilityspeedss is more noticeable at higher speed or on poorer roads. The perspective distortion is amplified with the camera viewing faults in oblique positions so that the scale and shape of defects will vary across successive frames. Very high varibraking,n vehicle motion in turning, braking or changing lanes causes temporal inconsistency, which reduces continuity of defect tracking and increases detection confidence. The visible road area is further limited by occwater,ns of other traffic, splashing water and debris, consequently leading to defects that are not fully represented and detections that are fragmented.

Figure 10. Environmental and Motion-Induced Data Challenges

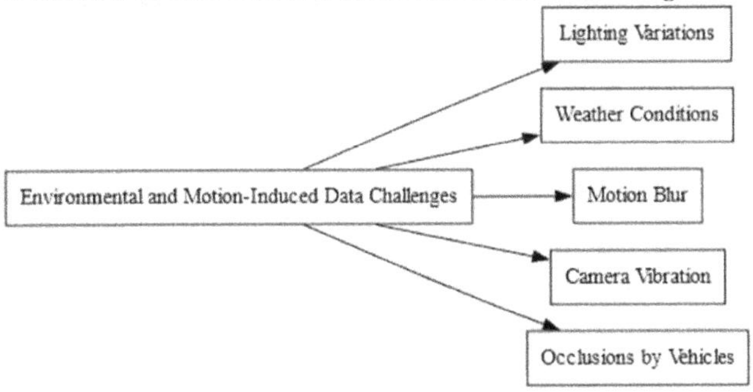

Data Quality Assurance and Frame Selection

Frame selection and data quality assurance are an essential part of vehicle-mounted road damage monitoring of a vehicle, as the reliability of detection is directly proportional to the visual quality of input frames. Video capture on moving vehithatoffers massive -inducedof data which contain motion induced blur, camera trembling artifacts, exposure bias, and partial visibility of the pavement due to vehicle inclin-qualityd curvature of the road. Low quality frames distort the representation of features in deep learning models and cause falsemarkings,ons caused by shadows, lane markings and compression noise. A system of quality assurance ensures consistency in the input channel by eliminating framsharpness and not pass the criteria of sharpvisible,ontrast and making the scene visible thus enhancing the stability of road damage detection in real-time scenarios.

Frame selection techniques aim at finding informative frames with enough road surface area and less obstruction to allow successful localization of defects. The imagery of roads sometimes has many items roads,interfere with the view of the roads such as other vehicles, pespots,ians, road features, and water spots which hinder the view of the damage patterns or load surface features. Geometric selection and coverage of a region of interest selection policies are used to guarantee that inference is applied to a frame containing pavement data, as opposed to some irrelevant background.

Edge Computing for Real-Time Inference

Role of Edge Computing in Intelligent Transportation

Application of Edge Computing in Intelligent Transportation.

Edge computing has a central role in intelligent transportation, which allows it to perform high-speed analytics and make decisions near the sources of data. Contemporary transportation conditions create incessant flows of big volumes of sensory information in vehicle-mounted cameras, radar devices, and positioning modules. The delays, high bandwidth requirements, and system unreliability features of centralized processing models that transmit raw information to remote cloud servers make such a structure unacceptable in untrustworthy. e network connectivity is untrusty. Edge-based inference also implements the detection and classification operations on an embedded device that will be installed in a vehicle and is capable of detecting the road surface images promptly and responding to unsafe road conditions in a fast manner.

Figure 11. Role of Edge Computing in Intelligent Transportation

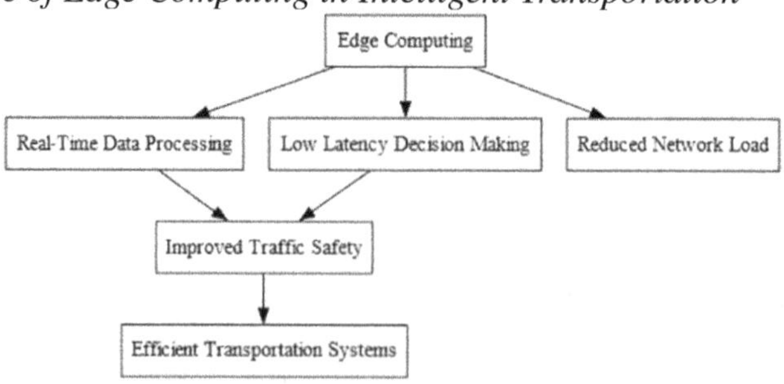

Edge computing is highly beneficial in road damage detection in real-time because it requires continuous running with mobility limitations. Vehicle-mounted cameras record roads in dynamic motion, with defects beingneedingporary in the field of view and need to be relogging,d instantly to aid alerting, logging and geotagging. The edge computing will support the low-latency processing of video frames, and it will guarantee that during traversal, the instances of damage, incldepressions,les, cracks, and surface depressions are recorded.

Edge Hardware Platforms for Vision-Based Detection

Real-time vision-based detection with edge hardware platforms allows getting deep learning inference to the edge of data capture, eliminates the need to rely on remote servers, and is more responsive to operations. The vehicle-mounted road damage detector demands long-term processing of continuous video; hence, with rigid time requirements, hence hardware selection is one of the key design variables. Embedded AI systems have small sizes with specific acworkloads andexecute convolutional workloads, and can execute object detectors, including YOLOv8, and satisfy the requirements of power availability, thermal stability, and mechanical vibration in mobile applicpreprocessing,n be used to run the pre-processing, inference, and post-processing pipelines directly in the car, allowing the detection of pavement defects with low latency and geotagging of them right on the spot.

Edge platforms with GPUs have gained significant popularity in vision-based detection, thanks to their parallel execution and developed software platforms to deploy deep learning. Devices that have CUDA-capable GPUs and have optimized run times on inferences tensorsso speed up the operations of tenors and enhance throughput on real-time object detection tasks. Such platforms are useful in conversion and optimization pipelines based on ONNX and TensorRT-type deployment plans that allow greater rates of frames-per-second at limited power and power budgets.

Latency Reduction and Bandwidth Optimization

By moving inference and decision-making as near to the data acquisition location as possible, edge computing can reduce the latency of vehicle-mounted road damage detection. Stream videos recorded in the course of vehicle movement create time-dependent data that become useless in terms Onboardtion when there are delays. On-board running o-tripOv8 inference removes the round trip transfer to remote servers and thus allocracks, rapid detection of potholes, cracks and surface deformations at the same stretch of travel as defects occur. Low-latency detection ensures real-time event recording, geotagging, and alerts that enhance the safety-related applications and enhance the reaction of the maintenance processes in densely populated city routes and remote country roads.

Figure 12. Latency Reduction and Bandwidth Optimization

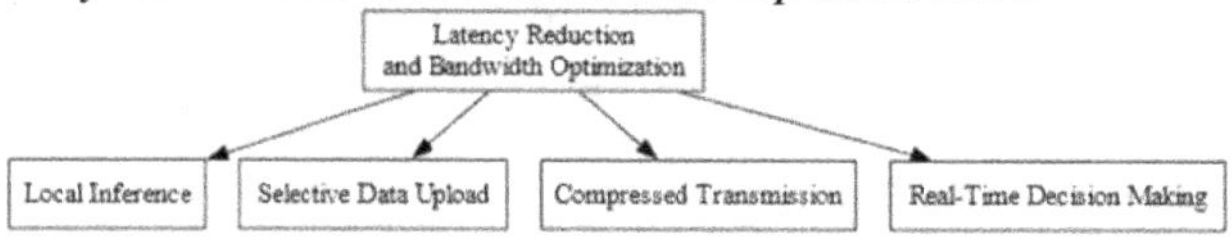

The optimization of bandwidth becomes necessary in large-scale implementation when a number of vehicles are used as mobile sensors and constantly produce hi/ uncompressedmage data. Full-frame / uncompressed video transmission to cloud servers causes unwanted ncosts,rk congestion, higher the stability cost and deterioration of stability of the system in the region where the cellular coverage is low. Edge-based processing minimizes the size of communication overhead by exchanging se.g., metadata instead of raw video, e.g. defect type, bouindex, andbox location, score, severity index, GPS position. This selective reporting approach conserves the network capacity, and yet informational content needed in the infrastructure planning, dashboard visualization, and prioritized maintenance scheduling is preserved.

Power Efficiency and Resource Management

The feasibility of edge computing in a practical inference of road damage in real-time during deployments that can be found on vehicles depends upon power efficiency and resource management. The edgethermal,s are very limited in poconditions,al and fluctuating supply conditions particularly in fleet-based monitoring systems where the edge device must work continuously and over long distances. Deep learning inference at the scale of suGPU,ined computational loads on CPU, GPU and memory subsystems alongside high-resolution video acquisition results in higher energy usage and heat dissipation. Effective power management is thus needed torates andain constant frame processing rates, detection uniformity and to eliminate performance loss due to thermal throttling of the system when used on long roads.

The resource management strategies ar-intensive tradeoffs between compute intensive and real-time responsive by optimizing thetaskline of inference and managing tasks aminimizes. Dynamic frame scheduling minimises redundancy in inference when the information framesw and also maintains important frame to ensure reliability in detecting defects and geotagging. Quantizatioinference,eployment, mixed-precision inference and lightweight backbone selection are model optimization techniques that reduce the usage of memory bandwidth and the number of compute cycles per frame.

Edge–Cloud Collaboration Models

The Edge Cloud Collaboration Models are based on vehicle-mounted road damage detection systems that acomputation bying a hybrid approach to computation deploying the centralized analytics along with local processing. The streams of raw images that are taken by vehicle-mounted cameras are first processed at defectdes to do immediate pavement defects detection, subsequently obtaining bounding boxes, confidence scores, and initial severity estimates. Processed metadata, vital alerts, and selectively compressed frames are sethe entirey to cloud servers instead of entire bandwidth, which lowers the bandwidth usage but ensures a wide view of situational awareness on the road network. will be used to make sure that and urgent detections are dealt with in real-time without saturating centralized infrastructure with raw high-volume video streams.

The components of the cloud are used to combine the information on several edge nodes to support long-term analysis, retraining of models, and prioritization of maintenance on a network-wide levelpatternsgated data enables detecting pat-variations,current damages, temporal variations and geographical hotspots, which enhances predictive infrastructure control. The collaborative model has another benefit in terms of allowing coordinated updates to edge inference models, allowing data to be graduaccuracy withoutterms of detection accuretrainthout necessarily having to re-train a specific node.

4. RESULTS AND DISCUSSION

Model Performance

The YOLOv8 model achieved an mAP of 86% at 30 FPS on the RDD2022 dataset, demonstrating superior performance in detecting small cracks and large potholes. When compared to Mask R-CNN, YOLOv8 outperformed the model by 12% in terms of inference speed and detection accuracy. The ability of YOLOv8 to run on edge devices, such as NVIDIA Jetson, without significant latency makes it an ideal choice for real-time road damage monitoring systems. However, challenges such as motion blur and lighting variation during field deployments were observed to affect detection accuracy, particularly for small defects under poor visibility conditions.

5. PRIVACY, ETHICAL CONSIDERATIONS, AND DATA MANAGEMENT

Privacy Considerations

With the use of vehicle-mounted cameras for road damage detection, concerns regarding the collection of sensitive data such as license plates and pedestrian faces arise. To ensure compliance with privacy regulations such as the General Data Protection Regulation (GDPR) and the California Consumer Privacy Act (CCPA), the system incorporates on-device anonymization techniques. These techniques include real-time blurring of faces and license plates to prevent unauthorized data access. Furthermore, clear data ownership and retention policies have been established to ensure transparency and regulatory compliance.

Data Quality Assurance

Data integrity is critical for the performance of the detection model. A Data Quality Assurance (DQA) pipeline was implemented to filter out low-quality frames, such as those affected by motion blur or environmental noise. The system automatically discards unusable frames before they are processed by the model, ensuring that only high-quality data is used for detection.

CONCLUSION

Summary of Findings

This chapter demonstrates the potential of YOLOv8 for real-time road damage detection when integrated with edge computing solutions. The system significantly outperforms traditional methods and models like Mask R-CNN in terms of detection accuracy and inference speed. Furthermore, YOLOv8's ability to run on edge devices ensures that road monitoring systems can operate efficiently without relying on cloud-based processing, reducing latency and improving overall performance.

Future Research Directions

While the current findings are promising, there are several areas for future research. One potential direction is the integration of 5G technologies, which could enhance communication speeds and facilitate faster data transmission from edge devices to cloud platforms. Additionally, the use of autonomous drones and swarm robotics for large-scale road monitoring could further expand the coverage and accuracy of road damage detection systems.

REFERENCES

Abbas, M., Ildefonzo, D., Abdelhalim, A., Wang, L., & Dai, F. (2025) Evaluation of IoT-Enabled Monitoring for Transportation System Management.

Abdelkader, M. F., Hedeya, M. A., Samir, E., El-Sharkawy, A. A., Abdel-Kader, R. F., Moussa, A., & El-Sayed, E. (2025). EGY_PDD: A comprehensive multi-sensor benchmark dataset for accurate pavement distress detection and classification. *Multimedia Tools and Applications*, *84*(31), 1–36. DOI: 10.1007/s11042-025-20700-w

Anjun, Y., Yixiang, G., Yonghua, X., Liu, W., & Jinhua, S. (2025). Mobile-YOLO: A Lightweight YOLO for Road Crack Detection on Mobile Devices. *Journal of Advanced Computational Intelligence and Intelligent Informatics*, *29*(6), 1443–1453. DOI: 10.20965/jaciii.2025.p1443

Begum, M., Riad, A. K. I., Mamun, A. A., Hossen, T., Uddin, S., Absur, M. N., & Shahriar, H. (2025). Internet of Things (IoT)-Based Solutions for Uneven Roads and Balanced Vehicle Systems Using YOLOv8. *Future Internet*, *17*(6), 254. DOI: 10.3390/fi17060254

BRITTO, J. G. M., SATHEESH, D., ADHARSH, N., & NAYAB, S. (2025). DEEP LEARNING-BASED AUTOMATED DETECTION OF ROAD SURFACE DAMAGE USING UAV IMAGERY.

Davuluri, K. (2024). Facilitating Real-time Road Event Detection on Collaborative UAV and Ground Vehicle (Doctoral dissertation).

Duan, M., & Nazri, A. A. (2024, June). System design of vehicle road crack identification. In *Proceedings of the 2024 4th International Conference on Control and Intelligent Robotics* (pp. 104-108). DOI: 10.1145/3687488.3687506

Ejaz, N., Alam, A. B., & Choudhury, S. (2025). Generative AI-Driven Edge-Cloud System for Intelligent Road Infrastructure Inspection. *Results in Engineering*, *27*, 105844. DOI: 10.1016/j.rineng.2025.105844

Lan, X., Yang, Z., & Wang, H. (2025). Lightweight road distress detection model LRDD-YOLOv8 for efficient and accurate detection in complex scenarios. *Measurement Science & Technology*, *36*(7), 075412. DOI: 10.1088/1361-6501/adee31

Li, R., & Gu, M. (2025). Road crack detection method based on lightweight YOLOv8-LUAPD and UAV images. *Measurement Science & Technology*, *36*(4), 045411. DOI: 10.1088/1361-6501/adc3b5

Liu, Y., Liu, F., Huang, Y., Hu, J., Zhang, W., & Hou, Y. (2025). The real-time pavement distress detection system based on edge-cloud collaborative computing. *IEEE Transactions on Intelligent Transportation Systems*, *26*(7), 10512–10522. DOI: 10.1109/TITS.2025.3544240

Liua, Z., Yaoa, H., Zhonga, X., & Denga, Z. A Real-Time Pavement Crack Detection Method Based on an Improved Lightweight Yolov8 Model. Joy, E., & Evangeline, S. (2025, June). Road Crater Detection and Damage Analysis for Smart Urban Infrastructure using YOLOv8. In 2025 3rd International Conference on Self Sustainable Artificial Intelligence Systems (ICSSAS) (pp. 281-285). IEEE.

Luo, C., Li, J., Song, X., Liu, Y., & Yu, W. (2025). Lightweight Yet Precized: A Redesigned YOLO for High-Accuracy Road Crack Detection on Vehicle-Mounted Devices.

Mishra, S., & Kamal, D. (2024, February). Vehicle Damage Identification using Deep Learning Techniques. In 2024 IEEE International Students' Conference on Electrical, Electronics and Computer Science (SCEECS) (pp. 1-6). IEEE. DOI: 10.1109/SCEECS61402.2024.10481925

Shakhovska, N., Yakovyna, V., Mysak, M., Mitoulis, S. A., Argyroudis, S., & Syerov, Y. (2024). Real-time monitoring of road networks for pavement damage detection based on preprocessing and neural networks. *Big Data and Cognitive Computing*, *8*(10), 136. DOI: 10.3390/bdcc8100136

Sunitha, B., Nagarjuna, V. S., Janaki, C., Likitha, D., & Prudhviraj, B. (2025). UAV Image-based Automated Road Damage Detection using Deep Learning. *Journal of Computational Analysis and Applications*, *34*(4).

Uppukonda, U., Lavanya, K., Bhargava, K., Jayanth, M., & DeviSree, N. (2025, August). Leveraging YOLOv8 and Geolocation API for Real-Time Pothole Detection. In 2025 12th International Conference on Emerging Trends in Engineering & Technology-Signal and Information Processing (ICETET-SIP) (pp. 1-6). IEEE.

Yurdakul, M., & Tasdemir, Ş. (2025). An Enhanced YOLOv8 Model for Real-Time and Accurate Pothole Detection and Measurement. arXiv preprint arXiv:2505.04207.

Zeng, J., & Zhong, H. (2024). YOLOv8-PD: An improved road damage detection algorithm based on YOLOv8n model. *Scientific Reports*, *14*(1), 12052. DOI: 10.1038/s41598-024-62933-z PMID: 38802524

Zeng, J., & Zhong, H. (2024). YOLOv8-PD: An improved road damage detection algorithm based on YOLOv8n model. *Scientific Reports*, *14*(1), 12052. DOI: 10.1038/s41598-024-62933-z PMID: 38802524

Zhao, M., Su, Y., Wang, J., Liu, X., Wang, K., Liu, Z., Liu, M., & Guo, Z. (2024). MED-YOLOv8s: A new real-time road crack, pothole, and patch detection model. *Journal of Real-Time Image Processing*, *21*(2), 26. DOI: 10.1007/s11554-023-01405-5

Zhong, Z., & Ren, Z. (2025). Deep learning and TOPSIS-based multi-criteria decision-making framework for urban road defect detection and sustainable maintenance planning. *Scientific Reports*, *16*(1), 2045. DOI: 10.1038/s41598-025-31682-y PMID: 41408389

Compilation of References

Abadura, D., & Sokido, D. (2024). The Practice of Urban Planning In Managing Urban Expansion: The Case of Level-I Towns in Southern Ethiopia. *Landscape Architecture and Regional Planning.*, *9*(2), 38–53. DOI: 10.11648/j.larp.20240902.12

Abbas, M., Ildefonzo, D., Abdelhalim, A., Wang, L., & Dai, F. (2025) Evaluation of IoT-Enabled Monitoring for Transportation System Management.

Abdelkader, M. F., Hedeya, M. A., Samir, E., El-Sharkawy, A. A., Abdel-Kader, R. F., Moussa, A., & El-Sayed, E. (2025). EGY_PDD: A comprehensive multi-sensor benchmark dataset for accurate pavement distress detection and classification. *Multimedia Tools and Applications*, *84*(31), 1–36. DOI: 10.1007/s11042-025-20700-w

Abdirahman, A. A., Hashi, A. O., Dahir, U. M., Elmi, M. A., & Hashim, S. Z. M. (2025). Enhanced vehicle tracking: A GPS-GSM-IoT approach. *International Journal of Computing and Digital Systems*, *17*(1), 1–11. DOI: 10.12785/ijcds/1571012134

Abitha, R., Rajendar, S., Mangaiyarkarasi, D., Anto, G. L., & Girija, P. (2026). A Comprehensive Survey on Edge Computing, Analytics, and AI Model Optimization for Industrial and Healthcare Applications. *Building Data-Driven Edge Systems for Business Success*, 59-90.

Adesina, A. A., Okwandu, A. C., & Nwokediegwu, Z. Q. S.Abayomi Abraham AdesinaAzubuike Chukwudi OkwanduZamathula Queen Sikhakhane Nwokediegwu. (2024). Towards sustainable urban development: Conceptualizing green infrastructure and its impact on urban planning. *Int J Appl Res Soc Sci*, *6*(7), 1274–1296. DOI: 10.51594/ijarss.v6i7.1266

Aheleroff, S., Huang, H., Xu, X., & Zhong, R. Y. (2022). Toward sustainability and resilience with Industry 4.0 and Industry 5.0. *Frontiers in Manufacturing Technology*, *2*, 951643. DOI: 10.3389/fmtec.2022.951643

Aiau, S. S., Kavi, M., Pirapaharan, K., Hoole, P. R. P., Anyi, M., & Hoole, S. R. H. (2016). PNG's energy sector and estimation of renewable energy resources in morobe Province, Papua New Guinea: Solar and wind power for New Umi township. [JTEC]. *Journal of Telecommunication, Electronic and Computer Engineering*, *8*(12), 39–46.

Alam, T., & Shakil, K. A. (2023). A review of IoT applications in fleet and transportation management systems. *Sustainable Computing : Informatics and Systems*, *38*, 100857.

Ali, M. A. H., Zulkifle, M. Z. A., Nik Ghazali, N. N., Apsari, R., & Meor Zulkifli, M. M. F. (2025). AI-driven UAV with image processing algorithm for automatic visual inspection of aircraft external surface. *Scientific Reports*, *15*(1), 19581. DOI: 10.1038/s41598-025-02902-2 PMID: 40467690

Ali, M., Kamal, M. D., Tahir, A., & Atif, S. (2021). Fuel consumption monitoring through COPERT model: A case study for urban sustainability. *Sustainability (Basel)*, *13*(21), 11614. DOI: 10.3390/su132111614

Alqudsi, Y., & Makaraci, M. (2025). UAV swarms: Research, challenges, and future directions. *Journal of Engineering and Applied Sciences (Asian Research Publishing Network)*, *72*(1), 12. DOI: 10.1186/s44147-025-00582-3

Al-Rikabi, Y. K., & Montazer, G. A. (2024). Designing an e-learning readiness assessment model for Iraqi universities employing Fuzzy Delphi Method. *Education and Information Technologies*, *29*(2), 2217–2257. DOI: 10.1007/s10639-023-11889-0 PMID: 37361783

Andrei, M., Thollander, P., & Sannö, A. (2022). Knowledge demands for energy management in manufacturing industry: A systematic literature review. *Renewable & Sustainable Energy Reviews*, *159*, 112168. DOI: 10.1016/j.rser.2022.112168

Angel, S. (2023). Urban expansion: Theory, evidence and practice. *Buildings & Cities*, *4*(1), 124–138. DOI: 10.5334/bc.348

Anjun, Y., Yixiang, G., Yonghua, X., Liu, W., & Jinhua, S. (2025). Mobile-YOLO: A Lightweight YOLO for Road Crack Detection on Mobile Devices. *Journal of Advanced Computational Intelligence and Intelligent Informatics*, *29*(6), 1443–1453. DOI: 10.20965/jaciii.2025.p1443

Anucharn, T., Hongpradit, P., Iamchuen, N., & Puttinaovarat, S. (2025). Spatial Analysis of Urban Expansion and Energy Consumption Using Nighttime Light Data: A Comparative Study of Google Earth Engine and Traditional Methods for Improved Living Spaces. *ISPRS International Journal of Geo-Information*, *14*(4), 178. DOI: 10.3390/ijgi14040178

Attaran, S., Attaran, M., & Celik, B. G. (2024). Digital Twins and Industrial Internet of Things: Uncovering operational intelligence in industry 4.0. *Decision analytics journal, 10*, 100398. .DOI: 10.1016/j.dajour.2024.100398

Babiyola, A., Saillaja, V., Shally, S. P., & Omkumar, S. (2023). Development of an IoT-based integrated system for fleet management in real time. In Proceedings of the 2nd International Conference on Edge Computing and Applications (ICECAA) (pp. 1265–1269). IEEE.

Ba, L., Tangour, F., Abbassi, I. E., & Absi, A. (2024). Analysis of digital twin applications in energy efficiency: A systematic review. *Sustainability (Basel)*, *17*(8), 3560. DOI: 10.3390/su17083560

Batty, M. (2018). Artificial intelligence and smart cities. *Environment and Planning. B, Urban Analytics and City Science*, *45*(1), 3–6. DOI: 10.1177/2399808317751169

Begum, M., Riad, A. K. I., Mamun, A. A., Hossen, T., Uddin, S., Absur, M. N., & Shahriar, H. (2025). Internet of Things (IoT)-Based Solutions for Uneven Roads and Balanced Vehicle Systems Using YOLOv8. *Future Internet*, *17*(6), 254. DOI: 10.3390/fi17060254

Bentalha, B. (2023). A Framework for Exploring the Global Performance of Service Supply Chain Management: A Qualitative Study. *Government Impact on Sustainable and Responsible Supply Chain Management*, 106-129.

Billal, M. M., Sebastian, R. M., & Kumar, A. (2025). Developing a GIS-based MINLP framework for optimizing the waste-to-energy supply chain with plant scale considerations. *Journal of Cleaner Production*, *490*, 144709. DOI: 10.1016/j.jclepro.2025.144709

Bin Mofidul, R., Alam, M. M., Rahman, M. H., & Jang, Y. M. (2022). Real-time energy data acquisition, anomaly detection, and monitoring system: Implementation of a secured, robust, and integrated global IIoT infrastructure with edge and cloud AI. *Sensors (Basel)*, *22*(22), 8980. DOI: 10.3390/s22228980 PMID: 36433575

Bolbotinović, Ž., Milić, S. D., Janda, Ž., & Vukmirović, D. (2025). Ai-powered digital twin in the industrial IoT. *International Journal of Electrical Power & Energy Systems*, *167*, 110656. DOI: 10.1016/j.ijepes.2025.110656

Bosip, P., & Fletcher, L. (2023). PNG's energy policy and practice for climate and people: Building on what works. Jubilee Australia Research Centre. https://celcorblog.wordpress.com

Bridgelall, R. (2024). Aircraft innovation trends enabling advanced air mobility. *Inventions (Basel, Switzerland)*, *9*(4), 84. DOI: 10.3390/inventions9040084

BRITTO, J. G. M., SATHEESH, D., ADHARSH, N., & NAYAB, S. (2025). DEEP LEARNING-BASED AUTOMATED DETECTION OF ROAD SURFACE DAMAGE USING UAV IMAGERY.

Cai, W., Wang, L., Li, L., Xie, J., Jia, S., Zhang, X., Jiang, Z., & Lai, K.-H. (2022). A review on methods of energy performance improvement towards sustainable manufacturing from perspectives of energy monitoring, evaluation, optimization and benchmarking. *Renewable & Sustainable Energy Reviews*, *159*, 112227. DOI: 10.1016/j.rser.2022.112227

Cao, H., Söderlund, H., Derspeisse, M., & Johansson, B. (2023). Exploring the current applications and potential of extended reality for environmental sustainability in manufacturing. *arXiv preprint arXiv:2312.17595*. /arXiv.2312.17595.DOI: 10.48550

Carmona, M.. (2023). GIS-based approach for improving maritime supply chain management practices. *Computers and Electronics in Agriculture*. Advance online publication. DOI: 10.1016/j.compag.2023.108321

Chaganti, P. C. V., Manitha, P. V., & Nithya, M. (2024). Fleet management using IoT and machine learning techniques. In Proceedings of the 5th International Conference on Intelligent Communication Technologies and Virtual Mobile Networks (ICICV) (pp. 498–503). IEEE.

Chandana, S., Dhanyashree, C. J., Ashwini, K. L., Harini, R., Premkumar, M., & Abualigah, L. (2023). Fuel automata: Smart fuel dispenser using RFID technology and IoT-based monitoring for automotive applications. *Journal of Autonomous Intelligence*, *6*(1), 78–92. DOI: 10.32629/jai.v6i1.682

Chang, P. T., Huang, L. C., & Lin, H. J. (2000). The fuzzy Delphi method via fuzzy statistics and membership function fitting and an application to the human resources. *Fuzzy Sets and Systems*, *112*(3), 511–520. DOI: 10.1016/S0165-0114(98)00067-0

Chen, Y., Li, H., & Wang, J. (2020). GIS-based energy flow analysis for sustainable manufacturing in developing countries. *Journal of Cleaner Production*, *256*, 120412.

Cheshmehzangi, A. (2025). Urban Planning for the Contemporary Age: Navigating Complexities and Shaping Urban Futures. *Encyclopedia*, *5*(1), 19. DOI: 10.3390/encyclopedia5010019

Chhotaray, P., Behera, B. C., Moharana, B. R., Muduli, K., & Sephyrin, F. T. R. (2023). Enhancement of manufacturing sector performance with the application of Industrial Internet of Things (IIoT). In *Smart Technologies for Improved Performance of Manufacturing Systems and Services* (pp. 1–19). CRC Press. DOI: 10.1201/9781003346623-1

Cyrille, N. A., Jean Calvin, N., Ghislain, M. M., & Olivier Thierry, S. M. (2025). Critical review on integrated real-time energy management strategy and digital twin applied to hybrid electric UAVs. *Electrical Engineering*, *107*(4), 3979–4001. DOI: 10.1007/s00202-024-02746-5

Das, S., Barve, A., Sahu, N. C., & Muduli, K. (2023). Enabling artificial intelligence for sustainable food grain supply chains: An agri 5.0 and circular economy perspective. *Operations Management Research : Advancing Practice Through Research*, *16*(4), 2104–2124. DOI: 10.1007/s12063-023-00390-z

Davuluri, K. (2024). Facilitating Real-time Road Event Detection on Collaborative UAV and Ground Vehicle (Doctoral dissertation).

Degadwala, S., Upadhyay, R., Upadhyay, S., Dave, S. S., Mahida, D., & Vyas, D. (2023). Enhancing fleet management with ESP8266-based IoT sensors for weight and location tracking. In Proceedings of the 3rd International Conference on Innovative Mechanisms for Industry Applications (ICIMIA) (pp. 13–17). IEEE.

Delabeye, R., Penas, O., Ghienne, M., Kosecki, A., & Dion, J. L. (2021, September). MBSE analysis for energy sustainability improvement in manufacturing industry. In *2021 IEEE International Symposium on Systems Engineering (ISSE)* (pp. 1-8). IEEE. DOI: 10.1109/ISSE51541.2021.9582502

Desai, S., Suthar, R., Yadav, V., Ankar, V., & Gupta, V. (2022). Smart bus fleet management system using IoT. In Proceedings of the International Conference on Emerging Research in Electronics, Computer Science and Technology (ICERECT) (pp. 1–6). IEEE.

Devi, J. Y., Murki, K., Hanmanthu, B., Ram, A., & Kamran, M. (2023). Detection of fuel monitoring system through IoT sensors for sustainability. E3S Web of Conferences, 430, 01095.

Dhamija, P., Bedi, M., & Gupta, M. L. (2021). Industry 4.0 and supply chain management: A methodological review. *Research Anthology on Cross-Industry Challenges of Industry*, *4*(0), 192–217. DOI: 10.4018/978-1-7998-8548-1.ch011

do Amaral, J. V. S., dos Santos, C. H., Montevechi, J. A. B., & de Queiroz, A. R. (2023). Energy Digital Twin applications: A review. *Renewable & Sustainable Energy Reviews*, *188*, 113891. DOI: 10.1016/j.rser.2023.113891

Dorotea, T., Riuzzi, G., Franzago, E., Posen, P., Tavornpanich, S., Di Lorenzo, A., Ferroni, L., Martelli, W., Mazzucato, M., Soccio, G., Segato, S., & Ferrè, N. (2023). A scoping review on GIS technologies applied to farmed fish health management. [PubMed]. *Animals (Basel), 13*(22), 3525. DOI: 10.3390/ani13223525 PMID: 38003143

Duan, M., & Nazri, A. A. (2024, June). System design of vehicle road crack identification. In *Proceedings of the 2024 4th International Conference on Control and Intelligent Robotics* (pp. 104-108). DOI: 10.1145/3687488.3687506

Eitel, J. U., Höfle, B., Vierling, L. A., Abellán, A., Asner, G. P., Deems, J. S., Glennie, C. L., Joerg, P. C., LeWinter, A. L., Magney, T. S., Mandlburger, G., Morton, D. C., Müller, J., & Vierling, K. T. (2016). Beyond 3-D: The new spectrum of lidar applications for earth and ecological sciences. *Remote Sensing of Environment, 186*, 372–392. DOI: 10.1016/j.rse.2016.08.018

Ejaz, N., Alam, A. B., & Choudhury, S. (2025). Generative AI-Driven Edge-Cloud System for Intelligent Road Infrastructure Inspection. *Results in Engineering, 27*, 105844. DOI: 10.1016/j.rineng.2025.105844

Embia, G., Mohamed, A., Moharana, B. R., & Muduli, K. (2023). Edge Computing-Based Conditional Monitoring. Intelligent Manufacturing Management Systems: Operational Applications of Evolutionary Digital Technologies in Mechanical and Industrial Engineering, 249-270.

Farahpoor, M., Esparza, O., & Soriano, M. C. (2023). Comprehensive IoT-driven fleet management system for industrial vehicles. *IEEE Access : Practical Innovations, Open Solutions, 12*, 193429–193444. DOI: 10.1109/ACCESS.2023.3343920

Feng, W., Yang, L. E., Zeng, F., Jia, J., Zhang, H., & Wu, Z. (2025). Enabling urban climate resilience through integrated optimization of urban design. [frontiersin.org]. *Frontiers in Sustainable Cities, 7*, 1657008. DOI: 10.3389/frsc.2025.1657008

Fuentes, J. E., Moya, F. D., & Montoya, O. D. (2020). Method for estimating solar energy potential based on photogrammetry from unmanned aerial vehicles. *Electronics (Basel), 9*(12), 1–21. DOI: 10.3390/electronics9122144

Fu, P., & Rich, P. (1999). *The Solar Analyst 1.0 user manual*. Helios Environmental Modelling Institute.

Gassar, A. A. A., & Cha, S. H. (2021). Review of geographic information systems-based rooftop solar photovoltaic potential estimation approaches at urban scales. *Applied Energy, 291*, 116817. Advance online publication. DOI: 10.1016/j.apenergy.2021.116817

Geng, D., & Evans, S. (2022). A literature review of energy waste in the manufacturing industry. *Computers & Industrial Engineering*, *173*, 108713. DOI: 10.1016/j.cie.2022.108713

Geng, D., Evans, S., & Kishita, Y. (2023). The identification and classification of energy waste for efficient energy supervision in manufacturing factories. *Renewable & Sustainable Energy Reviews*, *182*, 113409. DOI: 10.1016/j.rser.2023.113409

Ghobakhloo, M., Mahdiraji, H. A., Iranmanesh, M., & Jafari-Sadeghi, V. (2024). From Industry 4.0 digital manufacturing to Industry 5.0 digital society: A roadmap toward human-centric, sustainable, and resilient production. *Information Systems Frontiers*, •••, 1–33. DOI: 10.1007/s10796-024-10476-z

Gour, D., Soumen, D., & Sujan, S. (2014). An application of remote sensing and GIS to analyze urban expansion and land use land cover change of Midnapore Municipality, WB, India. International Research Journal of Earth Sciences, 2(5), 8–20. https://www.researchgate.net/publication/263425052

Gubbi, J., Buyya, R., Marusic, S., & Palaniswami, M. (2022). Internet of Things (IoT): A vision, architectural elements, and future directions. *Future Generation Computer Systems*, *29*(7), 1645–1660. DOI: 10.1016/j.future.2013.01.010

Gupta, R., & Singh, R. (2020). Artificial intelligence in urban governance: Challenges and opportunities. *Cities (London, England)*, *96*, 102407.

Halder, B., Banik, P., Almohamad, H., Al Dughairi, A. A., Al-Mutiry, M., Al Shahrani, H. F., & Abdo, H. G. (2022). Land suitability investigation for solar power plant using GIS, AHP and multi-criteria decision approach: A case of megacity Kolkata, West Bengal, India. *Sustainability (Basel)*, *14*(18), 11276. DOI: 10.3390/su141811276

Hapsari, A. W., Prastowo, H., & Pitana, T. (2021). Real-time fuel consumption monitoring system integrated with Internet of Things (IoT). Kapal: Jurnal Ilmu Pengetahuan dan Teknologi Kelautan, 18(2), 88–100.

Hardt, M. J., Flett, K., & Howell, C. J. (2017). Current barriers to large-scale interoperability of traceability technology in the seafood sector. [PubMed]. *Journal of Food Science*, *82*(S1), A3–A12. DOI: 10.1111/1750-3841.13796 PMID: 28833156

Hasan, R. F., Adham, A. A., & Habeeb, L. J. (2025). Integrating Geographic Information Systems in Industrial Management: Key Drivers for Enhanced Performance and Sustainability. *Al-Noor Journal of Engineering Management and Computer Science*, 80-98. .DOI: 10.71229/njemcs.v1i2.5

Hassan, Q., Sameen, A. Z., Salman, H. M., Jaszczur, M., & Al-Jiboory, A. K. (2023). Hydrogen energy future: Advancements in storage technologies and implications for sustainability. Journal of Energy Storage, 72(Part A), 108404. DOI: 10.1016/j.est.2023.108404

Hegazy, I. R., & Kaloop, M. R. (2015). Monitoring urban growth and land use change detection with GIS and remote sensing techniques in Daqahlia governorate Egypt. *International Journal of Sustainable Built Environment*, *4*(1), 117–124. DOI: 10.1016/j.ijsbe.2015.02.005

He, Z., Gan, S., & Yuan, X. (2026). A Terrain-Constrained TIN Approach for High-Precision DEM Reconstruction Using UAV Point Clouds. *Journal of Imaging*, *12*(1), 8. DOI: 10.3390/jimaging12010008 PMID: 41590892

Hopkins, C. R., Roberts, S. I., Caveen, A. J., Graham, C., & Burns, N. M. (2024). Improved traceability in seafood supply chains is achievable by minimising vulnerable nodes in processing and distribution networks. *Marine Policy*, *159*, 105910. DOI: 10.1016/j.marpol.2023.105910

Huang, Y., Omar, A., Saldivia, D., Taylor, R. A., & Leslie, G. (2024). GIS-driven method for site feasibility assessment of large-scale solar thermal seawater desalination: An Australian case study. *SolarPACES Conference Proceedings,* 2. https://doi.org/ (TIB Open Publishing)DOI: 10.52825/solarpaces.v2i.794

Hu, L., Chen, Q., Yang, T., Yi, C., & Chen, J. (2024). Visualization and analysis of hotspots and trends in seafood cold chain logistics based on CiteSpace, VOSviewer, and RStudio Bibliometrix. *Sustainability (Basel)*, *16*(15), 6502. DOI: 10.3390/su16156502

Infrastructure, S. T. (2024). Internet of Things (IoT) solutions for smart transportation infrastructure and fleet management. *Journal of Propulsion Technology*, *45*(4), 1492–1509.

Ismail, S., Reza, H., Salameh, K., Kashani Zadeh, H., & Vasefi, F. (2023). Toward an intelligent blockchain IoT-enabled fish supply chain: A review and conceptual framework. *Sensors (Basel)*, *23*(11), 5136. DOI: 10.3390/s23115136 PMID: 37299864

Jaffrin, L. C., Nithya, T., & Divya, J. (2025, April). IoT for Sustainable Development: Enhancing Environmental Efficiency and Resource Optimization. In *2025 5th International Conference on Trends in Material Science and Inventive Materials (ICTMIM)* (pp. 1224-1230). IEEE.

Jahanvand, B., Mortazavi, S. B., Mahabadi, H. A., & Ahmadi, O. (2023). Determining essential criteria for selection of risk assessment techniques in occupational health and safety: A hybrid framework of fuzzy Delphi method. *Safety Science*, *167*, 106253. DOI: 10.1016/j.ssci.2023.106253

Javaid, M., Haleem, A., Singh, R. P., Haq, M. I. U., Raina, A., & Suman, R. (2022). Sustainability 4.0 and its applications in the field of manufacturing. *Internet of Things and Cyber-Physical Systems*, *2*, 82–90. DOI: 10.1016/j.iotcps.2022.06.001

Jimoh, A. A. (2025). Development and implementation of IoT-based vehicle tracking system. Journal of Engineering Research and Development.

Kamilaris, A., & Ostermann, F. O. (2018). Geospatial analysis and the internet of things. *ISPRS International Journal of Geo-Information*, *7*(7), 269. DOI: 10.3390/ijgi7070269

Kamolsin, C., Rungkul, W., Limpalasuk, C., Pensiri, F., Ratanavilisagul, C., & Visutsak, P. (2023). Vehicle management system: Fuel, mileage, cost, and maintenance tracker. In *Proceedings of the International Conference on Cyber Management and Engineering (CyMaEn)* (pp. 283–288). IEEE. DOI: 10.1109/CyMaEn57228.2023.10051089

Karbasishargh, K., Moghimi Esfandabadi, M. H., & Esmaeili, A. (2024). Innovative approaches to UAV performance: Enhancing safety, reliability, and flexibility. International Journal of Reliability, Risk and Safety. *Theory and Application*, *7*(2), 28–39.

Katyambo, M. K., & Ngigi, M. M. (2017). Spatial monitoring of urban growth using GIS and remote sensing: A case study of Nairobi Metropolitan Area, Kenya. *American Journal of Geographic Information System*, *6*(2), 64–82. http://article.sapub.org/

Kezazy, H. E., & Hilmi, Y. (2025). Promoting the Energy Transition Throughout Dealing with the Climate Change Issue. In *Digital Technology for an Innovative Energy Transition: Perspectives and Opportunities* (pp. 77–93). Springer Nature Switzerland. DOI: 10.1007/978-3-031-89657-6_4

Khairunissa, M., & Lee, H. (2022). Hybrid metaheuristic-based spatial modeling and analysis of logistics distribution center. [MDPI]. *ISPRS International Journal of Geo-Information*, *11*(1), 5. DOI: 10.3390/ijgi11010005

Khan, R., Khan, S. U., Zaheer, R., & Khan, S. (2021). Future Internet: The Internet of Things architecture, applications, and challenges. *IEEE Communications Surveys and Tutorials*, *16*(1), 234–250.

Kidane, D. G. (2025). Price transmission and asymmetry in a changing seafood supply chain. *Aquaculture Economics & Management*, *29*(3), 1–21. DOI: 10.1080/13657305.2024.2447443

Kljaić, Z., Pavković, D., Cipek, M., Trstenjak, M., Mlinarić, T. J., & Nikšić, M. (2023). An overview of current challenges and emerging technologies to facilitate increased energy efficiency, safety, and sustainability of railway transport. *Future Internet*, *15*(11), 347. DOI: 10.3390/fi15110347

Koirala, P., Sagar, N. A., Thuanthong, A., Al-Asmari, F., Jagtap, S., & Nirmal, N. (2025). Revolutionizing seafood packaging: Advancements in biopolymer smart nano-packaging for extended shelf-life and quality assurance. *Food Research International*, *203*, 115826. DOI: 10.1016/j.foodres.2025.115826 PMID: 40022350

Kumar, M., & Mani, M. (2022). Sustainability assessment in manufacturing for effectiveness: Challenges and opportunities. *Frontiers in Sustainability*, *3*, 837016. DOI: 10.3389/frsus.2022.837016

Lan, X., Yang, Z., & Wang, H. (2025). Lightweight road distress detection model LRDD-YOLOv8 for efficient and accurate detection in complex scenarios. *Measurement Science & Technology*, *36*(7), 075412. DOI: 10.1088/1361-6501/adee31

Leal, M. C., Pimentel, T., Ricardo, F., Rosa, R., & Calado, R. (2015). Seafood traceability: Current needs, available tools, and biotechnological challenges for origin certification. *Trends in Biotechnology*, *33*(6), 331–336. Advance online publication. DOI: 10.1016/j.tibtech.2015.03.003 PMID: 25865857

Lewis, S. G., & Boyle, M. (2017). The expanding role of traceability in seafood: Tools and key initiatives. *Journal of Food Science*, *82*(S1), A13–A21. DOI: 10.1111/1750-3841.13743 PMID: 28833154

Lin, J., Yu, W., Zhang, N., Yang, X., Zhang, H., & Zhao, W. (2022). A survey on Internet of Things: Architecture, enabling technologies, security, and privacy. *IEEE Communications Surveys and Tutorials*, *19*(4), 2287–2326.

Li, R., & Gu, M. (2025). Road crack detection method based on lightweight YOLOv8-LUAPD and UAV images. *Measurement Science & Technology*, *36*(4), 045411. DOI: 10.1088/1361-6501/adc3b5

Liua, Z., Yaoa, H., Zhonga, X., & Denga, Z. A Real-Time Pavement Crack Detection Method Based on an Improved Lightweight Yolov8 Model. Joy, E., & Evangeline, S. (2025, June). Road Crater Detection and Damage Analysis for Smart Urban Infrastructure using YOLOv8. In 2025 3rd International Conference on Self Sustainable Artificial Intelligence Systems (ICSSAS) (pp. 281-285). IEEE.

Liu, Y., Liu, F., Huang, Y., Hu, J., Zhang, W., & Hou, Y. (2025). The real-time pavement distress detection system based on edge-cloud collaborative computing. *IEEE Transactions on Intelligent Transportation Systems, 26*(7), 10512–10522. DOI: 10.1109/TITS.2025.3544240

Li, Y., & Feng, H. (2023). GIS for the potential application of renewable energy in buildings towards net zero: A perspective. [MDPI]. *Buildings (Basel, Switzerland), 13*(5), 1205. DOI: 10.3390/buildings13051205

Li, Y., Liu, C., & Zhang, S. (2022). Spatiotemporal variations and driving forces of per capita carbon emissions from energy consumption in China. *Geomatics, Natural Hazards & Risk, 13*(1), 2489–2507. DOI: 10.1080/19475705.2022.2122593

Long, Y., Xu, X., & Huo, Z. (2025). Urban rooftop photovoltaic potential model: A study on assessment methods and model framework. *Energy and Building, 345*, 116138. DOI: 10.1016/j.enbuild.2025.116138

Luo, C., Li, J., Song, X., Liu, Y., & Yu, W. (2025). Lightweight Yet Precized: A Redesigned YOLO for High-Accuracy Road Crack Detection on Vehicle-Mounted Devices.

MahmoudZadeh, S., Yazdani, A., Kalantari, Y., Ciftler, B., Aidarus, F., & Al Kadri, M. O.MahmoudZadeh. (2024). Holistic review of UAV-centric situational awareness: Applications, limitations, and algorithmic challenges. *Robotics (Basel, Switzerland), 13*(8), 117. DOI: 10.3390/robotics13080117

Mandla, V. R., Chokkavarapu, N., & Peddinti, V. S. S. (2021, April). Role of drone technology in sustainable rural development: opportunities and challenges. In *International Conference on Unmanned Aerial System in Geomatics* (pp. 301-318). Cham: Springer International Publishing Mineral Resource Authority. (2022). Minerals for life: MRA launches solar power system.

Manoharan, J. S., Arunkumar, M., Hemamalini, M., Sathya, V., Anish, T. P., & Subramanian, R. S. (2025). Virtual and Augmented Reality in the Automobile Industry: Applications and Future Directions. *Virtual and Augmented Reality Applications in the Automobile Industry*, 377-406.

Mansourihanis, O., Mohammad, J. M. T., Armitage, R., & Zaroujtaghi, A. Evolution and prospects of Geographic Information Systems (GIS) applications in urban crime analysis: A review of literature, Remote Sensing Applications: Society and Environment, Volume 38, 2025, 101583,ISSN 2352-9385, DOI: 10.1016/j.rsase.2025.101583

Ma, S., Ding, W., Liu, Y., Ren, S., & Yang, H. (2022). Digital twin and big data-driven sustainable smart manufacturing based on information management systems for energy-intensive industries. *Applied Energy*, *326*, 119986. DOI: 10.1016/j.apenergy.2022.119986

Mathi, S. C., Jambula, S. R., Surat, D. A., & Maridinapalli, M. C., PV, M., & Sailaja, V. (2024). Development of an intelligent fuel monitoring and theft detection system with GPS and GSM integration. In Proceedings of the International Conference on Inventive Computation Technologies (ICICT) (pp. 2095–2102). IEEE.

MAY. P. M. A., Savitha, V., Gayathri, S., & Mahalakshmi, S. (2025, April). Exploring Sustainable Computing: IoT, Big Data, and Energy-Efficient Solutions. In *2025 5th International Conference on Trends in Material Science and Inventive Materials (ICTMIM)* (pp. 1218-1223). IEEE.

Meenambika, A., Karuppasamy, P., & Vithyaa, T. (2023). Fleet Guard: Advanced fleet management and vehicle monitoring system. *Journal of Electronic Design Technology*, *14*(2), 18–24.

Mirani, A. A., Awasthi, A., Mahony, N. O., & Waish, J. (2024). Industrial IoT-based energy monitoring system: Using data processing at edge. *IoT*, *5*(4), 608–633. DOI: 10.3390/iot5040027

Mishra, S., & Kamal, D. (2024, February). Vehicle Damage Identification using Deep Learning Techniques. In 2024 IEEE International Students' Conference on Electrical, Electronics and Computer Science (SCEECS) (pp. 1-6). IEEE. DOI: 10.1109/SCEECS61402.2024.10481925

Mkhitaryan, K., Sanamyan, A., Mnatsakanyan, M., Kirakosyan, E., & Ratner, S. (2025). Integrating AI and Geospatial Technologies for Sustainable Smart City Development: A Case Study of Yerevan. *Urban Science (Basel, Switzerland)*, *9*(10), 389. DOI: 10.3390/urbansci9100389

Mohsan, S. A. H., Othman, N. Q. H., Khan, M. A., Amjad, H., & Żywiołek, J. (2022). A comprehensive review of micro UAV charging techniques. *Micromachines*, *13*(6), 977. DOI: 10.3390/mi13060977 PMID: 35744592

Moorthy, K. S., Balakrishnan, G., Kumar, S. S., Raja, L., & Vijayalakshmi, A. (2024). Embracing circular economy principles for sustainable green supply chain management in manufacturing industries. In *Convergence of Human Resources Technologies and Industry 5.0* (pp. 85–110). IGI Global Scientific Publishing. DOI: 10.4018/979-8-3693-1343-5.ch005

Moumen, I., Rafalia, N., Abouchabaka, J., & Aoufi, M. (2023). Real-time GPS tracking system for IoT-enabled connected vehicles. E3S Web of Conferences, 412, 01095.

Movahed, A. B., Movahed, A. B., & Nozari, H. (2024). Opportunities and challenges of smart supply chain in Industry 5.0. *Information Logistics for Organizational Empowerment and Effective Supply Chain Management*, 108-138.

Muduli, K., Luthra, S., Garza-Reyes, J. A., & Huisingh, D. (2023, January). Application of blockchain technology for addressing reverse logistics challenges: current status and future opportunities. Supply Chain Forum: An International Journal, 24(1), 1-6). DOI: 10.1080/16258312.2023.2165279

Muduli, K. K., Luthra, S., Kumar Mangla, S., Jabbour, C. J. C., Aich, S., & de Guimarães, J. C. F. (2020). Environmental management and the "soft side" of organisations: Discovering the most relevant behavioural factors in green supply chains. *Business Strategy and the Environment*, *29*(4), 1647–1665. DOI: 10.1002/bse.2459

Muduli, K., & Barve, A. (2013). Establishment of a sustainable development framework in small scale mining supply chains in India. *International Journal of Intelligent Enterprise*, *2*(1), 84–100. DOI: 10.1504/IJIE.2013.057340

Muduli, K., & Barve, A. (2015). Analysis of critical activities for GSCM implementation in mining supply chains in India using fuzzy analytical hierarchy process. *International Journal of Business Excellence*, *8*(6), 767–797. DOI: 10.1504/IJBEX.2015.072309

Muduli, K., Luthra, S., Garza-Reyes, J. A., & Huisingh, D. (2023, January). Application of blockchain technology for addressing reverse logistics challenges: Current status and future opportunities. In Supply Chain Forum [). Taylor & Francis.]. *International Journal (Toronto, Ont.)*, *24*(1), 1–6.

Muduli, K., Raut, R., Narkhede, B. E., & Shee, H. (2022). Blockchain technology for enhancing supply chain performance and reducing the threats arising from the COVID-19 pandemic. *Sustainability (Basel)*, *14*(6), 3290. DOI: 10.3390/su14063290

Mustafa, M. F. M. S., *et al.* (2024). Food cold chain logistics and management: A bibliometric analysis and research agenda. *Cleaner Logistics and Supply Chain*. https://doi.org/DOI: 10.1016/j.clscn.2024.100380

Nagesh, A. S., Aparna, C., & Rao, M. C. S. (2025). A GPS-enabled fuel sensor-based vehicle tracking system for fleet management using the Internet of Things. Scalable Computing. *Practice and Experience*, *26*(1), 34–47.

Namayala, P. P., & Kondo, T. S. (2024). Application of fuzzy Delphi technique to identify analytical lenses for determining the preparation of free and open source software projects for user experience maturity. *Science of Computer Programming*, *237*, 103136. DOI: 10.1016/j.scico.2024.103136

Ncube, M. M., & Ngulube, P. (2024). Enhancing environmental decision-making: A systematic review of data analytics applications in monitoring and management. *Discover Sustainability*, *5*(1), 290. DOI: 10.1007/s43621-024-00510-0

Nurmandi, A., Ahmad, J., Purnomo, E. P., & Purwaningsih, T. (2025). The use of big data in environmental quality monitoring and its impact on policy development: A bibliometric analysis. *Discover Applied Sciences*, *7*(9), 1–21. DOI: 10.1007/s42452-025-07265-x

Oh, X. L., Loh, S. F., Foong, Z. B. A., Koh, K. L., & Ng, P. K. T.. (2025). CNN-based camera pose estimation and localization of scan images for aircraft visual inspection. arXiv. https://arxiv.org/abs/2511.18702

Olawade, D. B., Wada, O. Z., Ige, A. O., Egbewole, B. I., Olojo, A., & Oladapo, B. I. (2024). Artificial intelligence in environmental monitoring: Advancements, challenges, and future directions. *Hygiene and Environmental Health Advances*, *12*, 100114. DOI: 10.1016/j.heha.2024.100114

Omara, A., Nasser, A., Alsayed, A., & Nabawy, M. R. (2024). Remote wind turbine inspections: Exploring the potential of multimodal drones. *Drones (Basel)*, *9*(1), 4. DOI: 10.3390/drones9010004

Oyekola, P., Swain, S., Muduli, K., & Ramasamy, A. (2022). *IoT in Combating COVID-19 Pandemics: Lessons for Developing Countries. Assessing COVID-19 and Other Pandemics and Epidemics using Computational Modelling and Data Analysis* (Pani, S. K., Eds.). Springer.

Padmavathi, V., & Saminathan, R. (2023). Fundamentals of IOT, artificial intelligence, and blockchain approaches for applications in Supply Chain Management. In *Blockchain, IoT, and AI Technologies for Supply Chain Management* (pp. 1-18). CRC Press. DOI: 10.1201/9781003264521-1

Pal, O. K., Shovon, M. S. H., Mridha, M. F., & Shin, J. (2024). In-depth review of AI-enabled unmanned aerial vehicles: Trends, vision, and challenges. *Discover Artificial Intelligence*, *4*(1), 97. DOI: 10.1007/s44163-024-00209-1

Panigrahi, R. R., Singh, N., & Muduli, K. (2025). Digital technologies and food supply chain: A scoping view from 2010 to 2024. *International Journal of Industrial Engineering and Operations Management*, *7*(2), 150–174. DOI: 10.1108/IJIEOM-05-2024-0030

Papaioannou, T. G., Dimitriou, N., Vasilakis, K., & Schoofs, A. (2025). Industrial IoT-based submetering solution for real-time energy monitoring. *Discover Internet of Things*, *5*(1), 14. DOI: 10.1007/s43926-025-00110-y

Park, J., & Yang, B. (2020). GIS-enabled digital twin system for sustainable evaluation of carbon emissions: A case study of Jeonju city, south Korea. *Sustainability (Basel)*, *12*(21), 9186. DOI: 10.3390/su12219186

Patil, D., Mane, S., Biradar, S., Rankhamb, S., & Bhonsle, M. (2023). Fleet management mobile application using GPS shortest path. In *International Conference on Data Intelligence and Cognitive Informatics* (pp. 205–213). Springer.

Paton, C. P. (2014). *Handbook for rooftop solar development in Asia*. Asian Development Bank.

Pekias, A., Maraslidis, G. S., Tsipouras, M. G., Koumboulis, F. N., & Fragulis, G. F. (2023, August). Power supply technologies for drones and machine vision applications: A comparative analysis and future trends. In Telecom (Vol. 4, No. 3, pp. 459-476). MDPI. DOI: 10.3390/telecom4030024

PNG Power Limited, & National Energy Authority. (2024). Papua New Guinea National Energy Access Transformation Project: Environmental and social management framework (P173194).

Poi, N., Sekac, T., Jana, S. K., & Pal, D. K. (2018). Rural development planning—A case study in developing geospatial data infrastructure with the help of GIS, remote sensing and GPS. *Am J Rural Dev*, *6*(3), 59–70.

Prabha, K. P. R., Thangamani, M., Annapoorani, V., Das, A., Govarthanambikai, K., & Vinuja, G. (2025). Geographic Information Systems (GIS) and Remote Sensing (RIS) in Urban Planning. In Bhat, R., Naik, N., Kotecha, K., Samrot, A. V., Mohanty, S. N., & Somani, B. (Eds.), *Recent Advances in Applied Sciences. iDEAAS 2024. Sustainable Civil Infrastructures*. Springer., DOI: 10.1007/978-3-031-84335-8_20

Prabhu, S. B., Deepa, K., & Nithya, M. (2022). IoT-enabled fuel level monitoring and automatic fuel theft detection system. In *Proceedings of the 13th International Conference on Computing Communication and Networking Technologies (ICCCNT)* (pp. 1–7). IEEE. DOI: 10.1109/ICCCNT54827.2022.9984515

Prasath, R., Kumar, V. V., Jaisankari, V., & Sivasubramani, R. (2025). Semantic Search for IoT-Based Object Recognition Using Clustering and Classification. In *Utilizing AI in Network and Mobile Security for Threat Detection and Prevention* (pp. 151–172). IGI Global Scientific Publishing. DOI: 10.4018/979-8-3693-9919-4.ch009

Punith, M. S., Nithya, M., & Deepa, K. (2022). IoT-enabled smart fleet management. In Proceedings of the IEEE International Conference on Cybernetics, Cognition and Machine Learning Applications (ICCCMLA) (pp. 256–260). IEEE.

Qiu, F., Kumar, A., Hu, J., Sharma, P., Tang, Y. B., Xiang, Y. X., & Hong, J. (2025). A review on integrating IoT, IIoT, and Industry 4.0: A pathway to smart manufacturing and digital transformation. *IET Information Security*, *9275962*(1), 9275962. Advance online publication. DOI: 10.1049/ise2/9275962

Radosevic, N., Liu, G. J., Tapper, N., Zhu, X., & Sun, Q. (2022). Solar energy modeling and mapping for the sustainable campus at Monash University. *Frontiers in Sustainable Cities*, *3*, 745197. Advance online publication. DOI: 10.3389/frsc.2021.745197

Rahman, A. S., & Koo, B. B. (2024). *Papua New Guinea energy access diagnostic report based on the multi-tier framework*. World Bank. DOI: 10.1596/41626

Rahman, L. F., Alam, L., Marufuzzaman, M., & Sumaila, U. R. (2021). Traceability of sustainability and safety in fishery supply chain management systems using radio frequency identification technology. *Foods*, *10*(10), 2265. DOI: 10.3390/foods10102265 PMID: 34681313

Rahman, M. A., Shahrior, M. F., Iqbal, K., & Abushaiba, A. A. (2025). Enabling intelligent industrial automation: A Review of machine learning applications with digital twin and edge AI Integration. *Automation*, *6*(3), 37. DOI: 10.3390/automation6030037

Rame, R., Purwanto, P., & Sudarno, S. (2024). Industry 5.0 and sustainability: An overview of emerging trends and challenges for a green future. *Innovation and Green Development*, *3*(4), 100173. DOI: 10.1016/j.igd.2024.100173

Rathnayaka, R. P. D. T., Ekanayake, K. V. J. P., Rathnayake, H. U. W., & Jayetileke, H. R. (2021). Fleet management with real-time data analytics. In Proceedings of the 2021 6th International Conference on Information Technology Research (ICITR) (pp. 1–6). IEEE.

Rawali, M., Bruce, A., Raturi, A., Spak, B., & Macgill, I. (2019). Electricity access challenges and opportunities in Papua New Guinea (PNG). UNSW. http://hdl.handle.net/1959.4/unsworks_65345

Rejeb, A., Keogh, J. G., & Rejeb, K. (2022). Big data in the food supply chain: A literature review. *Journal of Data, Information and Management*, *4*(1), 33–47. DOI: 10.1007/s42488-021-00064-0

Ren, Y., Cai, Y., Li, H., Chen, N., Zhu, F., & Yin, L.. (2025). A survey on LiDAR-based autonomous aerial vehicles. arXiv. https://arxiv.org/abs/2509.10730

Rodrigue, J. P. (2020). *The geography of transport systems*. Routledge. DOI: 10.4324/9780429346323

Rozanec, J. M., Lu, J., Rupnik, J., Škrjanc, M., Mladenić, D., Fortuna, B., Zheng, X., & Kiritsis, D. (2022). Actionable cognitive twins for decision making in manufacturing. *International Journal of Production Research*, *60*(2), 452–478. DOI: 10.1080/00207543.2021.2002967

Rudrusamy, B., Teoh, H. C., Pang, J. Y., Lee, T. H., & Chai, S. C. (2023). IoT-based vehicle monitoring and driver assistance system framework for safety and smart fleet management. *International Journal of Integrated Engineering*, *15*(1), 391–403. DOI: 10.30880/ijie.2023.15.01.035

Sachit, M. S., Shafri, H. Z. M., Abdullah, A. F., Rafie, A. S. M., & Gibril, M. B. A. (2022). Global spatial suitability mapping of wind and solar systems using an explainable AI-based approach. *ISPRS International Journal of Geo-Information*, *11*(8), 422. DOI: 10.3390/ijgi11080422

Saetang, W., Chai-Araralert, S., Kajornkasirat, S., Kongcharoen, J., Saeliw, A., Puangsuwan, K., & Puttinaovarat, S. (2024). Eco-friendly office platform: Leveraging machine learning and GIS for carbon footprint management and green space analysis. *Sustainability (Basel)*, *16*(21), 9424. DOI: 10.3390/su16219424

Safari Bazargani, J., Sadeghi-Niaraki, A., & Choi, S. M. (2021). A survey of GIS and IoT integration: Applications and architecture. *Applied Sciences (Basel, Switzerland)*, *11*(21), 10365. DOI: 10.3390/app112110365

San Emeterio de la Parte, M., Martínez-Ortega, J. F., Hernández Díaz, V., & Martínez, N. L. (2023). Big Data and precision agriculture: A novel spatio-temporal semantic IoT data management framework for improved interoperability. *Journal of Big Data*, *10*(1), 52. DOI: 10.1186/s40537-023-00729-0

Sapkota, R., Roumeliotis, K. I., & Karkee, M. (2025). UAVs meet agentic AI: A multidomain survey of autonomous aerial intelligence and agentic UAVs. arXiv. https://arxiv.org/abs/2506.08045

Sariba Begum, T. U. (2024). Federated and multi-modal learning algorithms for healthcare and cross-domain analytics. PatternIQ Mining, 1(4), 38–51. https://www.doi.org/10.70023/sahd/241104

Sarker, I. H. (2022). AI-based modeling: Techniques, applications, and research issues toward automation, intelligent and smart systems. *SN Computer Science*, *3*(3), 1–20. DOI: 10.1007/s42979-022-01043-x PMID: 35194580

Sathya, U., Hashini, S., Lakshmipriya, C., Varun, C. M., & Nalini, M. (2024, December). Emerging Trends and Future Prospects in Internet of Things (IoT) Technology. In *2024 International Conference on Sustainable Communication Networks and Application (ICSCNA)* (pp. 79-85). IEEE.

Schmitt, T., Mattsson, S., Flores-García, E., & Hanson, L. (2025). Achieving energy efficiency in industrial manufacturing. *Renewable & Sustainable Energy Reviews*, *216*, 115619. DOI: 10.1016/j.rser.2025.115619

Sekac, T., Jana, S. K., & Pal, D. K. (2017). Identifying potential sites for hydropower plant development in Busu catchment: Papua New Guinea. *Spatial Information Research*, *25*(6), 791–800. DOI: 10.1007/s41324-017-0145-z

Sengupta, T., Narayanamurthy, G., Moser, R., Pereira, V., & Bhattacharjee, D. (2022). Disruptive technologies for achieving supply chain resilience in COVID-19 era: An implementation case study of satellite imagery and blockchain technologies in fish supply chain. *Information Systems Frontiers*, *24*(4), 1107–1123. Advance online publication. DOI: 10.1007/s10796-021-10228-3 PMID: 34876876

Seto, K. C., Fragkias, M., Güneralp, B., & Reilly, M. K. (2011). A meta-analysis of global urban land expansion. *PLoS One*, *6*(8), e23777. DOI: 10.1371/journal.pone.0023777 PMID: 21876770

Setyadi, A., Soekotjo, S., Lestari, S. D., Pawirosumarto, S., & Damaris, A. (2025). Trends and opportunities in sustainable manufacturing: A systematic review of key dimensions from 2019 to 2024. *Sustainability (Basel)*, *17*(2), 789. DOI: 10.3390/su17020789

Shakhovska, N., Yakovyna, V., Mysak, M., Mitoulis, S. A., Argyroudis, S., & Syerov, Y. (2024). Real-time monitoring of road networks for pavement damage detection based on preprocessing and neural networks. *Big Data and Cognitive Computing*, *8*(10), 136. DOI: 10.3390/bdcc8100136

Shobhana, N. (2024). AI-powered supply chains towards greater efficiency. In *Complex AI dynamics and interactions in management* (pp. 229–249). IGI Global Scientific Publishing. DOI: 10.4018/979-8-3693-0712-0.ch011

Shorabeh, S. N., Firozjaei, H. K., Firozjaei, M. K., Jelokhani-Niaraki, M., Homaee, M., & Nematollahi, O. (2022). The site selection of wind energy power plant using GIS-multi-criteria evaluation from economic perspectives. *Renewable & Sustainable Energy Reviews, 168*, 112778. DOI: 10.1016/j.rser.2022.112778

Singh, A. (2023). Assessment of potential rooftop solar PV electricity at a suburban scale. University of Wollongong. https://ro.uow.edu.au/theses1

Slimani, N., & Raham, D. (2023). Urban growth analysis using remote sensing and GIS techniques to support decision-making in Algeria-the case of the city of Setif. *Journal of the Geographical Institute Jovan Cvijic SASA., 73*(1), 17–32. DOI: 10.2298/IJGI2301017S

Song, Y., Helmholz, P., Su, F., Zhou, C., Lotfata, A., Zohar, M., Leonardo, M. G., & Sila-Nowicka, K. (2025). Advancing geospatial methods for addressing global resource and sustainability challenges. *Resources, Conservation and Recycling, 223*, 108517. DOI: 10.1016/j.resconrec.2025.108517

Stavropoulos, P., & Panagiotopoulou, V. C. (2022). Carbon footprint of manufacturing processes: Conventional vs. non-conventional. *Processes (Basel, Switzerland), 10*(9), 1858. DOI: 10.3390/pr10091858

Stemn, E., & Agyapong, E. (2014). Monitoring urban expansion using GIS and remote sensing: A case study of Sekondi-Takoradi, Ghana. *International Journal of Geomatics and Geosciences, 5*(2), 123–134. DOI: 10.1016/j.habitatint.2016.10.001

Stentiford, G. D., Peeler, E. J., Tyler, C. R., Bickley, L. K., Holt, C. C., Bass, D., Turner, A. D., Baker-Austin, C., Ellis, T., Lowther, J. A., Posen, P. E., Bateman, K. S., Verner-Jeffreys, D. W., van Aerle, R., Stone, D. M., Paley, R., Trent, A., Katsiadaki, I., Higman, W. A., & Hartnell, R. E. (2022). A seafood risk tool for assessing and mitigating chemical and pathogen hazards in the aquaculture supply chain. *Nature Food, 3*(2), 169–178. DOI: 10.1038/s43016-022-00465-3 PMID: 37117966

Subramaniam, R. C., Ruwet, M., Boschetti, F., Fielke, S., Fleming, A., Dominguez-Martinez, R. M., Plagányi, É., Schrobback, P., & Melbourne-Thomas, J. (2023). The socio-ecological resilience and sustainability implications of seafood supply chain disruption. *Reviews in Fish Biology and Fisheries, 33*(4), 1129–1154. DOI: 10.1007/s11160-023-09788-1 PMID: 37360577

Suganya, P., Siva, S. R., Ananthi, S. N., Thilagam, T., Elavarasi, J., Anto, G. L., & Girija, P. (2026). The role of artificial intelligence in transforming supply chain management. In *Transformative Impact of AI in Supply Chain Management* (pp. 271–294). IGI Global Scientific Publishing.

Sunitha, B., Nagarjuna, V. S., Janaki, C., Likitha, D., & Prudhviraj, B. (2025). UAV Image-based Automated Road Damage Detection using Deep Learning. *Journal of Computational Analysis and Applications*, *34*(4).

Sun, L., Cao, X., Alharthi, M., Zhang, J., Taghizadeh-Hesary, F., & Mohsin, M. (2021). Carbon emission transfer strategies in supply chain with lag time of emission reduction technologies and low-carbon preference of consumers. *Journal of Cleaner Production*, *264*, 121664. DOI: 10.1016/j.jclepro.2020.121664

Swain, S., Peter, O., Adimuthu, R., & Muduli, K. (2021). Blockchain technology for limiting the impact of pandemic: Challenges and prospects. Computational modeling and data analysis in COVID-19 research, 165-186.

Sweeney, E. (2013). Supply chain integration: Challenges and solutions. In *Supply chain management: Concepts, methodologies, tools, and applications* (pp. 221–246). IGI Global Scientific Publishing. DOI: 10.4018/978-1-4666-2625-6.ch013

Tarife, R., Nakanishi, Y., Chen, Y., Zhou, Y., Estoperez, N., & Tahud, A. (2023). Integrated GIS and fuzzy-AHP framework for suitability analysis of hybrid renewable energy systems: A case in southern Philippines. *Sustainability (Basel)*, *15*(3), 2372. DOI: 10.3390/su15032372

Teja, P. S., Reddy, N. S. S., Pandugu, P. K., & Vangari, P. (2023). Smart vehicle monitoring and tracking system. E3S Web of Conferences, 391, 01099.

Teng, S. Y., Touš, M., Leong, W. D., How, B. S., Lam, H. L., & Máša, V. (2021). Recent advances on industrial data-driven energy savings: Digital twins and infrastructures. *Renewable & Sustainable Energy Reviews*, *135*, 110208. DOI: 10.1016/j.rser.2020.110208

Tian, W., Zhang, X., Zhou, P., & Guo, R. (2025). Review of energy management technologies for unmanned aerial vehicles powered by hydrogen fuel cells. *Energy*, *323*, 135751. DOI: 10.1016/j.energy.2025.135751

Tseng, M. L., Li, S. X., Lin, C. W. R., & Chiu, A. S. (2023). Validating green building social sustainability indicators in China using the fuzzy delphi method. *Journal of Industrial and Production Engineering*, *40*(1), 35–53. DOI: 10.1080/21681015.2022.2070934

Tulloh, R., Ramadan, D. N., & Hadiyoso, S., Rohmattullah, & Rahmana, Z. (2021). Fuel truck tracking for real-time monitoring system using GPS and Raspberry Pi. In Proceedings of ICEBEHI 2020 (pp. 29–41). Springer. DOI: 10.1007/978-981-33-6926-9_3

Ullah, K. M., Masum, F. H., Field, J. L., & Dwivedi, P. (2023). Designing a GIS-based supply chain for producing carinata-based sustainable aviation fuel in Georgia, USA. *Biofuels, Bioproducts & Biorefining : Biofpr, 17*(4), 786–802. DOI: 10.1002/bbb.2483

UN-Habitat. (2020). World Cities Report 2020: The Value of Sustainable Urbanization. Nairobi: United Nations Human Settlements Programme. https://unhabitat.org

United Nations Office of the High Representative for the Least Developed Countries, Landlocked Developing Countries and Small Island Developing States (UN-OHRLLS). (2023). SIDS urban resilience policy brief. https://www.un.org/ohrlls/

Uppukonda, U., Lavanya, K., Bhargava, K., Jayanth, M., & DeviSree, N. (2025, August). Leveraging YOLOv8 and Geolocation API for Real-Time Pothole Detection. In 2025 12th International Conference on Emerging Trends in Engineering & Technology-Signal and Information Processing (ICETET-SIP) (pp. 1-6). IEEE.

Venketbabu, T., Kennady, R., Saranya, R., Rashmi, P., & Anto, G. L. (2026). Enhancing Data Analytics and Forecasting Education Through Machine Learning Algorithms. In *Integrating AI and Machine Learning into Business and Management Education* (pp. 247–270). IGI Global Scientific Publishing.

Walzberg, J., Lonca, G., Hanes, R. J., Eberle, A. L., Carpenter, A., & Heath, G. A. (2021). Do we need a new sustainability assessment method for the circular economy? A critical literature review. *Frontiers in Sustainability, 1*, 620047. DOI: 10.3389/frsus.2020.620047

Wang, Y.. (2020). Cloud cover challenges in tropical urban remote sensing: A meta-analysis. *Remote Sensing, 12*(5), 789. DOI: 10.3390/rs12050789

Wen, X., Cao, H., Hon, B., Chen, E., & Li, H. (2021). Energy value mapping: A novel lean method to integrate energy efficiency into production management. *Energy, 217*, 119353. DOI: 10.1016/j.energy.2020.119353

Wolfs, T. (2017). *A GIS-based approach to estimate the geographic potential for rooftop solar PV: Developing a database at building-level for Denmark.* Aalborg University.

Wongwan, W., Pleerux, N., Thanomsat, N., & Moukomla, S. (2023). Assessment of solar rooftop potential using a digital surface model based on a photogrammetric survey by an unmanned aerial vehicle. *Environment Asia, 16*(3), 68–81. DOI: 10.14456/ea.2023.36

World Bank. (2020). *Digital economy in Pacific Island countries: Opportunities and challenges.* World Bank Group.

Wu, S., Lv, M., Ning, Z., Guo, S., & Chen, Y. (2025). Advancements in energy management strategies for hydrogen fuel cell hybrid UAVs: Towards intelligent, sustainable, and autonomous flight systems. *Aerospace (Basel, Switzerland), 12*(12), 1097. DOI: 10.3390/aerospace12121097

Xu, M. (2025). Enhancing GIS Models for Sustainable Development in Human Settlements Using Intelligent IoT Infrastructure and Human-Machine Interaction. *Results in Engineering, 106713*, 106713. Advance online publication. DOI: 10.1016/j.rineng.2025.106713

Yamini, N., Kumar, R. B., Mohamed, J. S., & Iyengar, N. C. S. (2023). Smart IoT-based system for monitoring and detecting fuel theft and fuel indication for refilling process. *International Journal of Human Computations & Intelligence, 2*(5), 247–255.

Yao, H., & Zhou, Q. (2023). Research status and application of rooftop photovoltaic Generation Systems. *Cleaner Energy Systems, 5*, 100065. DOI: 10.1016/j.cles.2023.100065

Yin, H., Krayenhoff, E. S., Voogt, J., Heusinger, J., & Aliabadi, A. A. (2025). Simulation of Tilted Rooftop Photovoltaic Panels at City Scale: Novel Measurements, Model Development, and Application in WRF. *Journal of Geophysical Research: Atmospheres, 130*(9), e2024JD043133.

Yurdakul, M., & Tasdemir, Ş. (2025). An Enhanced YOLOv8 Model for Real-Time and Accurate Pothole Detection and Measurement. arXiv preprint arXiv:2505.04207.

Yurtay, Y. (2025). Carbon footprint management with Industry 4.0 technologies and ERP systems in sustainable manufacturing. *Applied Sciences (Basel, Switzerland), 15*(1), 480. DOI: 10.3390/app15010480

Yu, W., Patros, P., Young, B., Klinac, E., & Walmsley, T. G. (2022). Energy digital twin technology for industrial energy management: Classification, challenges and future. *Renewable & Sustainable Energy Reviews, 161*, 112407. DOI: 10.1016/j.rser.2022.112407

Zeng, J., & Zhong, H. (2024). YOLOv8-PD: An improved road damage detection algorithm based on YOLOv8n model. *Scientific Reports, 14*(1), 12052. DOI: 10.1038/s41598-024-62933-z PMID: 38802524

Zhang, H., Qi, R., Liu, Y., Wang, T., Zhong, F., Zhou, Q., & Wang, K. (2024). The spatial impact of digital economy on carbon emissions reduction: Evidence from 215 cities in China. *Frontiers in Environmental Science, 12*, 1370938. DOI: 10.3389/fenvs.2024.1370938

Zhang, Y., Chen, X., & Li, L. (2023). Intelligent fleet management using IoT and machine learning: A comprehensive review. *IEEE Internet of Things Journal*, *10*(5), 4123–4145.

Zhao, M., Su, Y., Wang, J., Liu, X., Wang, K., Liu, Z., Liu, M., & Guo, Z. (2024). MED-YOLOv8s: A new real-time road crack, pothole, and patch detection model. *Journal of Real-Time Image Processing*, *21*(2), 26. DOI: 10.1007/s11554-023-01405-5

Zhong, Z., & Ren, Z. (2025). Deep learning and TOPSIS-based multi-criteria decision-making framework for urban road defect detection and sustainable maintenance planning. *Scientific Reports*, *16*(1), 2045. DOI: 10.1038/s41598-025-31682-y PMID: 41408389

Zhou, K., Fu, C., & Yang, S. (2016). Big data driven smart energy management: From big data to big insights. *Renewable & Sustainable Energy Reviews*, *56*, 215–225. DOI: 10.1016/j.rser.2015.11.050

Ziya. (2025). UAV Autonomous Navigation Dataset. Retrieved from https://www.kaggle.com/datasets/ziya07/uav-autonomous-navigation-dataset

About the Contributors

Kamalakanta Muduli, is presently working as Associate Professor and Head of the School of Mechanical Engineering, Papua New Guinea University of Technology, Lae, Morobe Province, Papua New Guinea. He is also a Visiting Researcher at University of Derby, UK. He has obtained PhD from School of Mechanical Sciences, IIT Bhubaneswar, Orissa, India. Dr Muduli has over 20 years of academic experience in Universities in India and Papua New Guinea. Dr Muduli is a recipient of ERASMUS+ KA107 award provided by European Union. He has published 106 papers in peer reviewed international journals most of which are indexed in Clarivate analytics, Scopus and listed in ABDC and more than 37 papers in National and International Conferences. He has been also guest edited 10 Special Issues in journals and 19 books published by Springer, Taylor and Francis, MDPI, CRC Press, Wiley Scrivener and Apple Academic Press. Dr Muduli also has supervised four PhD students. His current research interest includes Manufacturing, Sustainable supply chain management, and Industry 4.0 applications in operations and supply chain management. Dr Muduli is a fellow of Institution of Engineers India. He is also a senior member of Indian Institution of Industrial Engineering, member of American Society of Mechanical Engineers(ASME) and Professional member of Industrial Engineering and Operations Management(IEOM) Society.

Sujoy Kumar Jana presently working as Professor and Course Coordinator, MSc in Remote Sensing and GIS at School of Surveying and Land Studies at the Papua New Guinea University of Technology, Papua New Guinea. He having 25 years of extensive work on research, teaching, training and administrative in the university system. He also consults on disaster risk governance and policy, vulnerability and risk assessment, land use land cover mapping, capacity development of decision-makers, river and coastal resilience, critical infrastructure functional risk and urban planning. He has more than 4 books and 50 papers in international and national peer-reviewed journals. His professional activities include roles as editor, editorial board member and reviewer of international journals.

Rashmi Prava Das did her PhD from CV Raman Global University. She did her Bachelors in Computer Science and Engineering from North Orissa University and Masters in Computer Science from Birla Institute of Technology, Meshra. She has around 12 years of teaching and research experience in various degree level engineering colleges in Odisha as lecturer, Senior lecturer and Assistant Professor. She had also supervised few master's thesis. She has published research papers in international journals, conferences and books as chapters.

Indrajit Pal is an Associate Professor and Chair in the Climate Change, Disaster and Urban Sustainability, Asian Institute of Technology, Thailand. Dr. Indrajit Pal is also an Adjunct Professor at Chulalongkorn University Social Research Institute, Thailand. He served as a faculty member at Centre for Disaster Management at Lal Bahadur Shastri National Academy of Administration, Mussoorie, India (A Premier National Institute for training Indian Administrative Services Officers). Dr. Pal holds doctoral and two master's degrees, e.g., Applied Geology and Sociology with urban sociology specialization. Indrajit having more than 16 years of experience on research, teaching, training, advocacy, consultancy primarily focused on Disaster Governance, Risk Management, Incident Command System, Disaster Risk Reduction, Hazard and Risk Assessment, Risk Perceptions, Participatory DRM, Public Health Risk, GIS & Remote Sensing in DRM, Climate Change Adaptation and disaster risk management. Indrajit is reviewer of more than ten international journals and has published number of books and research articles in refereed national and international journals. Indrajit is a member of Board of Directors for the Global Alliance of Disaster Research Institutes (GADRI), Japan

***

Duggirala Aravind is currently associated with the Department of Aviation, ASET College of Science and Technology, Chennai, India. He previously completed his Bachelor's degree in Electrical and Electronics Engineering at MITS University. His research interests include aviation operations, renewable energy technologies in air transport, Artificial Intelligence, Logistics and Business management. (aravindaset.edu@gmail.com).

Kasukurthi Aravind is an MBA student in Airlines and Airport Management at ASET College of Science and Technology, India. His interests include innovation, entrepreneurship, global business, and artificial intelligence.

R. Deepa is an Assistant Professor in the Department of Computing Technologies at SRM Institute of Science and Technology, Kattankulathur, where she teaches and mentors students in computer science and emerging technologies. SRM Institute of Science and Technology.She has published multiple research papers in national and international conferences and journals on topics including machine learning, data analytics, and NaturalLanguage Processing applications.

Deepak Gupta is a distinguished academician with over 20 years of teaching experience and expertise in Computer Science and Engineering. He holds a Ph.D. in Computer Engineering and has an extensive research portfolio, including many international journal papers, multiple conference proceedings, and several books. His research interests span machine learning, artificial intelligence, and data mining. Currently, Dr. Gupta serves as an Associate Professor and Head of Department of CSE (Emerging Branches) at ITM Gwalior, MP, India. He has also contributed significantly to the field through patents and copyrights and actively engages as a reviewer, conference chair and editor.

L.K Shoba received her B.E degree in Electronics and Communication Engineering from Noorul Islam college of Engineering, Kumaracoil, Manonmanium sundaranar university, India in 2003. She has received her M.Tech. degree in Information Technology from Madras Institute of Technology, Chennai, Anna University, India in 2009. She holds a Ph.D degree in Information and communication Engineering from Anna University, India in 2023. She is in teaching profession since June 2009. Currently she is working as Assistant Professor in the Department of Computing Technologies, SRM Institute of Science and Technology, kattankulathur, Her research area of interest is Artificial Intelligence and Machine Learning.

M. Gunasekar is an Assistant Professor in the Department of Computer Science and Business Systems. He is associated with Vel Tech Multi Tech Dr. Rangarajan Dr. Sakunthala Engineering College, Avadi, Chennai. His academic interests focus on computing technologies and their applications in business systems.

Dinesh M. G. is a faculty member at Easa College of Engineering and Technology, Coimbatore, Tamil Nadu. He specializes in engineering education and contributes to academic research, teaching, and student development. With a commitment to innovative learning, he actively engages in institutional growth and professional advancement.

Samrat Majumdar is a Research Scholar in the department of School of Mechanical Engineering of KIIT Deemed to be University.

Bikash Ranjan Moharana presently working as a Senior Lecturer in the School of Mechanical Engineering at Papua New Guinea University of Technology, Lae, Morobe Province, Papua New Guinea. He received his PhD in Department of Mechanical Engineering from the National Institute of Technology, Rourkela, India in 2018. He has published more than 25 papers in various peer-reviewed Scopus, and SCI-indexed journals and conference proceedings. He has also been editing books approved for publication by Taylor and Francis-CRC Press, and IGI Global. His main research interests focus on various fusion welding processes, sustainable manufacturing, and process optimization. He is a life member in the Institution of Engineers India, IIW, IWS, and ASME.

P. Ranjidha is an Assistant Professor in the Department of Artificial Intelligence and Data Science at Panimalar Engineering College. She is involved in teaching and mentoring students in AI- and data-driven technologies. Her academic interests focus on applying artificial intelligence and data science concepts to real-world problem solving.

Chiranth R is a technology enthusiast and researcher with interests in artificial intelligence, automation, and web technologies. He focuses on developing practical, scalable solutions that bridge academic research and real-world applications. His work includes AI-driven systems, intelligent automation workflows, and web-based platforms, with experience in CRM and enterprise-oriented applications. Chiranth actively engages in research, hackathons, and innovation-driven projects, aiming to contribute impactful and applied technological advancements.

R. Vinoth is an Associate Professor in the Department of Computer Science and Engineering at RMK College of Engineering and Technology, RSM Nagar, Puduvoyal, Thiruvallur District – 601206. He is engaged in teaching, research, and academic administration in core and emerging areas of computer science. His professional focus includes mentoring students and contributing to institutional academic and research initiatives.

Ravikumar R. N. is currently a Postdoctoral researcher at Universiti Malaysia Perlis, Malaysia. Also, working as an Associate Professor in the Department of Computer Engineering at Marwadi University, Rajkot, Gujarat. He has 14+ years of national and international teaching experience. He completed his PhD at Amity University, Rajasthan, M.E at VMRF Deemed University, B.E at Anna University, Chennai. He is the co-founder of the Research Activities Club at Marwadi University and trains students on research activities, which leads to the various publications. He has published numerous research articles in Scopus, SCI journals, Book chapters, and Conference papers. Also, he holds several patents and authored books. His specialization is ML, DL, Federated Learning, Recommendation Systems, and Optimization Techniques.

E. Eswara Reddy is currently working as a Professor in the School of EConomics and Commerce, CMR University Bangalore. He has done his M.Com from S.V University Tirupathi and full time Ph.D from S.K.University Anantapur. He has also done his MBA from IGNOU in 2003. He has 28 years of experience in teaching, research and administration in the field of Commerce and Management. He has published 32 Papers in Journals and magazines and attended 12 National and international conferences. He has successfully guided 4 Ph.D scholars in the School of Economics and Commerce CMR University and 5 scholars in the School of Commerce and Management Dravidian University. He has also guided 45 research scholars for M.Phil Programme in Alagappa University, S.V.University, S.K.Univeisty, Periyar Univesity, M.K,University Madurai,. He has adjudicated 25 Ph.D thesis of Kannur University, Bharathidasan University and Bharathiar University. He has written 2 Books on Accounting for Management and Indian financial system Publihsed by Thakur Publications. He is a member of Board of examiners of Kannur University Kannur, Bharathidasan Univeristy, Bharathiar Univeristy, Dr, MGR deemed University Chennai and Bharath academy of Higher education and Research Chennai. He is also the editorial board member of Vidyanikethan Journal of Management publihsed by Oxford group Bangalore.He has also delivered expert lecture in UGC Academic stafff college Goa University in the Referesher course conducted for degree colleges faculty.of Commerce and Management Orcid Id 0000-0003-0387-1437

Jeremiah Renagi is a postgraduate researcher at the Papua New Guinea University of Technology, School of Surveying and Land Studies. His research focuses on estimating rooftop solar photovoltaic potential using GIS, remote sensing, and low-cost UAV photogrammetry. His work aims to support sustainable energy planning and improve electricity reliability in Papua New Guinea aligned with interests in geospatial analysis, renewable energy mapping, and spatial data modeling.

S. Deepa is an Assistant Professor in the Department of Computer Science and Business Systems at R.M.D. Engineering College. She is actively involved in teaching, mentoring, and academic development in interdisciplinary computing domains. Her professional interests focus on bridging computer science fundamentals with business and industry-oriented applications.

S. Gunanandhini is working in the Department of Computer Science and Engineering at S.A Engineering College. She is engaged in teaching and academic activities. Her interests are aligned with core and emerging areas of computer science.

Arun Kumar Singh is working as an Asso. Professor in the Department of Mathematics and Computer Science (PNG University of Technology, PNG). He received Ph.D. in CS/IT under the guidance of Dr. Neelam Srivastava (IET Lucknow) and Dr. R. P. Agarwal (IIT Roorkee), M.Tech. (IT-WCC) degree in 2005 from IIIT-Allahabad under the guidance of Prof. M. Radha Krishnan and B.E. (ECE) degree in 2002 from Dr. B. R. Ambedkar University, Agra, India. His research interests are IoT, Machine Learning, Big Data, Network Management, Wireless networking, Social Networking and Mobile computing.

Deepak Singhal is currently working as Associate Professor in School of mechanical engineering, KIIT Deamed to be university, Bhubaneswar, Odisha. His research areas are remanufacturing, system modelling, supply chain management, project management and quality management.

N.V. Suresh, renowned academician from Chennai, Tamil Nadu. He completed his schooling in Chennai. He obtained his BA degree (2012) from D.G.Vaishnav College, Madras University. He joined the Bharathiar University received an MBA degree and PGDM. He Awarded Ph.D. in Management from University of Madras. Dr.N.V.Suresh joined then IIKM in June 2014 as an Assistant Professor (2014 to 2016). He moved to Remo International College in June, 2019 as an Assistant Professor and rose through the ranks to the position of Head of the Department in Aug 2020 to July 2021 then Shifted to Working Academy of Maritime Education and Training (Deemed to be University) Presently Working as Associate Professor and Vice-Principal in ASET College of Science and Technology from March 2022 to Till Date. He is the Visiting Faculty for Institute of Distance Education, University of Madras, Tamil Nadu Open University, Anna University and Directorate of Online Education SRM University for MBA from 2015 to till date. Dr.N.V. Suresh has carried out all his research work entirely in India and has published 30 research papers in international journals.

Susanta Tripathy is currently working as a Professor in the School of Mechanical Engineering. He is also serving as a Dean, School of Design, KIIT-DU.

Index

C

D

E

F

G

H

I

L

M

O

P

R

S

U

Y

www.ingramcontent.com/pod-product-compliance
Lightning Source LLC
LaVergne TN
LVHW060003031026
849161LV00025B/1349